Cellular biochemistry and physiology

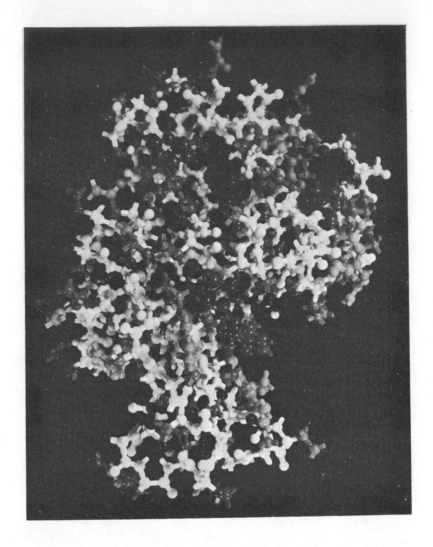

3-D model of the enzyme lysozyme, viewed down the z-axis, showing the deep cleft containing the 'active sites' which combine with the substrate (a polysaccharide-like substance, present in bacterial cell walls).

(*Acknowledgement:* Model built by Dr B. Nicholson, Department of Physiology & Biochemistry, Reading University, from push-fit plastic units (Labquip, Reading). Co-ordinates determined by Phillips et al. at the Royal Institution.)

Cellular biochemistry and physiology

N. A. Edwards and K. A. Hassall

Lecturer and Senior Lecturer, Department of Physiology
& Biochemistry, University of Reading

McGRAW-HILL · LONDON

New York · St Louis · San Francisco · Dusseldorf
Johannesburg · Kuala Lumpur · Mexico · Montreal · New
Delhi · Panama · Rio de Janeiro · Singapore · Sydney
Toronto

Published by
McGRAW-HILL Publishing Company Limited
MAIDENHEAD · BERKSHIRE · ENGLAND

07 094136 X

Printed by photo-lithography and made in Great Britain at the Pitman Press, Bath.

Preface

In the preface to a delightfully useful book on calculus first published in 1910,* Sylvanus Thompson, FRS, made comments which, if anything, have gained in significance over the years. The writers of textbooks, he said, 'seem to desire to impress you with their tremendous cleverness by going about it in the most difficult way. Being a remarkably stupid fellow . . . I beg to present to my fellow fools the parts that are not hard'. How poignantly these sentiments express some of the attempts to introduce the elements of biochemistry which have appeared in recent years!

It would be presumptuous of us to assume that we have everywhere succeeded where cleverer men have failed, but there is one factor overwhelmingly in our favour: we have lectured in Introductory Biochemistry to students reading a variety of disciplines and have additionally some knowledge of Advanced Level courses in chemistry and biology. We know what the average student is likely to know—and not to know— when he picks up with interest his first textbook on cellular biochemistry. We know from experience his disillusionment when he finds, as he sometimes does, that the first hundred pages are more akin to classical chemistry than to the subject he had enthusiastically anticipated. We know his frustration on discovering that, only too often, the essential reactions of biochemistry are later mentioned in rapid succession, without any serious attempt to explain why they occur, and often, in the cause of scientific accuracy, encompassed with so many qualifications, that the main theme is all but obscured.

We have attempted to overcome some of these difficulties in a variety of ways. Care is taken to explain reactions which perplex biology

---

* Sylvanus P. Thompson, FRS. *Calculus Made Easy.* MacMillan & Co., London (1910).

v

students; this is often done by the use of analogy or by comparison with similar reactions involving simpler compounds encountered at Advanced Level. Special emphasis is placed on the rationale underlying sequential reactions and energetic processes, even at the expense of a little teleological reasoning from time to time ('the next reaction involves the dephosphorylation of the hydroxyl group since the free hydroxyl group is to be removed in the reaction which follows'). Recognizing that our book will be used by students with somewhat different backgrounds, we have avoided giving an interminable introduction by providing references to well-known books for background reading on key non-biochemical topics (marked thus #) as well as providing, in Appendix 2, references for additional reading.

Almost anybody can write a text-book on his research subject, but the writer of an elementary book walks a precarious pathway. He must not give the impression that everything is universally cut-and-dried; nor must he obscure the woods with the trees. We have attempted to present our material in an interesting manner, based on sound educational principles, but with the minimal sacrifice of scientific accuracy. The reader is, however, warned that even the most sacred of scientific tenets amount at most to the temporary dogma of our era and are subject to modification in the light of further knowledge. In no subject is this more true than in the realm of so young a fledgling as biochemistry. The readers of our book will probably live long enough to see the subject soar to heights which no present writer can foresee.

Finally, we have tried above all to convey to our readers something of the fascination inherent in the subject, something of our own enthusiasm for it. We have not used the minimal number of words—a few lines of print may save a reader hours puzzling something out for himself. We will welcome suggestions for improvement. We have departed occasionally from accepted methods of presentation in an attempt to put first things first—even if those things are, very temporarily, somewhat unfamiliar. The fate of acetyl coenzyme A, for example, is of far greater importance in providing the cell with energy than the almost incidental processes of glycolysis and β-oxidation which convert carbohydrates and fats respectively to this important source of energy. Such an approach at once presents the compartmentation of the cell as a concept central to our approach and provides the biologist as well as the chemist with that pleasant psychological satisfaction of being on home ground which arises when a book, on a subject new to the reader, is opened for the first time.

The reader may well prefer to read the book twice, on two different intellectual wavelengths. For example, the beginner would probably find, during the first reading, that chapters 2 and 3 dealing with the basic properties of carbohydrates, nucleotides, and proteins were rather a strain on his memory. He would be advised *initially* to concentrate on getting a general overall impression rather than attempt to commit all the formulae to memory. When the reader subsequently encounters particular compounds in later chapters he should revise the earlier work. It will then be easier to appreciate the significance of the structure of the compounds by considering them in their biological context. Paraphrasing the words of Sylvanus Thompson: first of all, master the easy things and the things that interest you most, and the rest will follow. Remember, what two fools can do, so can another.

N. A. Edwards
K. A. Hassall

Contents

Preface   v

Chapter 1   *The Cellular Basis of Life*   1

  1.1 Why organisms are subdivided into cells   1
  1.2 Intracellular subdivision   2
  1.3 Structure of a generalized cell   3
  1.4 Ultracentrifugation   6
  1.5 Storage lipids and lipids of membranes   8
  1.6 Structure of lipoprotein membranes   17
  1.7 The plasmalemma, cell walls, cilia, and flagella   21
  1.8 The cytosol and cell vacuoles   24
  1.9 The nucleus   25
  1.10 Mitochondria   26
  1.11 Chloroplasts   28
  1.12 The endoplasmic reticulum   29

Chapter 2   *Proteins, Peptides, and Amino Acids*   32

  2.1 The structural versatility of proteins   32
  2.2 Structure of α-amino acids and the effect of pH   35
  2.3 The peptide bond   40
  2.4 Individual amino acid residues   41
  2.5 Some important reactions of amino acids   46
  2.6 The primary structure of representative peptides and proteins   48

	2.7	The three-dimensional structure of proteins	52
	2.8	Native and denatured proteins	58
Chapter	3	*Carbohydrates and Nucleotides*	60
	3.1	Classification and naming of carbohydrates	60
	3.2	Glucose—an aldohexose sugar	62
	3.3	Other aldoses	66
	3.4	Ketoses	69
	3.5	Derivatives of the monosaccharides	70
	3.6	Disaccharides	73
	3.7	Polysaccharides	75
	3.8	Nucleotides and nucleosides	79
	3.9	Dinucleotides	84
Chapter	4	*Enzymes*	87
	4.1	Enzymes as catalysts	87
	4.2	The specificity of enzymes	88
	4.3	Classification of enzymes	90
	4.4	The chemical and physical characteristics of enzymes	92
	4.5	How enzymes accelerate reactions	95
	4.6	Enzyme-substrate interaction and the Michaelis constant	99
	4.7	Inhibition of enzymes	104
	4.8	Enzymes acting in sequence	105
	4.9	Coenzymes and activators	106
Chapter	5	*Bioenergetics*	111
	5.1	Energy for the living cell	111
	5.2	Interconversion of energy	112
	5.3	Potential energy in physical and chemical systems	113
	5.4	The driving force of the living machine	117
	5.5	Free energy, enthalpy, and entropy	119
	5.6	Priming and coupling reactions	122
	5.7	Free energy, equilibria, and the equilibrium constant	125
	5.8	Electron flow: the relation of chemical to electrical potentials	128
	5.9	Steady states	129
	5.10	Adenosine triphosphate and other 'energy-rich' compounds	131

Chapter 6   *Mitochondrial Electron Transport and Oxidative*   136
                     Phosphorylation

6.1   The purpose of electron transport   136
6.2   NAD, the first electron acceptor   139
6.3   Flavin mononucleotide (FMN)   140
6.4   Ubiquinone   142
6.5   Cytochrome **b**   143
6.6   The other cytochromes   144
6.7   Succinate dehydrogenase   146
6.8   Formulae and names of some important inter-mediates   147
6.9   Redox potentials   151
6.10 Oxidative phosphorylation   154
6.11 The site of oxidative phosphorylation in the cell   156

Chapter 7   *The Tricarboxylic Acid Cycle*   159

7.1   Two important general reactions   159
7.2   The TCA cycle—its fuel and functions   161
7.3   Formation of citric acid   163
7.4   Intra-molecular rearrangement of citrate to iso-citrate   164
7.5   Isocitrate to α-ketoglutarate   165
7.6   The oxidative decarboxylation of α-ketoglutarate   165
7.7   Conversion of succinyl coenzyme A to succinic acid   169
7.8   Succinate to oxaloacetate   171
7.9   Energy captured by the cycle   172
7.10 The acids of the TCA cycle   173

Chapter 8   *Lipid Metabolism*   175

8.1   Hydrolysis of glycerides   176
8.2   The catabolism of long-chain fatty acids   177
8.3   The fatty acid oxidation spiral   178
8.4   The central metabolic position of acetyl coenzyme A   182
8.5   Nicotinamide adenine dinucleotide phosphate (NADP)   183
8.6   The biosynthesis of long-chain fatty acids   185
8.7   The biosynthesis of triglycerides   190
8.8   The biosynthesis of lecithin   192

Chapter 9   *Glycolysis and the Fate of Pyruvate*   193

    9.1   The overall position of glycolysis   193
    9.2   The four stages of glycolysis   196
    9.3   Formation of fructose-di-P   197
    9.4   Formation of triose phosphates   199
    9.5   Oxidation of glyceraldehyde-3-P   201
    9.6   Synthesis of pyruvate   203
    9.7   Oxidative decarboxylation of pyruvate to acetyl
          coenzyme A   205
    9.8   The fate of pyruvate under anaerobic conditions   206
    9.9   Energy capture resulting from glucose catabolism   209

Chapter 10   *Glycogenesis and Gluconeogenesis*   211

    10.1 Polysaccharide biosynthesis from glucose-1-P   213
    10.2 Branching enzyme   217
    10.3 Galactose and lactose metabolism   218
    10.4 Origin of oxaloacetate and its conversion to glucose-
          1-P   220
    10.5 Synthesis of oxaloacetate from amino acids, pyru-   223
          vate, and propionate
    10.6 The synthesis of oxaloacetate from acetyl coenzyme
          A   225
    10.7 A general outline of carbohydrate metabolism in
          green plants   229
    10.8 Regulation of blood glucose levels in the mammal   232

Chapter 11   *Photosynthesis and the Pentose-Phosphate Cycle*   234

    11.1 The overall process of photosynthesis   234
    11.2 The light reaction of photosynthesis   236
    11.3 The photosynthetic carbon cycle   241
    11.4 The pentose-phosphate, or the hexose-monophos-
          phate, pathway   249
    11.5 Differences between the pentose-phosphate and the
          main pathway of sugar degradation   252
    11.6 Utilization of sugars produced by the pentose-
          phosphate pathway   253
    11.7 Some processes requiring a supply of reduced
          NADP   255
    11.8 The microsomal electron transport system   258

Chapter 12   *Nitrogen Metabolism*   262

12.1   Nitrogen fixation   262
12.2   The synthesis of amino groups   267
12.3   General aspects of amino acid metabolism   270
12.4   Essential and non-essential amino acids   272
12.5   Glutamine, aspartic acid, and carbamyl phosphate   274
12.6   Nitrogen excretion in animals   276
12.7   The formation of urea in mammals and amphibia   279
12.8   Fate of amino-nitrogen in uricotelic animals   282
12.9   The biosynthesis of some important compounds containing nitrogen   285

Chapter 13   *Metabolism of the Carbon Skeletons of the Amino Acids*   291

13.1   The transfer of units containing only one carbon atom   293
13.2   The role of derivatives of vitamin B_6 in amino acid metabolism   297
13.3   The general pattern of amino acid synthesis and catabolism   298
13.4   The α-ketoglutarate family of amino acids   300
13.5   The 3-phosphoglycerate family of amino acids   303
13.6   The oxaloacetate family of amino acids   306
13.7   The pyruvate family of amino acids   308
13.8   The cyclic amino acids   309
13.9   The metabolism of histidine   313
13.10 Some important compounds synthesized from amino acids   313

Chapter 14   *Polynucleotides and Protein Synthesis*   324

14.1   The location and structure of the nucleic acids   324
14.2   The double helix of DNA   326
14.3   Types of RNA; their structure and function   331
14.4   The role of pyrophosphatases in the synthesis of nucleic acids and proteins   335
14.5   Origin of molecules of DNA and RNA   340
14.6   A model system for protein synthesis   343
14.7   Coding ratio and the genetic code   346
14.8   The structure of bacteria and viruses   348

Chapter 15   *Movement of Substances into and out of Cells*   352

15.1   Some general considerations   352
15.2   Osmosis and the uptake of water by cells   353
15.3   Osmotic forces in plant cells   355
15.4   Osmoregulation in animals   356
15.5   Intracellular inorganic cations and the sodium pump   357
15.6   Donnan equilibrium   360
15.7   Absorptive cells of the small intestine   361
15.8   Absorption of sugars and amino acids by cells of
       the intestinal epithelium   364
15.9   The digestion and absorption of glycerides   366
15.10 Michaelis kinetics applied to the transport of com-
       pounds into cells   369
15.11 Blood and interstitial fluid   370
15.12 The transport of oxygen into animal cells   372
15.13 Intracellular pH, and the passage of hydrogen ions
       and carbon dioxide out of tissue cells   376
15.14 How the erythrocyte participates in the transport of
       carbon dioxide and in buffering the blood   377

Chapter 16   *Excitable Cells of Nerve, Muscle, and Retina*   382

16.1   The resting potential of cells   382
16.2   The general structure of a nerve cell   386
16.3   The action potential of a nerve cell   387
16.4   How a nerve cell carries a message   391
16.5   Transmission of the nervous impulse from one cell
       to another   392
16.6   How a nerve stimulates a muscle   394
16.7   Structure of muscle   397
16.8   How muscular contraction is initiated   402
16.9   The chemistry of muscular contraction   402
16.10 Structure and function of the rod cell of the eye   406
16.11 Concluding remarks   410

Appendix 1   *Systematic Nomenclature of Enzymes*   411

Appendix 2   *References and books for wider reading*   417

Index   419

1

The cellular basis of life

1.1   Why organisms are subdivided into cells

The theory of evolution proposes that living systems evolve and develop
only in the direction which leads to their success in a particular environment.   Extinction is the ultimate fate of all organisms which fail to do
this.   Since almost all animals and plants are subdivided into small
independent functional units, or cells, it would appear that, as a biochemical entity, the cell is not only successful and efficient *in its own right* but
additionally, that the *condition of cellular subdivision* of large organisms
confers advantages to the organisms of overwhelming evolutionary
importance.

   Living organisms cannot exist as 'closed systems' independent of their
environment.   They must, of necessity, obtain chemical substances from
their surroundings and they must eliminate waste materials.   The passage
into and out of cells of such important substances as oxygen and carbon
dioxide occurs by the process of diffusion.   This commences at the outer
surface of unicellular or multicellular organisms but must also continue
within the organism unless alternative methods of conduction exist.
This raises an important issue, for diffusion is a relatively slow process
and is therefore only an efficient transport mechanism when the distances
involved are quite small.   A typical cell is usually of the order of 10–100
microns diameter (1 micron, μ, $= 10^{-3}$ mm), and so has a volume of
between 10^{-6} and 10^{-3} mm^3.   Diffusion through bulk phases of this
magnitude presents little or no problem, and it is by this means that
most substances move within bacteria, protozoa, and green algae.   But
higher plants and animals can have volumes many millions of times
greater than this, and were they not divided into sub-units it is difficult to

conceive how diffusion from the outside inwards could have been rapid enough to be compatible with evolutionary success.

A related consideration of very great importance is the possibility of cellular specialization, or division of labour, within a multicellular organism. If large organisms were not divided into cells, evolutionary specialization of parts of the organism for particular functions would, at best, have been much less profound than it has been in both the plant and animal kingdoms. Specialization of cells appears to have been an essential pre-requisite for the evolutionary experimentation which has led to the existence of so many and diverse forms of multicellular organisms.

1.2   Intracellular subdivision

The cell, far from being a homogeneous mass, is subdivided into compartments, an essential requirement if it is to perform the multitude of activities that life demands of it. The object of this chapter is to indicate, with minimal detail, the relation between some of these activities and the structural organization of the cell, thus setting the scene for the chapters which follow. The subject of cellular biochemistry and physiology loses much of its fascination if the cellular location of particular events is largely forgotten in favour of the intellectual exercise of writing formally correct (but biologically barren) sequences of chemical reactions.

At any moment of time numerous biochemical processes are taking place within an active cell, and each of these may involve a series of apparently complicated chemical reactions. From small and relatively simple substances, the cell builds many large molecules which are characteristic of living organisms and essential to their existence. These large molecules include *proteins*, *polysaccharides*, *nucleic acids*, and *lipids*. The sum total of all the synthetic processes of the cell is referred to as the cell's *anabolism* and the relatively simple substances which undergo the anabolic process of being built up to form a more complex molecule are known as *precursors* of the complex molecule.

The cell is not a static system, however, for, side by side with synthesis, cellular components are constantly being degraded to simpler compounds. All the chemical processes occurring in the cell are known collectively as the cell's *metabolism*, and a substance being changed in some way is called a *metabolite*. The breakdown of carbohydrates, fats, and proteins yields energy which the cell harnesses to drive its machinery. The breakdown processes are, in many ways, the antithesis of anabolism and are called *catabolic* processes; nevertheless, as later chapters will show, the set of chemical reactions, or *route*, leading to the synthesis of a substance is seldom exactly the same as that which leads to its destruction. It is not unusual to find that anabolic and catabolic processes are often separated spatially by being located in different parts of the cell, yet they are none the less interrelated, for anabolic activities are driven by the energy released from catabolism. In green cells of plants, light indirectly provides the energy for synthetic reactions through the process of photosynthesis.

The multitude of reactions occurring in the minute dimensions of the cell would more than tax the resources of the most elaborate chemical laboratory. The chemist can use one vessel for a reaction occurring in water and another for a reaction requiring a non-polar solvent such as benzene. Solid catalysts would probably be used in a third type of apparatus. The cell achieves similar 'compartmentation' by having an elaborate sub-structure; some parts of the cell are adapted for aqueous reactions, others accommodate surface reactions, and yet other regions provide a non-polar environment for essentially hydrophobic reactions.

The activities of the cell must be controlled so that they take place at a rate appropriate to the circumstances of the moment, for the cell must be able to adapt to changing environmental conditions. This applies equally to the rates of catabolism and anabolism, to the movement of potential metabolites within and between the various parts of the cell, and to the movement of materials between the cell and its surroundings.

1.3 Structure of a generalized cell

Most cells are clearly visible under a good light microscope, the resolving power of which is rather less than $0.4\ \mu$, and the classical microscopists were well aware that cytoplasm was by no means an amorphous jelly. They knew that, in addition to the nucleus (and, in plants, the chloroplasts) a variety of particulate structures of smaller size were present within the cell.

The advent of the electron microscope stimulated interest in the finer detail of cell structure and in the relation of such structure to function. It provides effective magnification of between 100,000 and 200,000 compared with only 1000 for the light microscope and the resolution ($1\ m\mu$, or 10^{-6} mm) reaches almost to molecular dimensions (the orbital diameter of a hydrogen atom is about $0.1\ m\mu$). Electron microscopy relies on the

The cellular basis of life

principle that some materials are more or less opaque to a beam of electrons so that differences in electron density of tissue particles are detected when penetrating electrons hit a fluorescent screen or photographic plate. To prepare material for electron microscopy it has to be dried and stained with electron-dense compounds such as osmium tetroxide or potassium permanganate and cut into very thin sections. This treatment is very 'unphysiological' so artefacts could well be introduced. Electron microscopes which give three-dimensional pictures have recently been introduced.

Some of our present ideas are possibly based on misconceptions resulting from the creation of artefacts, but, almost beyond question, many of the cellular sub-structures are either constructed from a particular type of membrane composed mainly of lipid and protein, or they are attached to such a membrane. Apart from electron microscopy, X-ray analyses of cell membranes, fractionation of the various parts of the cell, and studies of the electrical properties and permeability of cell surfaces all tend to indicate that membranes made of lipid and protein are the basic 'fabric of life', the material from which most of the subcellular particles or 'organelles' in all types of cells are constructed.

Figure 1.1 illustrates, in very diagrammatic form, some of the features of a non-specialized cell; animal cells typically lack a prominent cell wall, have no chloroplasts and usually have no well-defined vacuole. Within most animal cells the *nucleus* is the most prominent organelle, although in many plant cells the chloroplasts are also large. Mitochondria are just visible under the light microscope and these, like the chloroplasts, are largely composed of lipoprotein membranes. The *plasmalemma* or cell membrane, the *tonoplast*, which surrounds the vacuole, and the nuclear membrane are all more or less typical lipoprotein membranes.

The background fluid in which the organelles are suspended is known as the *cytosol*; it is composed largely of water in which ions of potassium, chloride, phosphate, and many others are to be found, together with traces of organic compounds, such as glucose and amino acids and colloidal solutions of certain proteins and enzymes. Electron micrographs reveal that a complicated set of branching lipoprotein membranes apparently divide the cytosol into numerous separate compartments. In cross section, these membranes resemble a network of intercommunicating tubules, and for this reason have been termed the *endoplasmic reticulum*. These membranes probably undergo quite rapid formation and dissolution and so are less permanent than the other membranes mentioned above, although it is probably true to say that no membrane is 'permanent' in the sense that it persists unaltered for the whole life of the cell.

Several enzymes, or enzyme systems, are closely associated with the membranes of the endoplasmic reticulum. In addition, numerous small spherical bodies, the *ribosomes*, are present in the cytosol and these tend, especially during periods of active protein synthesis, to become attached to the membranes of the reticulum.

One of the most remarkable aspects of cell biology is the similarity in the basic cell structure and the overall resemblance of the vast majority of biochemical reactions occurring in cells of many different kinds. A word of caution is, however, desirable, for detailed biochemical investigations have tended to consider the cells of a very few species of organisms—bacteria (especially *Escherichia coli*), yeasts, tissues of rats, mice and rabbits, and a few plants (e.g., *Chlorella* spp.) have provided much of the

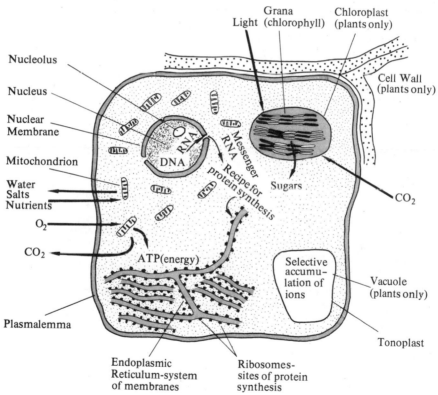

Fig. 1.1 Highly schematic structure of a generalized cell

The cellular basis of life

experimental material. Surprises may be awaiting the investigator who looks farther afield. Furthermore, bacterial cells are undoubtedly atypical in structure, lacking a definite nuclear area and lacking mitochondria. Nevertheless, bacterial cell organelles are based on lipoproteins, just as are those of larger organisms.

1.4 Ultracentrifugation

The location of various important cellular activities in certain parts of the cell (Table 1.1) often facilitates the investigation of these processes,

Table 1.1   Location of some important cellular activities

Location	Activity	Chapter
Nucleus	Genetic control of the proteins and enzymes of the cell	14
	Also synthesizes some essential cofactors	
Mitochondria	The inner membrane produces most of the ATP to drive anabolic processes.	6
	Other mitochondrial functions include:	
	(a) oxidation of acetyl coenzyme A with the formation of NADH	7
	(b) production of acetyl coenzyme A from fatty acids	8
	(c) production of acetyl coenzyme A from pyruvate	9
	(d) oxidation of glutamic acid to α-ketoglutarate	12
Chloroplasts	The seat of both light and dark reactions of photosynthesis (NADPH production and the reduction of CO_2 to fructose-6-P)	11
Ribosomes	Protein synthesis	14
Endoplasmic reticulum	(a) 'Rough' reticulum associated with ribosomes is probably concerned in the formation of despatching of proteins	14
	(b) 'Smooth' reticulum, ribosome-free, concerned with steps in the synthesis of lipids including long chain fatty acids and steroids, and with the processing of extraneous lipid-soluble molecules	8, 11
	(c) Golgi bodies and zymogen granules, associated with the intracellular storage of enzymes	1
Plasmalemma	Responsible for the uptake and elimination of water, certain ions, and organic molecules.	15
	Maintains a negative electrical charge inside nerve and muscle cells, a charge which enables conduction of impulses to occur.	16
Cytosol	(a) Seat of glycolysis by which carbohydrates are broken down to pyruvate	9
	(b) Hydrolysis of fats to glycerol and fatty acids	8
	(c) Locus of the enzymes of the pentose phosphate pathway, whose chief function is to produce NADPH for anabolic processes	11

for, by employing suitable techniques, it is frequently possible to separate the organelles so that they can be *studied in isolation*. Moreover, by adding cofactors normally provided by other cell fractions, a picture often emerges of the remarkable interdependence of the various parts of the cell.

If some suitable tissue or organ—liver, for instance—is ground up and then centrifuged, it is possible with suitable equipment to separate the various types of subcellular particles. The grinding-up process is called *homogenization*; a typical technique is to add to liver about twice its volume of 1·1 percent potassium chloride solution and to homogenize using the apparatus shown in Fig. 1.2. The 'pestle' is driven by means

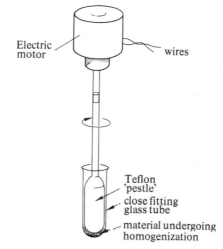

Electric motor

wires

Teflon 'pestle'

close fitting glass tube

material undergoing homogenization

(a) A Tissue Homogenizer

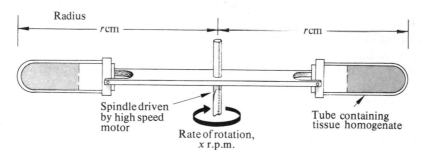

Radius

r cm

r cm

Spindle driven by high speed motor

Rate of rotation, *x* r.p.m.

Tube containing tissue homogenate

(b) Centrifugation

Relative centrifugal force $= (1118 \times 10^{-8} \times r \times x^2)g$

Fig. 1.2   Equipment used in cell fractionation

Table 1.2 Separation of particles by ultracentrifugation

Particle	Size	Centrifugation (minutes × g)
Nuclei	10 μ	10 × 600
Mitochondria	1–3 μ	10 × 7,000
Microsomes	100 mμ	60 × 105,000
Ribosomes	25 mμ	600 × 100,000

of an electric motor.   After homogenizing, incompletely homogenized material and debris (e.g., cell walls, in the case of plant homogenates) are removed by slow-speed centrifugation (about 1500 r.p.m.) using a conventional centrifuge.

To separate the cell organelles, a much more powerful centrifuge is necessary since the particles are much smaller—the smaller the particles, the greater the centrifugal force needed to bring about sedimentation. The special centrifuge is termed an *ultracentrifuge*; it can exert a centrifugal force exceeding 120,000 times the force of gravity, (g), when used at high speeds.   Centrifugal effectiveness is a function of the force applied in g and the time for which that force is applied; a relatively simple equation relates the force in g being applied to the number of revolutions per minute, shown on the dial of the centrifuge (Fig. 1.2).

Upon centrifuging for about 6000 g-minutes (i.e., acceleration in g multiplied by the time in minutes), the nuclei of the cells are thrown down.   The ultracentrifuge is then stopped, the supernatant solution poured into another centrifuge tube and then re-centrifuged at greater speeds.   At about 70,000 g-min, the mitochondria separate, whereupon the process is repeated.   Typical data are shown in Table 1.2.   The process of separation is termed differential centrifugation.   A point is eventually reached where the supernatant liquid is almost free of particulate structures and membranes and thus is known as the cytosol or soluble cell fraction.   In addition to water, ions, and small organic molecules it does, however, contain some substances in colloidal solution.   Preparative work of this kind is usually performed at low temperatures (e.g., 4°C) since this slows down biochemical reactions and helps to preserve the activity of the material.

1.5   Storage lipids and lipids of membranes

The lipoprotein membrane being the 'fabric' from which, in more or less modified form, most cell organelles are tailored, it is clearly to the structure of such a membrane that attention must first be drawn. And it is here that the first problem arises, for not only is it impossible to discuss simultaneously the lipids and proteins which are membrane sub-components, but it so happens that many, if not all, membranes are principally constructed from more complex lipids rather than from the simpler glycerol tri-esters with which most readers are likely to be familiar. It is proposed to solve the dilemma as follows. The present section will survey the chemical diversity of lipids, commencing with the simpler storage lipids, since such an account will also show why membranes are usually richer in lipids carrying charged or strongly polar groups; for the remainder of the chapter an elementary knowledge of the primary structure of proteins will be assumed, the subject being developed in chapter 2.

When non-polar liquids, such as hydrocarbon oils, octyl alcohol, or high molecular weight esters such as olive oil, are mixed with water, they form systems which separate into two layers with the denser liquid below the other. Upon shaking, a temporary emulsion is formed from which droplets of the disperse phase tend first to separate and then to coalesce to re-form a bi-phasal system. The principal reason for these changes is an energetic one; the non-polar long chains of covalently-linked carbon atoms are attracted towards similar chains in related molecules, but are repelled by the smaller, highly polar water molecules. If a soap or detergent is added to such an emulsion, the disperse phase is largely stabilized, since each droplet becomes surrounded by a sphere of electrical charges which repel those on nearby droplets (Fig. 1.3).

When a more complex lipid such as the phospholipid *lecithin* is shaken with water, the situation is comparable to the last one mentioned above, for the lecithin molecule automatically carries, according to the acidity of the medium, a positive or negative charge. These charged parts of the lecithin molecules are attracted to polar water molecules, while the non-polar parts of the molecules attempt to escape from the water by being attracted towards one another. The result is that the *hydrophilic* ('water-loving') and *hydrophobic* ('water-hating') parts of the molecules adopt a compromise by aggregating together in droplets or films with the charged groups on the outside in contact with the water and the non-polar parts directed away from the water.

As a group, lipids are substances more characterized by their physical properties than by their chemical structure. Of the several main groups of substances normally classified as lipids, the best known are the edible *fats and oils*. These are of great importance as sources of energy for the

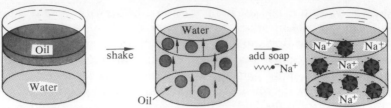

Neutral lipid + water. In the presence of soap, the charges on the carboxyl groups cause electrostatic repulsion of droplets which are tending to 'cream' owing to their lower density. In last diagram soap molecules not to scale

(a) Stabilization of lipid emulsion by soap

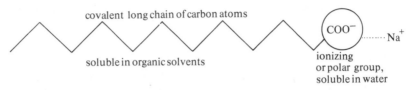

(b) An ampholytic molecule

Fig. 1.3   Emulsions of lipids in water

cell, but probably they are only rarely major components of lipoprotein membranes.   Of far greater importance in this structural sense are the *phospholipids*, even though much smaller total quantities of them are present in the cell.   Many other lipids are structurally quite unrelated to the fats and phospholipids, important examples being the *steroids* and the *carotenoids*.

All compounds classified as lipids share the physical characteristic of being soluble in organic solvents such as ether and hexane, but possess a low solubility in water.   For these reasons, they can be extracted from dried organic matter by dissolving them in petroleum 'ether' (a mixture of low boiling-point hydrocarbons) using a continuous extraction technique.   Such extracted compounds are essentially 'hydrophobic'— even though some of them do contain polar or even ionizing centres within their molecules.   They contrast in physico-chemical behaviour with the majority of highly polar or ionized substances such as low molecular weight acids, polyhydric alcohols, and most inorganic salts, which are 'hydrophilic' in character.   'Like tends to dissolve like' summarizes much of what is known of the solubility characteristics of the majority of solutes and solvents.

The edible *fats and oils* must not be confused with the hydrocarbon oils.   They are *esters*, formed when carboxylic acids unite with the trihydric alcohol, glycerol (Fig. 1.4a). Each of the three hydroxyl groups of the glycerol can form an ester by reaction with the carboxyl group of an acid, and, although mono- and diglycerides occur, most of the fats and oils present as major food reserves are, in fact, triglycerides. The three hydroxyl groups of glycerol are frequently esterified by union with different fatty acids to give mixed triglycerides such as that shown in Fig. 1.4d.

Two aspects of these glycerides are worthy of note. While short-chain fatty acids are occasionally present in one or more of the three possible

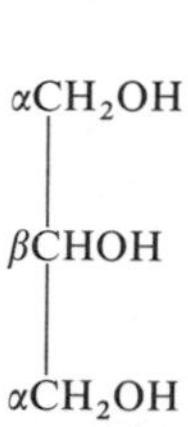

αCH_2OH

$\beta CHOH$

αCH_2OH

(a) *Glycerol*

$$CH_2-O-\overset{\overset{O}{\|}}{C}-(CH_2)_{16}-CH_3$$

$$CH-O-\overset{\overset{O}{\|}}{C}-(CH_2)_{16}-CH_3$$

$$CH_2-O-\overset{\overset{O}{\|}}{C}-(CH_2)_{16}-CH_3$$

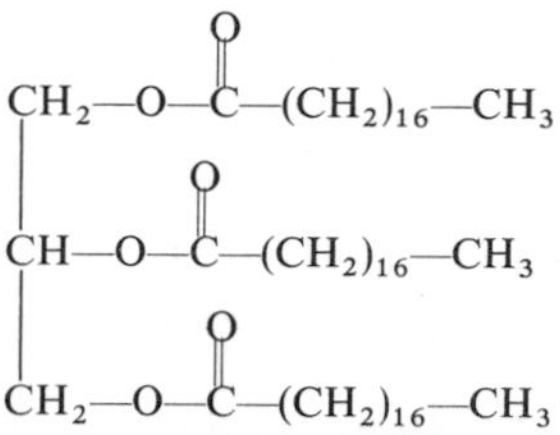

(b) *A Triglyceride.* In this example, *tristearin*, each hydroxyl group of glycerol is esterified by stearic acid.

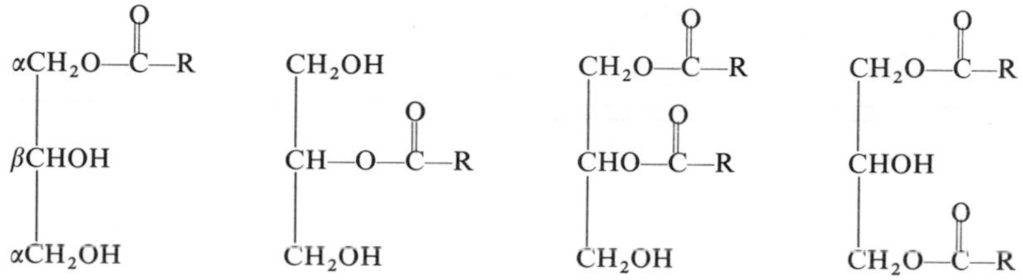

$\alpha CH_2O-\overset{\overset{O}{\|}}{C}-R$

$\beta CHOH$

αCH_2OH

α-monoglyceride

CH_2OH

$CH-O-\overset{\overset{O}{\|}}{C}-R$

CH_2OH

β-monoglyceride

$CH_2O-\overset{\overset{O}{\|}}{C}-R$

$CHO-\overset{\overset{O}{\|}}{C}-R$

CH_2OH

α,β-diglyceride

$CH_2O-\overset{\overset{O}{\|}}{C}-R$

$CHOH$

$CH_2O-\overset{\overset{O}{\|}}{C}-R$

α,α-diglyceride

(c) *Mono- and diglycerides*

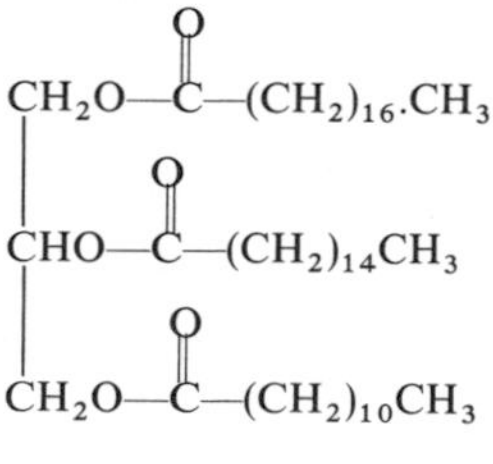

$$CH_2O-\overset{\overset{O}{\|}}{C}-(CH_2)_{16}.CH_3$$

$$CHO-\overset{\overset{O}{\|}}{C}-(CH_2)_{14}CH_3$$

$$CH_2O-\overset{\overset{O}{\|}}{C}-(CH_2)_{10}CH_3$$

(d) *A 'mixed' triglyceride.* In this example glycerol is esterified with stearic, palmitic, and dodecanoic acids

Fig. 1.4 Structure of glycerol and of glycerides

The cellular basis of life

positions (e.g., butyric acid is present in butter fat), the acids present are typically long-chain acids; it is their long, zig-zag paraffin-like chain composed of covalently-linked carbon atoms which gives the edible fats and oils their predominantly hydrophobic, lipid-soluble characteristics. Furthermore, nearly all the fatty acids occurring in cells, whether present there in the form of esters or as uncombined acids, contain an even *total* number of carbon atoms. It will be shown in chapter 8 that this is an inevitable consequence of the way their biosynthesis is achieved by the combination of acetic acid units.

Rather more than twenty 'fatty acids' occur in various fats and oils, some common examples being given in Table 1.3. The term 'fatty acid' is used here in a biological rather than a strictly chemical sense, for by no means all the acids present in edible fats and oils are alkanoic acids of the series $C_nH_{2n+1}COOH$. Some, for example, contain one or more double bonds, and several more complex acids are also known. Amongst

Table 1.3   Some important fatty acids present in cell lipids

A. Saturated fatty acids

$H_3C.(CH_2)_2.COOH$	Butyric acid
$H_3C.(CH_2)_4.COOH$	Caproic acid
$H_3C.(CH_2)_6.COOH$	Octanoic acid
$H_3C.(CH_2)_8.COOH$	Decanoic acid
$H_3C.(CH_2)_{10}.COOH$	Dodecanoic acid
$H_3C.(CH_2)_{12}.COOH$	Tetradecanoic acid
$H_3C.(CH_2)_{14}.COOH$	Palmitic (or, hexadecanoic) acid
$H_3C.(CH_2)_{16}.COOH$	Stearic (or, octadecanoic) acid
$H_3C.(CH_2)_{18}.COOH$	Arachidic acid
$H_3C.(CH_2)_{24}.COOH$	Cerotic acid

B. Unsaturated fatty acids

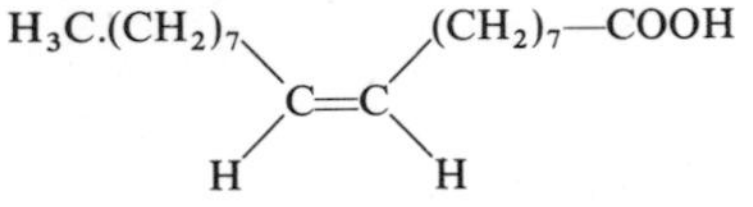

$$H_3C.(CH_2)_7 - \underset{H}{C} = \underset{H}{C} - (CH_2)_7 - COOH \qquad \text{Oleic acid}$$

$$H_3C.(CH_2)_4 - \underset{H}{C} = \underset{H}{C} - CH_2 - \underset{H}{C} = \underset{H}{C} - (CH_2)_7 - COOH \qquad \text{Linoleic acid}$$

$$H_3C.(CH_2)_4 - \underset{H}{C} = \underset{H}{C} - CH_2 - \underset{H}{C} = \underset{H}{C} - CH_2 - \underset{H}{C} = \underset{H}{C} - CH_2 - \underset{H}{C} = \underset{H}{C} - (CH_2)_3COOH \qquad \text{Arachidonic acid}$$

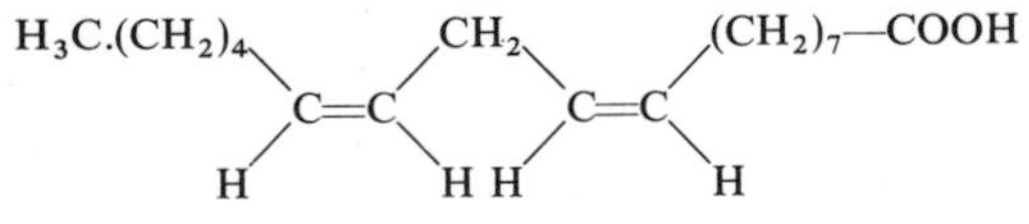

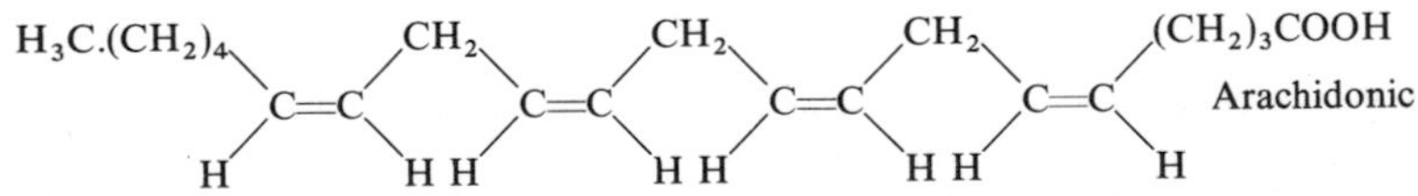

the commonest of the acids present in such materials as mutton fat, olive oil, and palm oil are *palmitic* acid, *stearic* acid, and the unsaturated *oleic* acid. Tristearin (Fig. 1.4b) is a typical example of an 'un-mixed' triglyceride, produced when each hydroxyl group of the glycerol forms an ester linkage with the same fatty acid.

Glycerides are hydrolysed or *saponified* commercially by boiling them with a solution of sodium or potassium hydroxide, the products being glycerol and soap, (the salts of the fatty acids). The enzyme *lipase* is an organic catalyst which brings about a rather similar hydrolysis *in vivo*, the hydrolysis being a preliminary step leading to the eventual liberation of a large amount of energy locked up within the glyceride molecules. Fats and oils are therefore valuable stores of energy and they are deposited particularly abundantly in the adipose tissues of animals and within the seeds of certain plants. However, almost all cells contain droplets of oil emulsified in the cytoplasm; such droplets stain red with the lipid-soluble dye, Sudan III.

Phospholipids are chemically related to the fats and oils, but their physical properties are in many ways very different and they fulfil a quite different function within cells. If one of the α-hydroxyl groups of glycerol is esterified with phosphoric acid instead of with a carboxylic acid, the phospho-diglyceride so formed is called *phosphatidic acid*. Although not formed in precisely this way within the cell, phosphatidic acid can be regarded as the precursor of most of the important phospholipids (Fig. 1.5a). It is of interest that of the two possible stereoisomers the L-form greatly predominates in living organisms. It is also noteworthy that the carboxylic acid residue attached at the β-position is frequently derived from an unsaturated acid.

The three main types of phospholipids in cells have a nitrogen-containing base linked to the phosphate group of L-α-phosphatidic acid. As Fig. 1.5b shows, all three bases commonly present in phospholipids are not only nitrogen compounds but also contain a hydroxyl group, and, in each case, it is the hydroxyl group which unites with the phosphoric acid residue. *Lecithin* is the best known of the many phospholipids in the cell; it has the base *choline* united to the phosphoric acid moiety of phosphatidic acid and it is therefore also known as phosphatidyl choline. The other two common bases, *ethanolamine* and the amino acid L-*serine*, give rise respectively to phosphatidyl ethanolamine and phosphatidyl serine.

An important characteristic of phospholipids in which they differ from ordinary triglycerides is that the phosphate moiety and (when present)

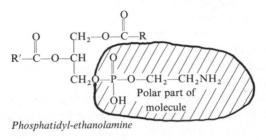

$$R'\!-\!\overset{\overset{\displaystyle O}{\|}}{C}\!-\!O\!-\!^\beta C.H \qquad ^\alpha CH_2\!-\!O\!-\!\overset{\overset{\displaystyle O}{\|}}{C}\!-\!R$$

$R'\!-\!C\!-\!O\!-\!\beta C.H$

↑
often an
unsaturated
fatty acid

$^\alpha CH_2\!-\!O\!-\!\overset{\displaystyle O}{P}\!-\!OH$

$\underbrace{\hspace{2cm}}_{\text{phosphate; polar part of molecule}}$
OH

(a) L-α-*phosphatidic acid*

(b) *Nitrogen bases present in phospholipids*

$HO\!-\!CH_2\!-\!CH_2\!-\!NH_2$
ethanolamine

$HO\!-\!CH_2\!-\!CH_2\!-\!\overset{+}{N}(CH_3)_3$
choline

$R'\!-\!\overset{\overset{\displaystyle O}{\|}}{C}\!-\!O\!-\!CH \qquad CH_2\!-\!O\!-\!\overset{\overset{\displaystyle O}{\|}}{C}\!-\!R$

$CH_2O\!-\!\overset{\overset{\displaystyle O}{\|}}{P}\!-\!O\!-\!CH_2\!-\!CH_2NH_2$
Polar part of
OH   molecule

Phosphatidyl-ethanolamine

$R'\!-\!\overset{\overset{\displaystyle O}{\|}}{C}\!-\!O\!-\!CH \qquad CH_2\!-\!O\!-\!\overset{\overset{\displaystyle O}{\|}}{C}\!-\!R$

$CH_2O\!-\!\overset{\overset{\displaystyle O}{\|}}{P}\!-\!O\!-\!CH_2\!-\!CH_2\!-\!\overset{+}{N}(CH_3)_3$
OH

Phosphatidyl-choline (= *lecithin*)

$R'\!-\!\overset{\overset{\displaystyle O}{\|}}{C}\!-\!O\!-\!CH \qquad CH_2O\!-\!\overset{\overset{\displaystyle O}{\|}}{C}\!-\!R$

$CH_2O\!-\!\overset{\overset{\displaystyle O}{\|}}{P}\!-\!O\!-\!CH_2\!-\!CHNH_2$
OH   COOH

Phosphatidyl-serine

(c) *Three common phospholipids derived from phosphatidic acid*

Fig. 1.5 Formulae of some phospholipids and their nitrogen bases (unionized forms except for choline) (R, R': long covalent carbon chains, non-polar)

the nitrogen bases are *ionized* and hence carry electrical charges. Consequently, whereas one part of the molecule is non-polar and has the solubility characteristics of paraffin-like compounds, the other part of the molecule is attracted to polar solvents such as water. In other words, one section of the molecule is hydrophobic and another is hydrophilic. As will be seen in the next section, it is this attribute of phospholipid molecules which, more than any other, renders them important components of lipoprotein membranes.

Lipid solvents extract a variety of other substances from dried cellular material. Amongst these are certain vitamins (e.g., vitamins A and D), various plant pigments and the steroids. However, from the point of view of the principal purpose of this chapter—a consideration of the structure of typical cell membranes—only the steroids need an introduction at this stage.

Cholesterol (Fig. 1.6a) is a typical example of a steroid, and it occurs in considerable quantities in most cells. At first sight its structure may look rather forbidding, yet those aspects of its chemistry which are relevant to lipoprotein structure and function are relatively simple. It consists essentially of four rings of carbon atoms linked together. Of these, three are six-membered rings, and one is a five-membered ring. Cholesterol contains one double bond, though, like other steroids, it is essentially a saturated compound and the molecule is, in consequence, not aromatic in character. Most steroids have methyl groups attached at positions 10 and 13, and either a keto- or a hydroxyl group at position 3. The cholesterol present in cell membranes is often esterified at the hydroxyl group in position 3. Cholesterol itself has a long side chain at position 17, but many other steroids have shorter substituents and in some a side chain is absent.

Several of the hormones present in higher animals are steroids. Cortisol (Fig. 1.6b), for example, is secreted by the mammalian adrenal cortex. Other examples are the male sex hormone, testosterone, and the female hormone, oestrogen. Certain important poisons, including digitalis and ouabain (Fig. 1.6c) are also based on the steroid structure.

Bile, a secretion of the liver, contains acids which possess the sterol nucleus. An example is cholic acid (Fig. 1.6d). These steroid acids have a carboxyl group at the end of the side chain at position 17. This group is usually conjugated with the amino acid, glycine, or with the sulphonic acid, taurine, to form glycocholic acid and taurocholic acid respectively. The bile acids play an important role in vertebrate digestion, for they assist in the emulsification of lipids in the small intestine.

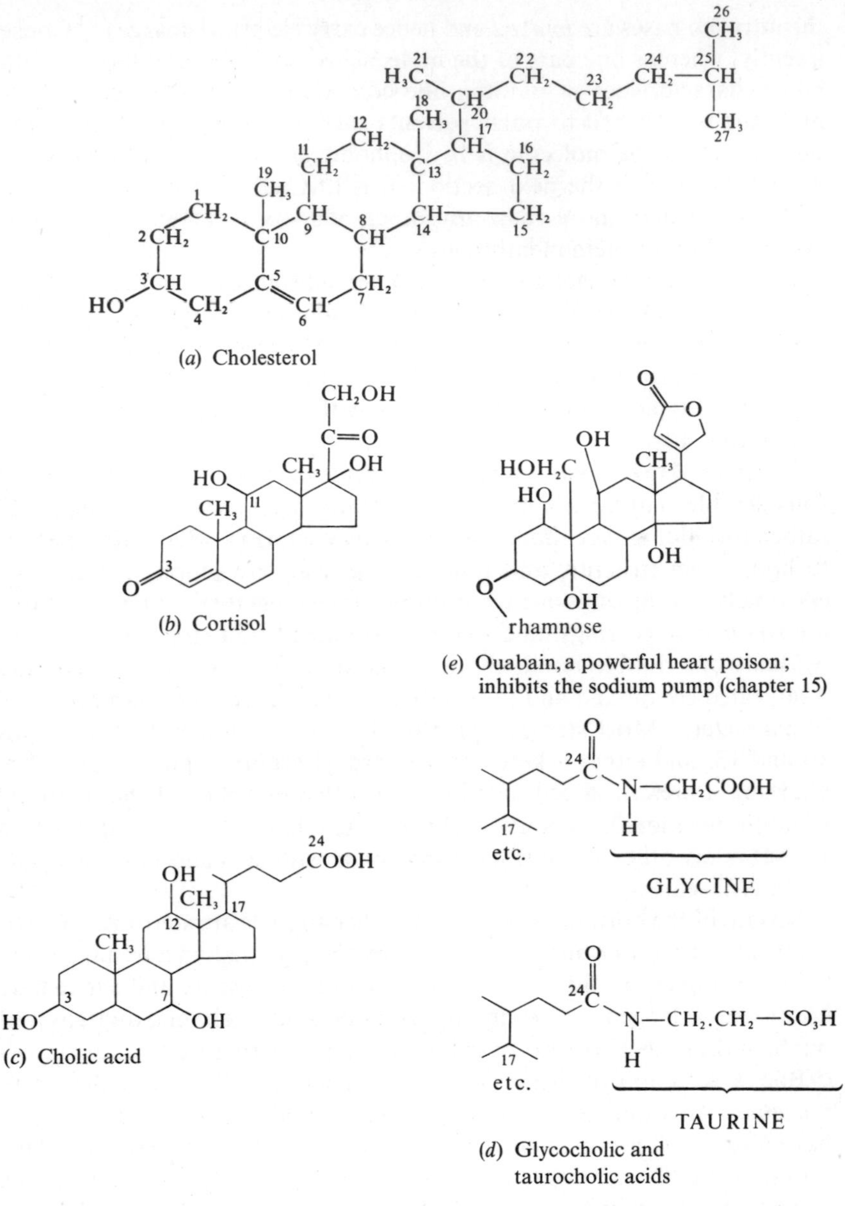

(a) Cholesterol

(b) Cortisol

rhamnose

(e) Ouabain, a powerful heart poison; inhibits the sodium pump (chapter 15)

(c) Cholic acid

GLYCINE

TAURINE

(d) Glycocholic and taurocholic acids

Fig. 1.6 Structure of cholesterol and other representative steriods. (The carboxyl group of cholic acid is usually conjugated to glycine or to taurine as shown in (d))

1.6   Structure of lipoprotein membranes

As was seen in the previous section, when a phospholipid such as lecithin is shaken with water, the non-polar parts of the molecules aggregate together but the charged parts are attached to the polar water molecules. In consequence, a configuration readily adopted by phospholipid molecules in contact with water is one comprising two layers with the charged groups directed outwards and the hydrocarbon chains remaining in close contact in the inner part of the membrane (Fig. 1.7a).

In cells, however, the situation is invariably complicated by the simultaneous presence of protein.   In the lipoprotein membranes of cells, it is found that the polar groups of the lipids are not in direct contact with the water, but, instead, the lipid sandwich is coated, probably automatically, with a thin coating of protein.   It would appear that the charged side-groups of some amino acid residues in proteins arrange themselves in close proximity to the charged groups of the phospholipids, thus interpolating longitudinally-arranged protein molecules between the lipid and the aqueous environment.   The result is a lipoprotein double layer: protein-lipid-lipid-protein.   Further evidence suggests that the protein on one side of the lipoprotein double layer, or 'membrane', may be different in character from that on the other.   At least in some membranes, the protein on one side of the double layer possibly has polysaccharide associated with it.

Electron micrographs of many cell membranes show two electron-dense layers of about 2 mμ thickness.   These border a less dense region which is approximately 3·5 mμ thick.   The outer layers are thought to be the protein film, the inner region of low electron density being that occupied by lipids.   Calculations of the molecular dimensions of lipids and proteins confirm that this suggested structure could well fit the electron micrograph measurements.

Figure 1.8 suggests, in a very diagrammatic way, a possible structure for a typical lipoprotein membrane of a cell.   This is similar to a model first proposed by Danielli and Davson.   A membrane should not be regarded as a static structure but as a dynamic system whose constituents are constantly undergoing slight rearrangement.   The protein and lipid components probably differ at various parts of the surface—in other words, although the transverse sections of the membrane may look superficially identical, the three-dimensional structure may best be regarded as a mosaic built up of separate but intimately-linked lipoprotein components.   Some workers have gone further, postulating that all the

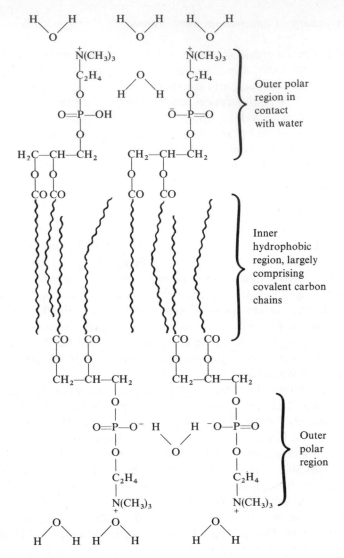

Fig. 1.7 Lipid associations
((a) Possible arrangement of lecithin film in water)

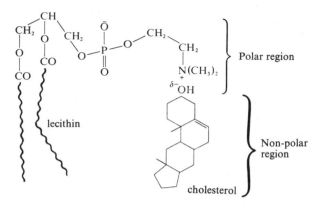

((b) Association between cholesterol and lecithin)

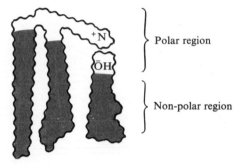

((c) Simplified representation of lecithin-cholesterol complex as used in Fig. 1.8)

lipoprotein membranes of the cell are constructed from separate 'blocks', or repeating units, of molecular weight between 10^5 and 10^6 which are united together in a planar fashion.

It is not known in what way variation in membrane structure leads to dissimilarities in membrane characteristics in different cells and, very probably, in different parts of one particular membrane in a single cell. Even less can we conceive how the nucleus 'arranges' for a particular protein to be present on a particular cell membrane for although the plasmalemma, endoplasmic reticulum, and mitochondria all have basically similar membranes, nevertheless each has unique biochemical attributes. Similarly, considerable variation appears to exist in the detailed structure of the lipid layers occupying the centre of different membranes. Cholesterol is very frequently present and one consequence of its presence

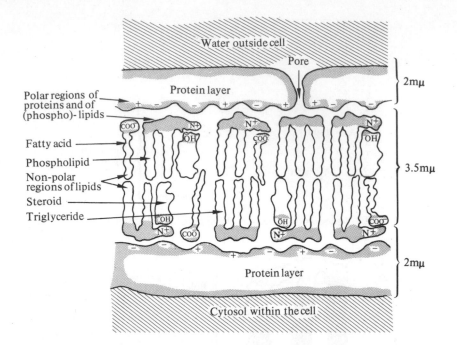

Fig. 1.8 A model of the lipoprotein membrane of cells

is that it tends to make molecules of phospholipid pack more closely together. It has been suggested that the cholesterol molecules arrange themselves amongst the non-polar carbon chains of the fatty acids of the phospholipids, the hydroxyl group of the cholesterol being associated with the charged nitrogen atom of the phospholipid (Fig. 1.7b,c). In different lipoprotein membranes the molecular proportions of different kinds of lipids vary considerably. In the plasma membrane of the intestinal mucosa, for example, phospholipid and cholesterol molecules are present in approximately equimolecular proportions, while, in liver mitochondria and endoplasmic reticulum, phospholipid molecules outnumber those of cholesterol by eight to one. Presumably, the three-pronged unit (Fig. 1.7c) normally formed by the association of one molecule of phospholipid and one of a sterol, is capable of occasional replacement by a single triple-pronged molecule of a triglyceride.

1.7 The plasmalemma, cell walls, cilia, and flagella

The boundary of the active cytoplasm in cells is known as the plasma-lemma. In plants, and occasionally in other organisms, this is surrounded in turn by a much more conspicuous cell wall. The existence of a specific surface membrane had been suspected long before the electron microscope was invented because of the characteristics of the cell boundary. For example, it was inferred that the plasmalemma was likely to contain lipid because many non-polar compounds, e.g., chloroform and higher alcohols, are able to enter erythrocytes (red blood cells) with ease. Such compounds probably gain access to cells by virtue of their solubility in the double layer of lipid molecules in the lipoprotein membrane (Fig. 1.8). The cell surface has, furthermore, a high electrical resistance, a fact which is compatible with the existence of a non-polar region in the membrane. On the other hand, the surface tension between a cell and its surroundings is low, an observation consistent with the presence of a superficial layer of a polar protein but inconsistent with lipid being in direct contact with the water.

Electron micrographs of material from a very wide range of cells show the plasmalemma to consist of a typical double layer of electron dense material surrounding a less dense (and probably lipid) inner layer. Chemical analysis of red cell ghosts (mammalian erythrocytes from which the cell contents have been removed by lysis to leave only the outer cell surface) reveals a ratio of about 1 part by weight of lipid to 1·7 parts by weight of protein in this particular membrane.

It is probable that the plasmalemma is typically a continuous structure around the cell with very few, if any, pores completely penetrating it. Needless to say, the plasmalemma is a structure of great importance to the cell since substances entering or leaving the cell must pass through it. The membrane is *semi-permeable* for while water can move into, or out of, the cell, solutes usually have difficulty in passing through it unless highly selective transport mechanisms exist to facilitate their movement. The importance of selective permeability and some of the problems it raises, are considered in chapter 15, but an example will illustrate the type of phenomenon which is frequently encountered. The vertebrate intestinal mucosal cells have a great ability to absorb D-glucose, yet the closely-related D-mannose is only slowly absorbed. In this case, a specific protein probably acts as a *carrier* which facilitates the uptake of glucose. Whether proteins of this type are actually the structural protein of the membrane or some other more mobile protein is still uncertain.

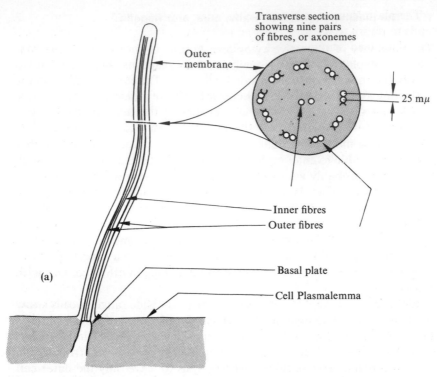

Transverse section showing nine pairs of fibres, or axonemes

Outer membrane

25 mμ

Inner fibres

Outer fibres

Basal plate

Cell Plasmalemma

(a)

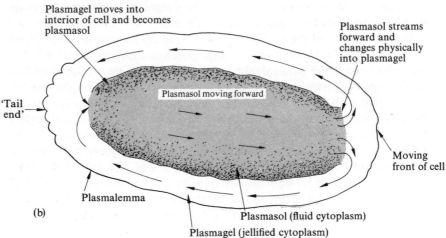

Plasmagel moves into interior of cell and becomes plasmasol

Plasmasol streams forward and changes physically into plasmagel

Plasmasol moving forward

'Tail end'

Moving front of cell

Plasmalemma

(b)

Plasmasol (fluid cytoplasm)

Plasmagel (jellified cytoplasm)

Fig. 1.9 (a) Structure of a typical cilium
             (b) Streaming of cytoplasm in amoeboid movement

The plasmalemma is a delicate membrane which, when it is distorted, tends to return to its normal shape by elastic recoil. If it is slightly torn, the edges tend automatically to flow together, probably reflecting the tendency for phospholipids to aggregate in ordered layers. However, the mechanical strength of cells is limited, and they are frequently strengthened by external materials; plant cells, in particular, are characterized by the presence of a cell wall. Cells of green plants secrete fibrous polysaccharides such as cellulose while the cell wall of fungi is frequently composed of chitin (chapter 3). Other substances such as lignin and pectin are often also present. Usually the cell wall is freely permeable to water and solutes but impregnation with lignins or waxes may render the walls impermeable causing the death of the cell. Sometimes specific areas are left unlignified and these *pits* serve to allow free passage of materials. In the xylem (wood) of most higher plants the cells are devoid of cell contents and only the wall remains, thus providing a mechanism for the effective conduction of water and ions from roots to the leaves.

In some types of cell present in organisms from almost every plant and animal phylum, the plasmalemma and certain underlying structures are modified to form projections, known as flagella and cilia, which, by waving or undulating, are often able to induce the locomotion of cells. This, at least, is their primitive function; their presence in certain protozoa and algae is well known, but it must not be overlooked that spermatozoa of higher animals also possess flagella. Furthermore, even when tissue structure precludes active movement of cells, ciliated cells are frequently to be found, and then they serve some more advanced function. They are necessary for the feeding process of many filter-feeders, including shellfish of many kinds. In mammals, they line the bronchioles and waft mucus out of the lungs. This is a cleansing mechanism, for the mucus carries with it particles of dust which have managed to enter; it is of interest that one of the adverse effects of nicotine is that it paralyses these cilia.

Electron microscopy has shown that no fundamental difference exists between a flagellum and a cilium. Each is surrounded by a membrane which is continuous with the plasmalemma of the main part of the cell. Twenty fibres, or *axonemes*, always arranged in a similar fashion no matter what the origin of the flagellum, run along the length of the structure. In cross section, it is found that a central pair of axonemes are surrounded by nine outer pairs. Each fibre is about 25 mμ in diameter, the electron-dense area being on the outside (Fig. 1.9a).

1.8 The cytosol and cell vacuoles

The cytosol is the aqueous medium which lies between the intracellular organelles. In many cells it contains suspended particles of carbohydrate storage materials such as glycogen in animal cells, and starch in plants. Several proteins are also present in the cytosol as colloidal solutions; amongst these proteins are the enzymes which catabolize glycogen and starch to form pyruvate and others associated with certain other metabolic systems listed in Table 1.1.

Every cell organelle is highly dependent on the activity of each of the others. For example, the cytosol provides the mitochondria with pyruvate. The mitochondria then use energy liberated when pyruvate is degraded to build up part of the molecule of adenosine triphosphate (ATP), an important substance to which constant reference will be made in later chapters. The ATP travels through the cytosol to other cellular organelles and so provides them with the *energy* they need if they are to fulfil their several and varied functions. For example, ATP provides the energy for the active uptake of many substances through the plasmalemma, and is also the ultimate energy source which motivates cilia and flagella. Anabolic activities, including the vital process of protein synthesis which occurs at the ribosomes, also require energy in the form of ATP.

The streaming of cytoplasm within cells is a very common feature in animals and plants. It may serve to transport large, and therefore relatively indiffusible particles, around the cell. Protoplasmic streaming occurs particularly obviously in amoeboid cells (e.g., amoeba, lymphocytes) which move by a streaming of the cytoplasm (Fig. 1.9b). It has been suggested that the process involves the modification of the molecules of protein from an extended to a folded shape, ATP somehow providing the energy. Phagocytosis, an ingestion process observed in amoeba and in many cells of higher animals, is probably brought about in a similar manner.

The cytosol of many plant cells contains vacuoles bounded by a membrane, the *tonoplast*. This membrane probably has a similar structure to the plasmalemma and may, indeed, arise from it by a process of invagination. Within the vacuoles, plant cells store water, ions, and food materials such as soluble sugars. They may possibly also act as receptacles for harmful cellular waste products. The tonoplast, like the plasmalemma, is probably selectively permeable, but it seems to possess a more developed mechanism for the uptake and retention of certain ions.

1.9 The nucleus

All active plant and animal cells contain a nucleus which is surrounded by a membrane whose dimensions and properties strongly suggest that it has a lipoprotein structure closely resembling that described in section 1.6, although some electron micrographs of the nuclear membrane suggest that it may be penetrated by pores of some 5–10 mμ diameter. If these pores really exist in the intact cell, they may allow large molecules such as RNA (ribonucleic acid) to move from the nucleus into the cytosol.

Within the nuclear membrane is a jelly-like fluid (a gel) composed largely of water.   The non-aqueous contents comprise some 70 % protein, 10 % DNA (deoxyribonucleic acid) as well as lipid and RNA.   The chromosomes are not visible in non-dividing cells but, nevertheless, the essential part of their chemical structure undoubtedly persists during the resting phase.   From a genetic point of view, the most important nuclear material is the DNA present in the chromosomes, for it is a carrier of information (chapter 14), the molecular structure of DNA providing a 'recipe' for the proteins produced by the cell.   This is one reason why the nucleus can be regarded as the control centre of the cell.   Almost all the chemical reactions the cell performs are catalysed by enzymes (all of which are proteins or polypeptides) and their structure is indirectly decided by the precise composition of the DNA in the nucleus.   The nucleus, therefore, plays a vital, if indirect, role in all the biochemical processes which go on in the cell.

The DNA passes its information about prospective protein structure to the ribosomes, the principal site of protein synthesis (some proteins are also made in the nucleus).   Since DNA itself does not leave the chromosomes, the message is carried by an intermediary, messenger RNA.   Both types of *nucleic acid*, RNA and DNA, are composed of *nucleotide bases* united with the sugar, *ribose* (in *ribo*nucleic acid, RNA), or the sugar-like *deoxyribose* (in *deoxyribo*nucleic acid, DNA).   In addition, both nucleic acids contain phosphate groups which, at physiological pH, are negatively charged.   The negative charges on these phosphate residues are attracted to the positive charges on certain basic proteins called histones, which are present in the nucleus; the resulting rather loosely-bound complex is called a *nucleoprotein*.

The nucleus contains a number of enzymes absent from the rest of the cell, or which are in short supply there.   In particular, those enzymes concerned with the synthesis of nucleotide bases and of a substance termed NAD (nicotinamide adenine dinucleotide) are located in the nucleus. Numerous oxidation and reduction reactions throughout the cell can only occur if the enzyme responsible for the reaction is itself accompanied by NAD and this substance is therefore known as a *coenzyme*. *Oxidation* reactions—the removal of electrons from compounds—play a particularly vital role in cellular chemistry, the NAD being important because the electrons removed from the substance being oxidized are frequently passed to NAD. The significance of this process will be explained in later chapters; for the present, it is important to recognize that, since NAD is largely synthesized in the nucleus, many oxidation and reduction reactions throughout the cell are dependent upon the nucleus for this essential coenzyme.

The nucleus again illustrates the interdependence among the cell organelles. The nucleus supplies the rest of the cell with NAD, with certain other coenzymes, and with messenger RNA. In return, the nucleus receives numerous metabolites from the cytoplasm, including sugars and amino acids; it also receives, in the form of ATP, the energy which is absolutely essential if it is to perform its principal activities, almost all of which consume energy.

1.10 Mitochondria

The shape, size, and number of these bodies vary greatly in different cells. They may be spherical or elongated in form, their shape varying with their metabolic activity. There may be only a few per cell, or, as in an active liver cell, there may be up to a thousand. Typically, they are just visible under the light microscope, but their elaborate membranous substructure was only revealed by electron micrographs. Each mitochondrion is constructed from two lipoprotein membranes (each a more or less typical protein-lipid-lipid-protein double layer) the inner one of which is convoluted to form folds or cristae which project into a central aqueous cavity (Fig. 1.10). The outer membrane is not convoluted and forms a skin-like limiting layer to the mitochondrion. Between the outer and inner membranes is a second mitochondrial cavity, which, like the central cavity, appears transparent in electron micrographs and is probably aqueous.

The outer membrane of the mitochondrion is structurally and physiologically very different from the inner membrane. Unlike the inner membrane, which is only permeable to relatively small molecules, the outer membrane is permeable to molecules with molecular weights of up to about 10,000, and many of the enzymes associated with it take part in the synthesis of phospholipids. In both these respects, the outer membrane closely resembles the membranes of the endoplasmic reticulum (section 1.12). The space between the outer and inner membranes appears to contain few enzymes.

Under aerobic conditions, pyruvate produced in the cytosol by the catabolism of carbohydrates finds its way into the central cavity of the mitochondrion, where it is changed to an important complex containing acetate and termed *acetyl coenzyme A*. Similarly, fatty acids, liberated from fats by hydrolysis in the cytosol, penetrate the outer mitochondrial membrane and are assisted across the inner membrane probably by union with a carrier; enzymes present in the central cavity subsequently metabolize these acids, again yielding acetate in the form of acetyl coenzyme A.

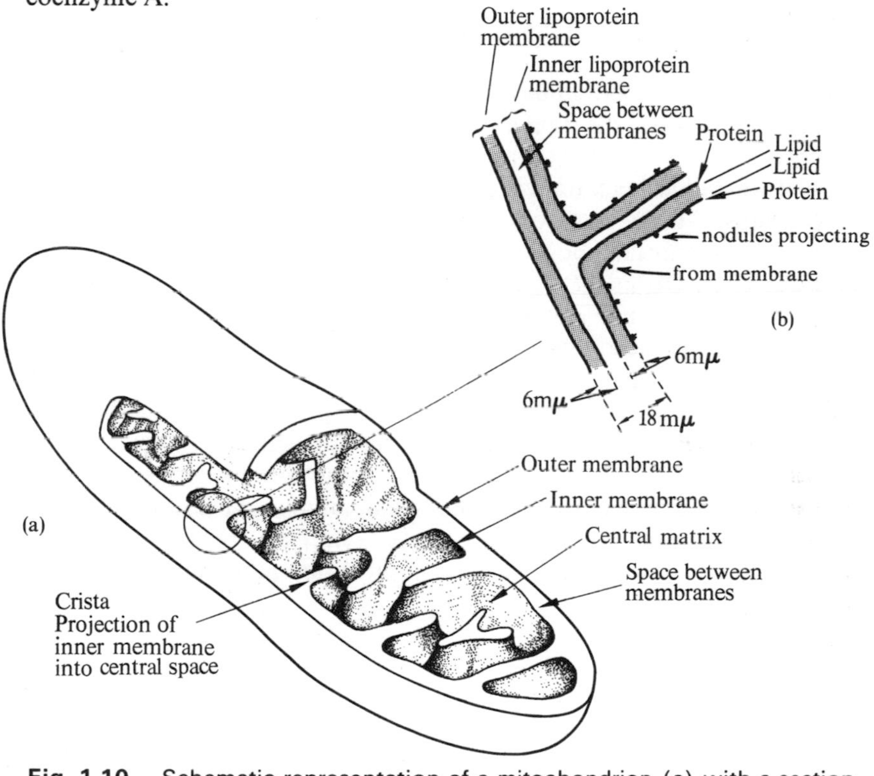

Fig. 1.10   Schematic representation of a mitochondrion (a) with a section cut away to reveal the interior and (b) enlarged to show detail of inner and outer membranes.

A set of enzymes present for the most part in aqueous solution in the central cavity (though at least one of them is firmly bound to the inner membrane), oxidizes acetyl coenzyme A to carbon dioxide and water. These enzymes bring about a series of reactions known collectively as the Tricarboxylic Acid Cycle (TCA cycle, chapter 7). This cycle is important because it results in the reduction of the electron carrier NAD, which was mentioned in section 1.9. The electrons which reduce NAD originate indirectly from the acetyl part of acetyl coenzyme A and hence, even more indirectly, from carbohydrates and fats. Reduced NAD formed in the mitochondrion is, in a sense, the cell's major driving force, for it delivers electrons to a special series of electron carriers situated in the inner membrane. The movement of the electrons along these carriers generates ATP by reactions known as *oxidative phosphorylation* (chapter 6). Since most of the cell's ATP is produced in this way, the mitochondrion has been appropriately named the 'power house' of the cell.

The inner mitochondrial membrane is an excellent example of a highly organized structure, the design of which specially fits it for a particular function. It is held together by a system of phospholipids and structural protein which has been found to contain amino acids rich in non-polar side groups. These are weakly bound to the non-polar parts of the lipids. Four phospholipoprotein particles, known as electron transport particles I, II, III, and IV (chapter 6), are closely associated with one another in the inner mitochondrial membrane. Electrons initially derived from the acetyl part of acetyl coenzyme A pass systematically from one particle to another, so that the electrons 'fall' from a high energy level to a lower one; as they do so, *energy is tapped off and used to make ATP*. When the electrons finally reach particle IV, they combine with molecular oxygen and hydrogen ions to produce water.

Bacterial cells, unlike those of plants and animals, lack mitochondria. However, it is probable that they possess electron transport particles. In some cases, these are associated with the outer membrane of the cell.

1.11   Chloroplasts

The chloroplasts are the seat of the most fundamental of all biochemical reactions for it is here that the machinery exists for trapping light energy and transforming it into chemical energy. In part, this chemical energy is in the form of ATP, but reducing power is also created which eventually reduces carbon dioxide to the level of sugar. These, and substances derived from them, act in the first instance as fuels for driving the life processes of plants, but it must not be overlooked that in the end all animals depend upon plants for their existence.

In the green cells of plants, chloroplasts are often the largest and most characteristic organelles. Sometimes there may be one or a few, but often there may be a hundred or more in one cell; like mitochondria, their position is not rigidly fixed, and they often migrate around the cell. Their shape and size differ in different species, especially in algae, but in higher plants they are typically between 0·5 and 2·0 μ thick and up to 10 μ diameter. On a dry weight basis, some 50% of the chloroplast is protein, 35% is lipid of various kinds, and 7% comprises pigments. The pigments (except in photosynthetic bacteria) almost invariable include chlorophyll **a**, but the nature and the amounts of accessory pigments differ considerably in organisms from different parts of the plant kingdom.

Chloroplasts, like most other subcellular particles, are composed of lipoprotein membranes. These are typically stacked together rather like a pile of coins, to provide chloroplastic sub-units termed *grana*. When light strikes the chloroplast, a series of reactions is initiated which leads to the production of ATP and of reducing power. The latter is, in fact, not reduced NAD, but reduced NAD phosphate (NADPH). ATP and NADPH in turn provide the energy which drives what must undoubtedly be regarded as the most important anabolic process in nature—the conversion of carbon dioxide to sugars. Furthermore, from the sugars and their derivatives, arising from photosynthesis, the carbon skeletons of all the other major components of the plant are ultimately synthesized.

1.12   The endoplasmic reticulum

The membranes of the reticulum, like other cell membranes, are probably typical lipoprotein double layers. In some cells, the nuclear membrane is connected to adjacent endoplasmic reticular membranes. Sometimes parts of the reticulum appear to be in the form of tubules, many of which are somewhat distended inside at what is presumed to be the internal lipid layer.

In certain areas of the cell, spherical particles of about 10 mμ diameter are attached to the membrane. These particles, the ribosomes, contain some phospholipid but are mainly protein and RNA. As mentioned above, messenger RNA passes from the nucleus via the cytosol to the ribosomes where the message is deciphered. With the co-operation of a number of other constituents of the cytoplasm, including the energy of ATP, amino acids are 'zipped' together in an order or sequence dictated by the structure of the messenger RNA and hence, ultimately, by the structure of the DNA in the nucleus.

Some regions of the endoplasmic reticulum lack ribosomes and in consequence appear to have a smooth surface (Fig. 1.11). It is here that

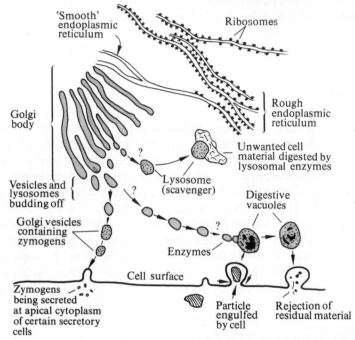

Fig. 1.11 A possible relationship among the rough and smooth endoplasmic reticulum and some associated bodies in various cells. (Much of the arrangement is still hypothetical)

the synthesis of cell lipids probably takes place—steroids, phospholipids, and fatty acids are built up from simpler precursors. Their synthesis probably takes place in a largely non-polar environment.

Besides its two anabolic functions (the synthesis of proteins and lipids) the endoplasmic reticulum is associated with the storage of enzymes in an inactive form. In some cells, and especially well developed in cells whose main function is secretion, there are structures called *Golgi bodies* (Fig. 1.11). These look in electron microscope pictures like pieces of smooth endoplasmic reticulum stacked together. Here inactive enzyme precursors known as *zymogens* accumulate in vacuoles whose walls possibly arise from the reticulum. In the pancreas, for example, the zymogens trypsinogen and chymotrypsinogen are stored in vesicles. The zymogens are modified after leaving the cell, becoming the active enzymes, trypsin, and chymotrypsin.

A second site of enzyme storage, particularly abundant in phagocytic cells (e.g., blood leucocytes) are vesicles known as *lysosomes*. These may form by the 'budding off' of pieces of endoplasmic reticulum and they contain several powerful enzymes capable of breaking down cellular components. These enzymes include RNA-ase and DNA-ase which degrade the nucleic acids; cathepsins, which are proteolytic enzymes; glycosidases which hydrolyse polysaccharides; and acid phosphatases, which hydrolyse organic phosphate esters. In phagocytic cells, foreign particles (e.g., bacteria) are probably engulfed by an invagination of the plasmalemma so that they are eventually held in a vacuole. The enzymes of the lysosome may then be discharged into this vacuole where the particle is destroyed by the action of the enzymes. Normally the enzymes in the lysosomes do not attack essential cell components, presumably because they are kept physically separated from them by the membrane of the vesicle. Sometimes, however, as when starvation occurs, the enzymes may be released in the cell in such a way as to hydrolyse some parts of it to provide food materials which enable the organism to survive longer.

In the last few years the role of the endoplasmic reticulum in attacking foreign molecules has received much attention. Whether the system responsible is also involved in some steps in the synthesis of various complex natural molecules is not clear. It is, however, known that certain types of foreign molecules, and especially molecules of lipid-soluble substances, can undergo oxidation, hydroxylation, and other types of metabolic alteration within the cells. The system responsible for these changes is largely associated with the 'smooth' endoplasmic reticulum. Needless to say, this protective mechanism of the cell is of fundamental importance, for it not only provides a mechanism for eliminating dangerous foreign substances which have entered the cell by accident, but it sometimes also attacks such foreign molecules as drugs and antibiotics which have been introduced into the cell for therapeutic purposes.

Proteins, peptides, and amino acids

2.1 The structural versatility of proteins

Proteins are of great interest and importance, for they not only play a
key role in forming and maintaining the *structure* of the cell, but, as
enzymes, many of them are indispensable catalysts of chemical reactions
upon which the life of the cell depends.

Every species of animal and plant possesses thousands of different
proteins and in many higher animals each individual has numerous
characteristic proteins not shared by any other—except perhaps by an
identical twin. Nevertheless, all proteins share a basic and relatively
simple construction, possessing one or more chains of *amino acids* linked
together in a particular order. Some twenty different amino acids can
be present in a protein and any one of them may occur many times.
However, for a particular individual protein the number of amino acids
present, and the order in which they are arranged is a definite characteristic
of the protein and is known as its *primary structure*. This contrasts
with some other large biological molecules, amongst them those of
starch and glycogen, where molecules differ slightly in their detailed
structure and size (p. 78).

When amino acids are linked together in chains as they are in peptides
and proteins, they are, of course, present in a chemically combined form,
each amino acid unit being linked to its neighbour by a peptide linkage
(p. 40). Each of these combined amino acid units is termed an amino
acid *residue*.

Proteins often have very high molecular weights—the muscle protein,
actomyosin, for instance, has a molecular weight of about 14 million.
Peptides are basically similar to proteins, but have smaller molecules,
the molecular weight seldom exceeding 5000. In consequence of their

large molecular weight, proteins form colloidal rather than true solutions. Thus the colloidal micelles# disperse light shone into the solution (the Tyndall effect) and they carry electrical charges. When subjected to very great centrifugal force in an ultracentrifuge, large colloidal particles sediment faster than smaller ones—this use of the ultracentrifuge is merely an extension of the technique mentioned on p. 8. Differential centrifugation can, in fact, be employed as a technique for the determination of molecular (or micellar) weights of proteins.

Another property of colloids# can be used to separate proteins from smaller molecules or ions. The technique, termed *dialysis*, can perhaps be thought of as a sieving procedure, and an example will best illustrate its use. Blood plasma is an aqueous fluid containing a number of dissolved proteins as well as many inorganic ions, and especially ions of sodium, potassium, chloride and bicarbonate. If plasma is placed in a cellulose membrane sac which in turn is suspended in a beaker of water (Fig. 2.1a), the salt ions diffuse through the membrane leaving behind the protein molecules which are too large to penetrate the small pores in the cellulose membrane.

Mixtures of proteins such as are present in plasma can be separated by *electrophoresis*. A simple example of this valuable technique employs a strip of filter paper of about 4 × 30 cm. A thin line of some 20 μl of plasma is carefully pipetted 10 cm from one end of the paper which is soaked in buffer (pH 8·5). The paper is supported by two glass plates (Fig. 2.1b) and each end dips into the buffer. Electrodes connected to a d.c. source produce a potential of 4 V per cm of paper length. This potential causes the charged plasma proteins to move along the paper, each protein being attracted to the oppositely charged electrode. After some 20 hours the paper is removed from the apparatus and the position taken up by the various proteins is located by staining the paper with a dye such as Amido Black. As the figure indicates, albumen moves the furthest, and the plasma globulins are separated into fractions known as α, β, and γ globulins.

Physical studies of proteins, including X-ray crystallography, have indicated that very frequently the chain of amino acids comprising their structure is folded in a definite pattern in space rather than having a

---

Appendix 2 contains references to books which explain terms marked by this sign. As explained in the Preface it has been necessary to assume that the majority of readers are familiar with certain background chemical terms, and to ask the remainder to spend a few minutes with the recommended book(s) before continuing.

Proteins, peptides, and amino acids

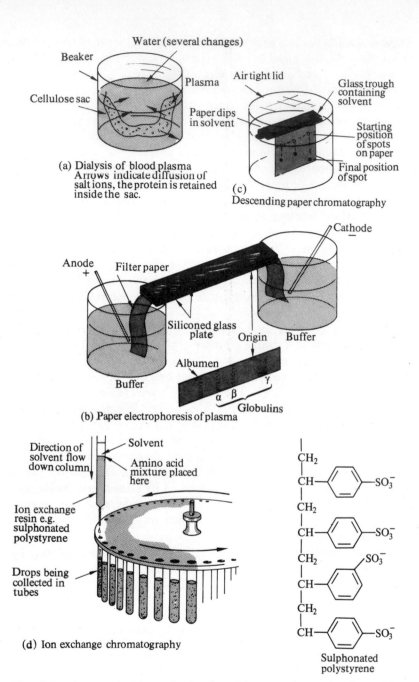

Fig. 2.1   Some valuable methods of studying proteins and amino acids

random disposition. The possibility of determining the three-dimensional shape of a protein molecule is of great interest because it promises to give biologists an insight into the fundamental relationship between sub-microscopic structure and the function of proteins in living cells. This is illustrated in its simplest form by the *fibrous proteins* present in feathers, hair, and horn, and which are collectively known as keratin. The molecules of keratin are long and threadlike, as are those of other fibrous proteins such as myosin, the muscle protein, and collagen, the protein of cartilage. The molecules of *globular proteins*, on the other hand, are more or less spherical, the peptide chains being wound around themselves in a complicated and characteristic fashion. Most enzymes are globular proteins, although myosin is a notable exception.

Proteins are often classified into two groups, the *simple proteins* which comprise amino acids only, and the *conjugated proteins*, the structure of which includes a component which is not an amino acid. Two types of conjugated proteins were mentioned in the previous chapter—*nucleoproteins*, which consist of protein linked to a nucleic acid, and *lipoproteins*, in which the protein is associated with lipid. *Glycoproteins* (mucoproteins) possess a carbohydrate as an integral part of their structure. Many enzymes are conjugated proteins and in such cases the non-amino acid moiety is often vital to the catalytic process; it is then called a prosthetic (helping) group.

2.2 Structure of α-amino acids and the effect of pH

The fundamental units from which proteins and peptides are built are carboxylic acids with an amino group attached to the carbon atom immediately adjacent to the carboxyl group. They are therefore known as α-amino acids. Alanine, for example, is derived theoretically from propionic acid:

$$H_3C-\overset{\alpha}{C}H_2-COOH \qquad\qquad H_3C-CH(NH_2)COOH$$
Propionic acid   Alanine

Each α-amino acid has the general formula $R.CH(NH_2)COOH$ where R can be one of about twenty possible side groups. These are discussed in section 2.4. For alanine R— is the methyl group.

Since typical α-amino acids have four different groups attached to the α-carbon atom, each exists in two stereoisomeric forms. The one exception is glycine, for which R = H. These are called Dextro (D) and Laevo (L) forms according to their configuration—the meaning of the symbols D and L is considered in relation to carbohydrates (p. 61). The two forms of alanine are written as follows; all that need be noted at this point is that by convention, L-alanine has the amino-group on the left when the carboxyl group is written at the top.

$$
\begin{array}{ccc}
\text{COOH} & \text{COOH} & \text{COOH} \\
| & | & | \\
H_2N—C—H & H—C—NH_2 & H—C—H \\
| & | & | \\
CH_3 & CH_3 & NH_2 \\
\text{L-Alanine} & \text{D-Alanine} & \text{Glycine (not optically active)}
\end{array}
$$

It can be shown easily with models that L and D forms are not super-imposable (and therefore are different) while at the same time they are related as an object to its mirror image. In all plants, animals and micro-organisms the α-amino acids found in proteins have the L-configuration.   D-amino acids do occur, however, in certain bacterial cell walls.

Many compounds encountered in relation to cell metabolism contain more than one sort of chemically reactive group.   The α-amino acids, for example, have an amino and a carboxyl group.   It is useful to remember that, in general, the various reactive groups present in such molecules can, in the first instance, be assumed to act independently of one another.   Thus carboxyl groups in amino acids for the most part react similarly to that of acetic acid, ionizing, forming esters and amides and so on.   Similarly, amino groups in amino acids can accept a proton $(—NH_2 \xrightarrow{H^+} —NH_3{}^+)$ or undergo acetylation.   As will be seen later, many amino acids contain other active centres; hydroxyl groups, like the hydroxyl groups in aliphatic alcohols, can form esters or ethers, while those attached to aromatic rings behave like those of phenol.   On the other hand, it must be remembered that this generalization is only a useful guide, and in more complex cases one reactive group can to some extent influence another, especially if they are situated near to one another in the molecule.

In aqueous solution, the α-amino acids are ionized to form what is sometimes called an *internal salt* or *Zwitterion*.

$$
\begin{array}{ccc}
R & R & R \\
| & | & | \\
H_2N—C—H \rightleftharpoons & H_2N—C—H & \rightleftharpoons H_3\overset{+}{N}—C—H \\
| & | & | \\
\text{COOH} & \text{COO}^- + H^+ & \text{COO}^- \\
\text{Amino acid,} & & \text{Internal salt} \\
\text{—covalent form} & &
\end{array}
$$

The carboxyl group loses a proton, becoming negatively charged, while the amino group acquires a positive charge by accepting a proton; the close proximity of the carboxyl group to the amino group enhances this ionization process and causes amino acids to have a high dipole moment which makes most of them very soluble in water. Their solubility also depends upon the nature of their side groups (R). Thus the α-amino acid, glutamic acid (—R = —CH$_2$CH$_2$COOH), is more soluble in water than valine (—R = —CH(CH$_3$)$_2$) because the former has an additional ionizing centre while valine has a non-polar and therefore hydrophobic side chain.

The properties of amino acids, and indeed of proteins too, are greatly influenced by the hydrogen ion concentration of the solution in which they find themselves. When in acid solution, they tend to take up hydrogen ions, becoming positively charged, but in alkaline solution they lose protons and by doing so they acquire a negative charge:

$$\overset{+}{H_3N}-\underset{\underset{COOH}{|}}{\overset{\overset{R}{|}}{C}}-H \underset{-H^+}{\overset{+H^+}{\rightleftarrows}} \overset{+}{H_3N}-\underset{\underset{COO^-}{|}}{\overset{\overset{R}{|}}{C}}-H \underset{-H^+}{\overset{+H^+}{\rightleftarrows}} H_2N-\underset{\underset{COO^-}{|}}{\overset{\overset{R}{|}}{C}}-H$$

High hydrogen ion concentration (low pH)—net positive charge	At the isoelectric point there is no net charge on the molecule	Low hydrogen ion concentration (high pH)—net negative charge

The degree of acidity or alkalinity of solutions is often expressed, not as the hydrogen ion concentration itself, but as pH. This is the negative logarithm (to base 10) of the hydrogen ion concentration in gram ions per litre. Just as the mass in grams of 6 × 10^{23} molecules is called the gram molecule or *mole*, so the mass of 6 × 10^{23} ions is called the gram ion. Thus one gram ion of hydrogen (H$^+$) is 1 g but one gram ion of hydroxyl (OH$^-$) is 17 g. For strong acids and alkalis, but not for weak ones, the pH can, to a close approximation, be calculated from the molar concentration. Thus M/100 HCl contains 0·01 or 10^{-2} gram ions H$^+$ per litre and the solution has a pH of 2. Similarly M/100 H$_2$SO$_4$, since it is dibasic, contains 0·02 gram ions H$^+$ per litre. It has a pH of 1·699 because

$$\log_{10} 0·02 = \overline{2}·301 = -2 + 0·301$$
$$= -1·699$$

and therefore $-\log 0·02$ is $+1·699$.

Pure distilled water contains equal numbers of hydrogen and hydroxyl ions (10^{-7} gram ions of each per litre) and so it has a pH of 7. Thus acidic solutions have a pH lower than 7 and alkaline solutions a pH greater than 7. It should not be overlooked that the pH scale is logarithmic, so that a tenfold change in acidity is represented by a pH change of 1·0. The pH of solutions is readily measured using a pH meter and a *glass electrode*.

For each amino acid and for most proteins there is a particular hydrogen ion concentration where the negative charges are exactly balanced by the positive charges. At this pH, which is known as the *isoelectric point*, they do not move when subjected to electrophoresis and possess their minimum solubility. This is often useful because it allows the precipitation of a particular amino acid or protein from a mixture by adjustment of the pH. In practice, various other factors such as temperature and inorganic ion concentration have to be appropriately adjusted at the same time.

Since amino acids and proteins readily accept or release hydrogen ions—depending on the pH of the solution—they readily act as *buffers*, tending to resist large changes in pH when acid is added or removed from the solution. In cells and in blood plasma, proteins play a vital role in buffering changes in pH which would otherwise occur (chapter 15, p. 376). This is important because the activity of many enzymes is greatly affected by pH (chapter 4).

A *buffer*[*] is typically a solution containing a weak acid, HA, together with a salt of that weak acid which, being highly ionized, provides anions of the acid, A^-. The dissociation of the weak acid can be written in the general form:

$$HA \rightleftharpoons H^+ + A^- \qquad (2.1)$$

The hydrogen ion concentration of a buffer solution depends on three quantities, the concentrations of the undissociated acid and of the anion, $[HA]$ and $[A^-]$, and also the dissociation constant of the acid, K_a where

$$K_a = \frac{[H^+][A^-]}{[HA]} \qquad (2.2)$$

The pH of the solution is related to these three quantities. By rearranging equation (2.2) and taking logarithms:

$$\log H^+ = \log K_a + \log [HA]/[A^-]$$

or $\qquad pH = pK_a + \log [A^-]/[HA] \qquad (2.3)$

The last equation is known as the Henderson-Hasselbalch equation; in it pK_a is the negative logarithm of K_a and the final term results from the fact that $-\log \dfrac{[HA]}{[A^-]}$ is equal to $+\log \dfrac{[A^-]}{[HA]}$.

Figure 2.2 illustrates how a typical buffer works. It is regarded as two reservoirs, one of acid HA and the other of its anion A^-. The relative size of the reservoirs depends on the number of moles of acid and salt of the acid used to make the buffer. On adding extraneous hydrogen ions to the buffer, these unite with A^- ions to form HA molecules, thus increasing the size of the HA reservoir very slightly $(+\Delta HA)$ and decreasing the reservoir of A^- correspondingly $(-\Delta A^-)$. Conversely, on adding extraneous hydroxyl ions, these unite with hydrogen ions to give neutral water molecules, thus disturbing the equilibrium shown in equation (2.1) so that the reservoir of HA molecules becomes slightly depleted and that of A^- ions becomes slightly larger. So long as

$$\frac{[A^- - \Delta A^-]}{[HA + \Delta HA]} \quad \text{or} \quad \frac{[A^- + \Delta A^-]}{[HA - \Delta HA]}$$

are not too different from the original quotient $[A^-]/[HA]$, the pH will stay nearly constant.

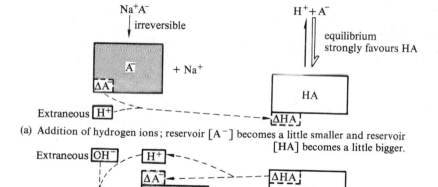

(a) Addition of hydrogen ions; reservoir $[A^-]$ becomes a little smaller and reservoir $[HA]$ becomes a little bigger.

(b) Addition of hydroxyl ions. By uniting with H^+ ions, the equilibrium $HA \rightleftharpoons H^+ + A^-$ is disturbed. Hence reservoir $[HA]$ gets smaller and reservoir $[A^-]$ bigger.

Fig. 2.2 Diagram showing how buffering involves small changes in the sizes of two reservoirs, one of anions (A^-) and the other of undissociated molecules (HA)

Proteins, peptides, and amino acids

2.3 The peptide bond

The linkage of amino acids to form peptide bonds is represented in its simplest form in Fig. 2.3. Note that in this example three amino acids

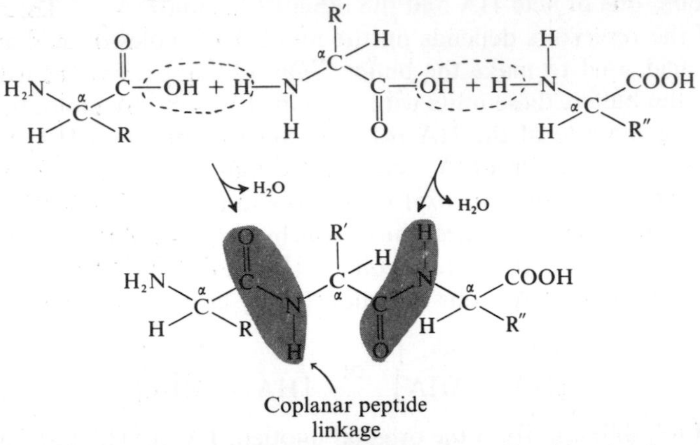

Coplanar peptide
linkage

Fig. 2.3 The formation of the peptide linkage. (In this simple example, a tripeptide is formed by the elimination of two water molecules from three amino acids molecules)

combine with the loss of two water molecules, forming a tripeptide containing two peptide linkages. The prefix *tri-* indicates the presence of three amino acid residues. The tripeptide, like most proteins and peptides, has a free amino group at one end and a carboxyl group at the other; these will be ionized to an extent which is determined by the pH.

The peptide linkage provides a particularly stable bond between the nitrogen atom of one amino acid and the carbon atom of the carboxyl group of the neighbouring amino acid. Its stability can probably be explained by the fact that it is best represented as a hybrid between two forms:

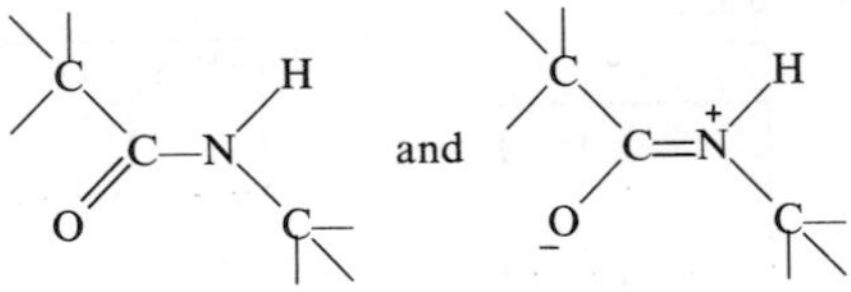

The double-bond character of the linkage between the carbon atom of the carbonyl group and the nitrogen atom makes the peptide linkage coplanar (Fig. 2.3), i.e., all the atoms lie in one plane, just as they do in ethylene. The bonds attached to the α-carbon atoms, however, have the normal tetrahedral arrangement. These features of the peptide bond confer upon proteins and peptides their stability of structure (section 2.7). Moreover, if the amino acid side groups are not to interfere with one another—i.e., show steric hindrance—the α-carbon atoms on either side of the peptide bond must normally have these groups arranged *trans* with respect to each other.

The great stability of the peptide bond can be illustrated by the fact that vigorous boiling of proteins in acid (6M HCl) or alkali (3M NaOH) for many hours is often necessary to hydrolyse them completely. Proteolytic enzymes achieve the same result at body temperature and a physiological pH (chapter 4).

Some proteins contain as many as a thousand peptide linkages firmly holding the α-amino acid residues together. Indirectly, the peptide bond is also responsible for maintaining the secondary structure of the protein chain—as will be explained later, the chain of amino acid residues is often wound or folded in a more or less complicated manner. The contribution of the peptide bond is explained by the fact that there is a tendency for the carbonyl *oxygen* atom of one peptide linkage to share its electrons with the *hydrogen* atom attached to the nitrogen of another. The *hydrogen bond* so formed is a type of weak electrostatic attraction and its existence is usually indicated by a dotted line:

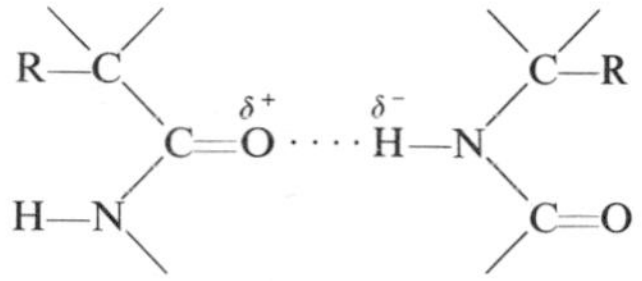

Proteins, as will be seen in section 2.7, very frequently have a considerable length of their amino acid chains arranged in the form of a helix, and the hydrogen bonding described above plays a major role in maintaining this form of secondary structure. However, other forms of bridge or linkage between amino acid residues within a peptide chain or in two different peptide chains also contribute to the spatial geometry or topography of the protein molecule. These bridges result from interactions between the side groups of certain amino acids and are described in the next section.

2.4 Individual amino acid residues

The side groups of the α-amino acids (Fig. 2.4) present in a protein largely determine its three-dimensional structure and its properties. In general,

Proteins, peptides, and amino acids

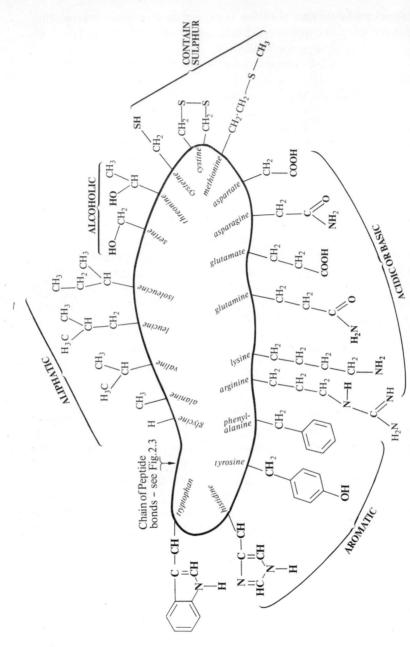

Fig. 2.4 The side groups of the common amino acids. (They can be envisaged as hanging from a common thread, the chain of peptide bonds. The groups in heavy type are probably particularly active)

side groups can be divided into five types—although some amino acids have side groups which could be classified in more than one way:

(a) Non-polar
(b) Alcoholic
(c) Those containing sulphur
(d) Acidic and basic
(e) Homo- or heterocyclic

There are, in addition, (f) two *imino* acids present in many proteins.

(a) Amino acids with non-polar side groups

The side groups of the non-polar amino acids (L-leucine, L-isoleucine, L-valine, and L-phenylalanine) are hydrophobic and tend to be attracted to other non-polar compounds. This attractive force helps to maintain the three-dimensional structure of proteins and can result in the formation of weak *non-polar bridges* (Fig. 2.5a). This type of attraction is similar to that which occurs between paraffin molecules.

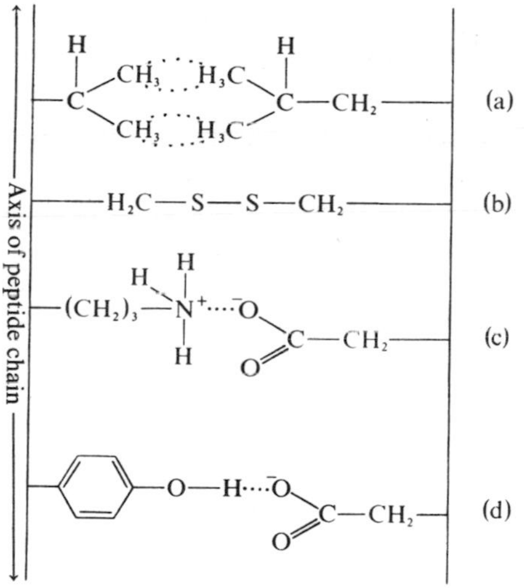

Fig. 2.5 Four types of bridge linking peptide chains together
(a) non-polar between valine and leucine
(b) disulphide bridge formed by a trans-chain cystine molecule
(c) ionic bond between lysine and aspartate
(d) hydrogen bond between tyrosine and aspartate

Proteins, peptides, and amino acids

(b) Amino acids with non-aromatic hydroxyl groups

The hydroxyl group of the alcoholic α-amino acids, L-serine, and L-threonine, is important in many proteins since it is able to take part in ester or ether linkages. Such linkages often attach the protein to a prosthetic group, and in the case of some enzymes, they may bind reactants to the catalytically active part of the enzyme. Sometimes, a phosphate group is attached to the hydroxyl group of serine. This happens, for example, to some of the serine residues present in casein, a protein present in milk and cheese.

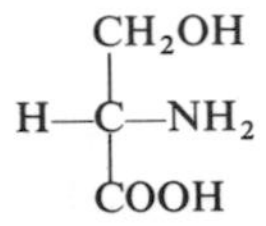

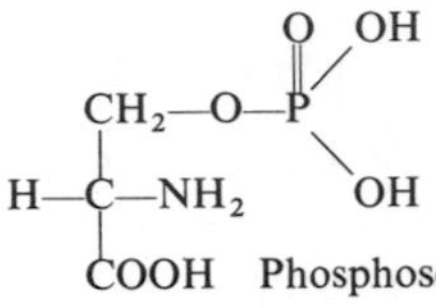

(c) Amino acids with side groups containing sulphur

Cysteine, one of the amino acids which contains sulphur, resembles serine except that it possesses a *sulphydryl group* instead of hydroxyl. It often acts as a reducing agent, for the hydrogen atom of the —SH group is more readily lost than that of the hydroxyl group (compare, in inorganic chemistry, the properties of H_2S and H_2O). In a number of enzymes the sulphur atom of cysteine is important because it provides a point of attachment for the reactants—an example is illustrated on p. 201.

The sulphydryl groups of cysteine residues are frequently involved in maintaining the three-dimensional structure of proteins. Two cysteine residues can link a pair of polypeptide chains together, or form a link between two parts of a single chain. The linkage, known as a *disulphide bridge* (Fig. 2.5b) actually involves an oxidative change which results in the formation of the 'double' or di-α-amino acid, L-cystine:

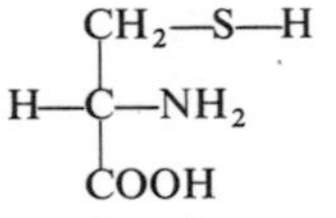

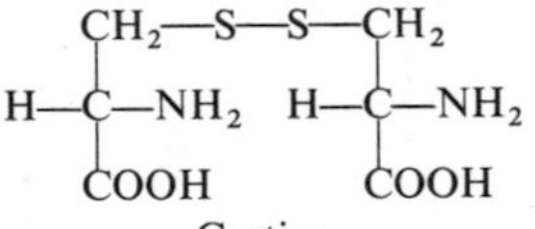

(d) Di-basic and di-acidic amino acids

The buffering power conferred by the ionizing terminal groups of peptide chains has been mentioned, but most of the buffering power of proteins results from the side groups of the di-basic and di-acidic amino acids.

A lysine residue, which can take up a proton, is an example of the first, and aspartic acid, which can lose a proton, is an example of the second.

$$\diagdown\!\!\!-\ CH_2.CH_2.CH_2.CH_2.NH_2 \quad + H^+ \rightleftharpoons\!\!-\ (CH_2)_3.CH_2\overset{+}{N}H_3$$

Lysine side group

Axis of peptide chain

$$\diagdown\!\!\!-\ CH_2-COOH \qquad\qquad -H^+ \rightleftharpoons\!\!-\ CH_2.COO^-$$

Aspartic acid side group

The side groups of these acidic and basic amino acids enable some enzymes to combine with their substrates. In addition, they are important in maintaining the three-dimensional structure of many proteins by forming *ionic* or *salt bridges* (Fig. 2.5c). An ionic bridge involves an electrostatic attraction between a positively charged and a negatively charged side group.

By contrast to the simple aliphatic amino acids, the side groups of the acidic and basic amino acid residues are attracted to water owing to their ionic nature. There is evidence to suggest that in some proteins (e.g., haemoglobin) the interior of the molecule contains numerous hydrophobic amino acid side groups (derived from simple aliphatic amino acids and phenylalanine), whereas the outer surface, in contact with water under physiological conditions, is surrounded by many hydrophilic acidic and basic amino acid residues.

(e) Amino acids with side groups containing homocyclic or heterocyclic rings

The four amino acids, L-tryptophan, L-histidine, L-tyrosine, and L-phenylalanine have side groups containing either benzene or heterocyclic rings (Fig. 2.4). In histidine and tryptophan, as in the case of the homocyclic compounds, it is not possible to ascribe to the bonds of the rings either double-bond or single-bond characteristics, since each is a *hybrid* of the two types. Other compounds containing these so-called 'aromatic' bonds will be encountered in later chapters—they include several vitamins and the bases of the nucleic acids.

The hydroxyl group of tyrosine is weakly acidic, like that of phenol, and is able to take part in hydrogen bonding, forming links between amino acid side chains (Fig. 2.5d). Furthermore, the 'aromatic' amino acids often fulfil an important function by holding substrates or prosthetic groups in position—this is particularly true of the *imidazole* ring of histidine. In myoglobin, for example, a ferrous ion is held in the centre of a porphyrin ring and, in addition, it is also linked to the nitrogen atom of two histidine residues (p. 144).

(f) The imino acids

L-Proline and L-hydroxyproline are present together with true α-amino acids in many proteins and especially in the protein, collagen, a constituent of cartilage.   They are *imino* acids, but clearly have a close resemblance to the α-amino acids, with which they form peptide bonds in the normal way.

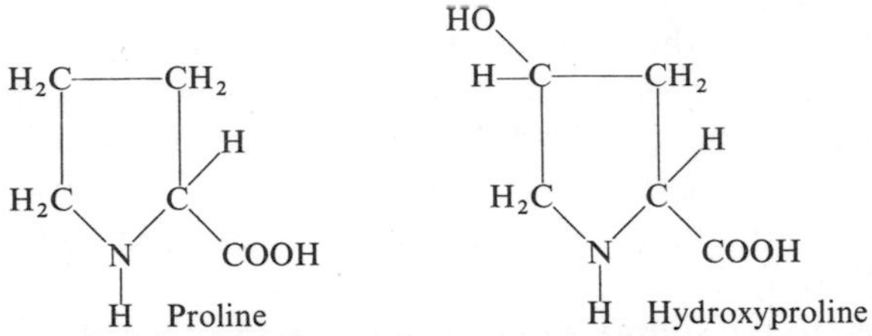

Proline   Hydroxyproline

However, because of the imino group, the α-carbon atoms are part of a five-membered heterocyclic ring, and, in consequence, the bond angles of peptide linkages containing them are somewhat different from the values characteristic of the true amino acids.   The result is that imino acids tend to produce a somewhat distorted or 'kinked' peptide union, and is one reason why many protein molecules have a complex molecular topography.

2.5   Some important reactions of amino acids

Amino acids undergo several reactions which are important in relation to the qualitative and quantitative investigation of proteins, and tests for proteins nearly always depend ultimately upon the reactions of the amino acid residues they contain.

(a) The Biuret test

Any compound with two or more peptide linkages close together responds to this test, so although proteins give a positive reaction, it is not entirely specific for proteins.   When copper sulphate in alkaline solution is added to a protein solution, a violet-purple colour is produced owing to the formation of a copper complex.

(b) The ninhydrin test

This test works best with free amino groups so a protein is usually first hydrolysed by boiling with hydrochloric acid.   When a dilute solution of triketohydrindene (ninhydrin) is warmed with amino acids, a purple colour is produced as a result of the following reaction:

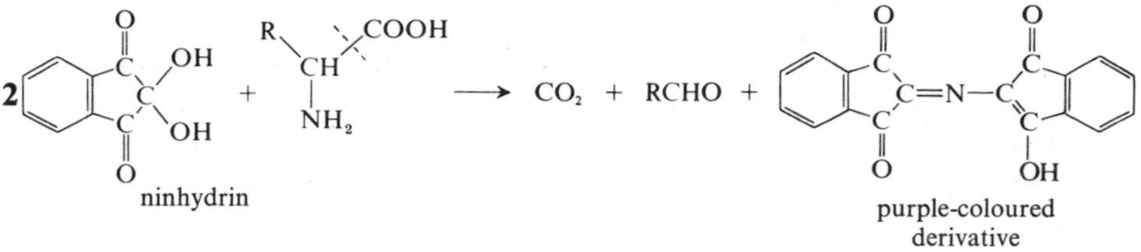

This reaction is particularly valuable for measuring the amounts of individual amino acids in proteins following the chromatographic separation of protein hydrolyzates (see (d) below).

(c) Tests for individual amino acids

Several tests enable the presence of specific amino acids in a protein to be confirmed. Some of these are summarized in Table 2.1. They

Table 2.1   Some tests for specific amino acids

Name	Reagent	
Hopkins-Cole test	Glyoxylic acid, or formaldehyde in conc. H_2SO_4	Violet colour indicates the presence of *tryptophan*
Sakaguchi test	α-Naphthol in sodium hypochlorite solution	Red colour indicates presence of *arginine*
Millon test	Mercury dissolved in conc. HNO_3	Red colour indicates presence of *tyrosine*
Xanthoproteic test	Boil with conc. HNO_3	Yellow colour indicates presence of *phenylalanine* or *tyrosine*
Sulphur test	Alkaline lead acetate	Black precipitate (PbS) indicates presence of *methionine* or *cystine* or *cysteine*
Folin and Ciocalteau test	Phosphomolybdo-tungstic acid	Blue colour indicates presence of *tyrosine* (the basis of an important quantitative test for proteins—Lowry's method)

are often named after their discoverers; details are available in practical manuals.

(d) Chromatography

This is a valuable technique for identifying several amino acids in a mixture such as a protein hydrolysate. There are many procedures

Proteins, peptides, and amino acids

available but one of the simplest is illustrated in Fig. 2.1c. Small spots of a solution of the amino acids are placed on a strip of filter paper and dried. The paper is then placed in a glass tank supported at one end in a glass trough containing a solvent—e.g., an appropriate mixture of butanol, acetic acid, and water. This flows slowly down the paper from the trough and as it does so it carries the amino acids with it. However, the distance which each amino acid moves in a fixed time differs according to its solubility in the solvent mixture. After a suitable period of time, the paper is removed from the tank, dried, and the final position of the amino acids can be located by spraying the paper with 1 % ninhydrin solution. On warming, purple spots appear at positions on the paper characteristic of the amino acids present in the original solution.

(e) Ion exchange chromatography

This technique, unlike paper chromatography, is suited for *quantitative* estimation of the amino acids present in a mixture. A suitable ion exchange resin such as sulphonated polystyrene is placed in a tall glass column (Fig. 2.1d) filled with dilute acid buffer. The mixture of amino acids is placed at the top of the column in a buffer of pH 3. Solvent is run slowly down the column, the pH being gradually increased. The basic amino acids such as lysine, asparagine, and glutamine, being positively charged at low pH, are attracted to negatively charged groups on the resin. Conversely, acidic amino acids have no affinity for the resin and pass fairly rapidly down the column with the solvent. Somewhat later, the non-polar amino acids emerge from the column and finally the basic amino acids are eluted from the resin. As solvent, containing dissolved amino acids, drips from the bottom of the column it is collected in test tubes in fractions of about 1 ml. When the ninhydrin reagent is added to the tubes it is possible to estimate colorimetrically every amino acid present, and since each passes down the column at a characteristic rate, the total quantity of the individual amino acids in the original mixture can be calculated. This technique has now reached a highly automated stage of development.

2.6   The primary structure of representative peptides and proteins

It is possible to determine the precise sequence of amino acids in a particular peptide or protein, but the task is time-consuming and arduous, the difficulty increasing rapidly with the molecular weights of the compounds under study. As a result, the structures of many peptides have been worked out, but so far the primary structures of less than twenty proteins have been completely elucidated. The method, pioneered by Sanger, depends on the step-by-step hydrolysis of the molecule, and, since various hydrolytic enzymes are usually employed, it will be referred to again later (p. 110). Essentially, however, enzymatic or chemical hydrolysis breaks the peptide chains into smaller lengths, each of which is characterized after separation and further degradation. Finally, the whole structure is worked out in a way reminiscent of the assembly of the pieces of a jig-saw puzzle.

The supreme test of the correctness of structural deductions so arrived at would be complete synthesis of the peptide or protein by purely chemical means, starting with free amino acids. Needless to say, this has so far only been achieved for a few relatively simple peptides and proteins—insulin, for example, has been synthesized by Chinese workers. In the present section, the structures of two related peptides, oxytocin and vasopressin, and of the protein, insulin, are briefly described.

Oxytocin and vasopressin

Oxytocin and vasopressin are two important hormones produced by the posterior pituitary gland of higher animals. The structure of oxytocin extracted from the pituitary gland of cattle is shown in Fig. 2.6. Both

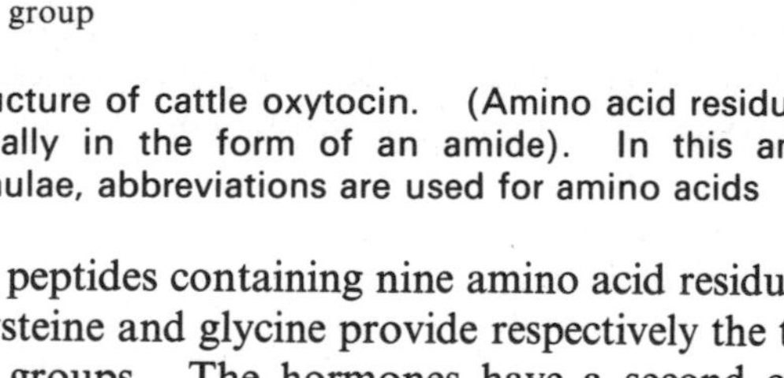

Fig. 2.6 Structure of cattle oxytocin. (Amino acid residue 9 (glycine) is actually in the form of an amide). In this and several later formulae, abbreviations are used for amino acids

hormones are peptides containing nine amino acid residues and, in both substances, cysteine and glycine provide respectively the terminal amino and carboxyl groups. The hormones have a second cysteine residue and this forms a disulphide bridge with the first, so giving the molecules the shape of a ring with a tail. Actually, the terminal glycine does not have a free carboxyl group but is in the form of its amide.

One function of oxytocin is to stimulate the release of milk in lactating mammals, while the major role of vasopressin is to reduce the volume of urine produced by the kidneys. Yet, despite these very different principal functions, it is a remarkable fact that *vasopressins* in several different mammals, including man, have almost the same structure as the *oxytocin* of cattle given in Fig. 2.6, the main differences relating to the amino acid residues in positions 3 and 6. It is also of great interest that vasopressin has varied slightly in composition during the evolutionary process (hence the plural form, vasopressins, used above) yet the different molecular species have the same anti-diuretic effect in different mammals. Vasopressin is also known as ADH, or anti-diuretic hormone.

Insulin

Insulin, which has a molecule of such a size that it can be regarded as a large polypeptide or a very small protein, is a vertebrate hormone secreted by the pancreas. Its main action is to control animals' carbohydrate metabolism (p. 232). It was the first protein to have its primary structure fully worked out. It is now known that the true molecular weight is 5733, although its tendency to dimerize led initially to estimates of about twice this value. Sanger showed that insulin consists of two peptide chains, one containing 21 amino acid residues (chain A) and a second with 30 residues (chain B). These are linked by two disulphide bridges (Fig. 2.7), a third cystine bridge occurring within the A-chain itself.

Insulins from a variety of mammals have now been studied and, in general, have been found to vary only in the amino acid composition of the A-chain at positions 8, 9, and 10. Insulins from man and rabbit are exceptional; in man, the terminal amino acid residue of the B-chain is threonine instead of alanine. Nevertheless—for the treatment of diabetes mellitus—bovine insulin is a very effective hormone in man, a fact which again indicates that similar or identical physiological responses can sometimes be evoked by protein-like substances of somewhat different composition.

Although insulin comprises two peptide chains, such evidence as is available suggests that the majority of proteins consist of only one long chain. For example, both bovine ribonuclease (RNA-ase, mol. wt. 12,700) and the proteolytic enzyme, papain (mol. wt. 20,700), consist of a single chain of amino acid residues; RNA-ase contains 124 of them and papain contains 198. Papain is of interest, for it was the first plant

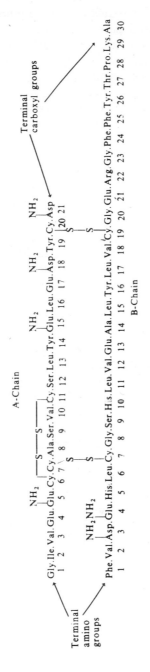

Fig. 2.7 The primary structure of insulin

Proteins, peptides, and amino acids

protein to have the complete sequence of its amino acid residues determined. Both RNA-ase and papain contain three disulphide bridges linking different parts of the same chain; clearly, such bridges, by limiting random kinetic twisting and turning of the molecule, give it a certain spatial pattern (as will be seen later, spatial arrangement also exists for other reasons). Such *secondary* structure is more conspicuous in many other more complex proteins, as the next section will show.

2.7 The three-dimensional structure of proteins

Having described the primary structure of several peptides and proteins it is important to emphasize that this, in itself, provides very little information about the overall shape of the molecule. The primary structure of a straw hat is a number of lengths of straw; only when the straw is folded and twisted in a definite manner does a three-dimensional structure of the hat become a recognizable entity. An understanding of this important aspect of protein chemistry has only just begun.

The major tool for tracing the shape the peptide chains of proteins follow is X-ray crystallography, aided by the computer. The work is very complex and time-consuming and relatively few proteins have so far been investigated; already, however, some important generalizations can be made about the factors which cause particular protein molecules to tend to assume one particular shape.

The three-dimensional topography of a protein is usually considered to result from the existence of structural factors of three types, although precise definition of the boundaries of each of these types is perhaps undesirable at this stage, for the subject is changing so rapidly. *Secondary structure* is largely a description of the hydrogen bonding between peptide linkages. When this occurs between peptide linkages within a *single* chain, this often leads to a remarkable spatial arrangement termed the α-helix. *Inter*molecular hydrogen bonding, between peptide linkages in *different* chains, occurs in some fibrous proteins, and leads to an arrangement known as pleated sheets. Sometimes secondary structure, as such, is absent, but disulphide, ionic, or other bridges within one chain produce a characteristic shape. *Tertiary structure* is a superimposed complication, for it is often found that chains, stabilized in some particular shape by hydrogen bonding, are themselves arranged in space in a specific pattern or 'conformation'. As an example, the α-helix may itself be wound into a helix, rather as a helical curtain spring could be wound around a broom-handle.

Secondary structure

Several types of secondary structure exist in various proteins, but limitations of space preclude discussion in any detail of more than one of them. X-ray crystallographic studies have shown that a group of animal fibrous proteins (keratin, myosin, elastin, and fibrin) have a regular repeating pattern along the length of the fibre. From a knowledge of the parameters of the repeating units, and using known values for bond lengths and bond angles in the peptide bond, Pauling was able to construct a model, called the α-*helix*, to describe the position of the amino acid residues in the fibre.

In the α-helix (Fig. 2.8) each peptide bond is planar (p. 54) and each carbonyl group is hydrogen-bonded to the amide group of the third amino acid from it along the chain. In Fig. 2.8, the hydrogen bonding is shown by dashed lines. For every 10 turns of the helix there are 36 amino acid residues with their side groups (R) projecting outwards from the axis of the helix. The hydrogen bonding greatly stabilizes the structure, the distance between the oxygen and nitrogen atoms being almost ideal for a strong hydrogen bond (0·286 mμ). Theoretically, there are

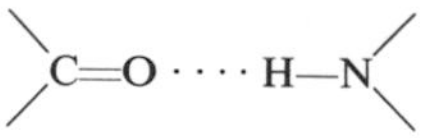

$$\diagdown\!\!\diagup C{=}O \cdots H{-}N\diagup\!\!\diagdown$$

two possible forms of the α-helix—the right-handed form illustrated in the figure, and a left-handed form which is almost, but not quite, the mirror image of the other. The right-handed form is the most probable, because it results in less steric hindrance between the side groups (R) and other neighbouring atoms.

Sometimes *intermolecular* hydrogen bonding results in pleated sheets, two conformations being possible. In the first, the linked peptide chains run in the same direction

$$\text{—C—CO—NH—C—}\; ; \text{—C—CO—NH—C—}$$

and in the second they face in opposite directions

$$\text{—C—CO—NH—C—}\; ; \text{—C—NH—CO—C—}.$$

The arrangements are called respectively parallel and antiparallel pleated sheets. One such conformation is illustrated in Fig. 2.9.

Tertiary structure

The α-helix only partially describes the structure of fibrous proteins. In hair, for example, the basic structure consists of numerous α-helices twisted together rather like the threads in a rope. This illustrates what

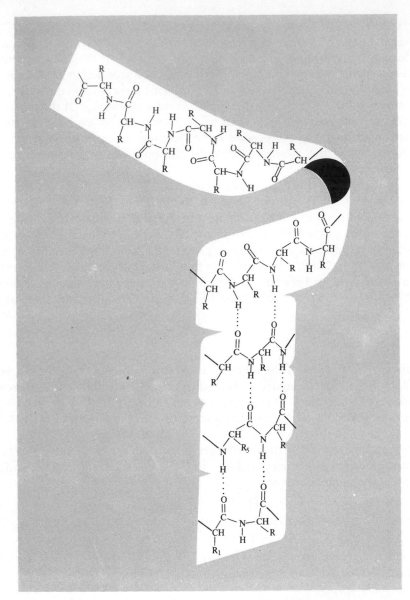

Fig. 2.8 The α-helix. (A very common arrangement for amino acid residues in proteins. The configuration of the atoms taking part in the peptide bonds of the right-handed form of the helix are shown, hydrogen bonds being indicated by dotted lines)

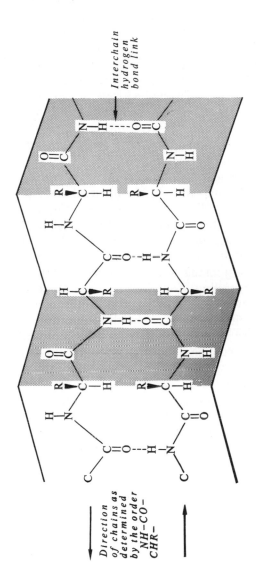

Fig. 2.9 Antiparallel pleated sheet structure between two peptide chains

Interchain hydrogen bond link

Direction of chains as determined by the order NH–CO–CHR–

Proteins, peptides, and amino acids

is understood by the term 'tertiary structure', this particular—and rather simple—example being referred to as a *super helix*. But perhaps the most important aspect of the α-helix is that it probably forms the basis upon which the tertiary structure of many non-fibrous (i.e., globular) proteins is patterned. The α-helix can be imagined to twist and turn to provide the overall shape of the molecule—although, in many proteins, parts of the α-helix may be 'unwound' over some part of the total length of the peptide chain. In the crystalline state, for example, about 50% of the insulin molecule is estimated to be in the α-helical form; what form the molecule takes in solution is less certain.

Since the α-helical structure is absent in some regions of many molecules, yet each molecule possesses, under constant environmental conditions, a more or less permanent and unique shape to which its biological characteristics are intimately related, it is apparent that types of bonding, other than hydrogen bonding associated with peptide linkages, must help to stabilize the conformation. Inter-chain linking via disulphide bridges and the other bridging devices illustrated in Fig. 2.5 are undoubtedly involved in different proteins; in addition, shape is partly determined by the nature and positioning of any prosthetic groups which may be present and by the distorting effect of imino acid linkages (p. 46).

After more than twenty years of elaborate investigations, Perutz has achieved a very detailed analysis of the three-dimensional structure of haemoglobin (mol. wt. 67,000), the globular protein present in red blood cells. Each molecule contains four large protein chains; two of these are called α-chains (a term which must not be confused with α-helix) and the other two are called β-chains. The α-chains contain 141 amino acid residues and the β-chains 146 so the whole molecule contains 574 amino acid residues. Within the folded structure of each α- and β-chain is a protohaem prosthetic group and therefore the haemoglobin molecule contains four such groups. The structure of protohaem is considered later (Fig. 6.6).

It is not proposed to discuss the molecule of haemoglobin in more detail, because the essential features of its structure are more easily described by reference to a second globular protein, myoglobin. This protein occurs in muscle, acting as a temporary oxygen store—it is particularly abundant in marine mammals such as seals and whales, but also occurs in man and other mammals. Myoglobin contains 153 amino acid residues and the chain has been found by Kendrew to have almost the same three-dimensional structure as each of the four peptide chains of haemoglobin.

Figure 2.10 provides a somewhat simplified impression of the myoglobin molecule. Some 75% of the structure is arranged in the α-helical form—there are, in fact, eight separate sections of α-helix, separated by regions where the helical structure is absent. Thus the molecule, which contains a protohaem group, has a definite, folded tertiary structure. The amino acids in the interior of the molecule have, for the most part, hydrophobic

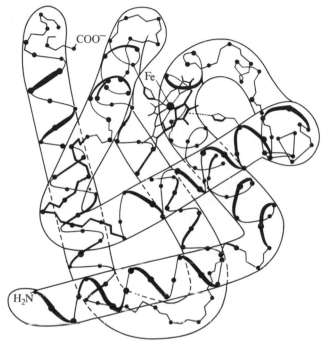

Fig. 2.10   The structure of myoglobin.   The ferrous ion is in the centre of a protohaem group

side groups which play an important role in maintaining the tertiary structure by non-polar bridges. Conversely, hydrophilic amino acid side-groups tend to be on the outside of the molecule and interact with the aqueous environment.

It is of particular interest to speculate that, despite the complexity of protein topography and the likelihood that no two proteins have molecules which are exactly the same shape, nevertheless, the tertiary structure of a particular protein may be an inevitable consequence of the sequence of amino acid residues in its primary structure. If this is so, tertiary structure may be created by successive approximations, but—in the constant

Proteins, peptides, and amino acids

environment of the cytosol—with ultimate certainty, at the moment of protein synthesis at the ribosomes (ch. 14). For, if the ribosome is imagined as moving in a series of jerks along the messenger RNA, attaching one amino acid residue at each jerk, an automatic folding might be anticipated, a folding which, at any instant, will be determined by the attempt of all the *hydrophobic* side chains present at that instant to escape from an aqueous environment by moving to the inside of the molecule, and of the *hydrophilic* groups both to form polar links with one another and to orientate themselves outwards. Knowing the sequence of amino acid residues in a particular protein, a computer may shortly be able to predict the magnitude of the forces leading to those successive approximations, and hence make it possible to put this supposition to the test.

Quaternary structure

Sometimes, molecules of proteins group together, held by more or less strong physical or chemical forces, to form a new structural unit with a geometry of its own and often with physiological properties different from those of the component proteins acting in isolation. Such a 'multimolecular molecule' clearly possesses a biochemical importance in its own right. Haemoglobin is an excellent example of a protein in which quaternary structure is important (pp. 56, 373).

2.8 Native and denatured proteins

It is known that the structure and function of proteins are intimately related, and a simple way to demonstrate the importance of the structure of active, *native* proteins is to *denature* them. There are several ways in which this can be done, but, in all cases, more or less disorganization of their three-dimensional structure occurs, and this leads to an alteration in their physical, chemical, and biological characteristics. This is particularly apparent when the denatured protein is an enzyme, for when enzymes are denatured they lose their ability to catalyse reactions. Similarly, when muscle proteins are denatured, they are unable to contract.

Heating is perhaps the method most frequently used for denaturing proteins. When an egg is boiled, for example, the soluble, almost colourless albumen becomes solid, insoluble, and white. In the laboratory, strong acids, alkalis, or oxidizing agents are all frequently used to denature proteins; their action results in the disruption of the various bridges which maintain the protein's three-dimensional shape, or in the formation of un-natural bridges. Salts of certain heavy metals such as mercury and tungsten (as tungstate) are also effective denaturing agents. In most cases, denaturation is irreversible because once the original molecular shape has been lost there is only a small probability that its unique original shape will be readopted. It follows, therefore, that to prepare a sample of native protein, the procedure must be carefully controlled so that the protein is not denatured.

3

Carbohydrates and nucleotides

Carbohydrates, lipids, and proteins are the most abundant of the organic compounds present in cells. This chapter is primarily concerned not with the chemistry, but with the range of molecular structure to be found amongst the carbohydrates and in a group of compounds, the nucleotides, which may be regarded as derivatives of carbohydrates. Nucleotides show great biological versatility, and illustrate—as do the carbohydrates themselves—that nature, having 'chosen' a particular molecular pattern, often adapts it so that various members of the group fulfil several unrelated biological functions.

3.1 Classification and naming of carbohydrates

The chemical compounds known collectively as carbohydrates[#] include such important materials as sugars, starch, glycogen, and cellulose. Some carbohydrates serve directly or indirectly as *sources of energy* in both plants and animals; others are components of the walls of plant cells or help to *provide rigidity* to the exoskeleton of certain invertebrate animals. They fall into three main groups, the monosaccharides, the disaccharides, and the polysaccharides, and since the last two groups are derivatives of the former, it is to the monosaccharides that attention must first be drawn.

Monosaccharides are aldehydes or ketones which also contain primary and (usually) secondary alcohol groups; they are all soluble in water and most of them are sweet. They can be represented by the following group formulae:

$$
\begin{array}{ccc}
\underset{\displaystyle C}{\overset{\displaystyle O\diagdown\quad\diagup H}{}} & \underset{\displaystyle C=O}{\overset{\displaystyle CH_2OH}{}} & \\
(CHOH)_n & (CHOH)_n & C_xH_{2x}O_x \\
CH_2OH & CH_2OH & \text{General formula} \\
\text{Aldehyde sugar} & \text{Ketone sugar} &
\end{array}
$$

Most carbohydrates are distinguished by the suffix *-ose*, the aldehyde sugars being described as *aldoses* and ketone sugars as *ketoses*. In almost all monosaccharides the carbon atoms are joined together in an unbranched chain, and the various members of the group are distinguished from one another by naming according to the number of carbon atoms present in the molecule. The commonest monosaccharides contain six, five, or three carbon atoms in their molecules and are named *hexoses*, *pentoses*, and *trioses*, though sugars with four and seven carbon atoms are occasionally important. Glucose is an aldo-hexose, fructose a keto-hexose, ribose an aldo-pentose, and glyceraldehyde is an aldo-triose.

The optical rotation observed when solutions of sugars are placed in a polarimeter may be dextro (d, or +) or laevo (l, or −). The simplest aldose, glyceraldehyde, is always the dextro-rotatory isomer when isolated from living cells, although an unnatural laevo-form can, of course, be synthesized. Conventionally—there is no other justification—dextro-glyceraldehyde is written on paper with the hydrogen atom of the asymmetric carbon atom on the *left* when the aldehyde group is on *top*:

$$
\begin{array}{ccc}
CHO & CHO & CHO \\
H\!-\!\underset{|}{C}\!-\!OH & (CHOH)_n & HO\!-\!\underset{|}{C}\!-\!H \\
CH_2OH & H\!-\!\underset{|}{C}\!-\!OH & CH_2OH \\
 & CH_2OH &
\end{array}
$$

(+)-Glyceraldehyde, the parent substance for the D- family of sugars.

All natural aldose sugars are of this type —i.e., D- family.

(−)-Glyceraldehyde, from which unnatural L- family sugars can be made.

Virtually all the monosaccharides in nature can be synthesized from D(+)-glyceraldehyde by a series of reactions which elongates the carbon chain. Thus they form a chemical family in the same way that the amino acids present in proteins are all related to L-alanine (chapter 2). However, the L-amino acids are not all *laevo*-rotatory and many monosaccharides differ from D-glyceraldehyde by showing laevo- instead of dextro-rotation. Thus D-glucose is dextro-rotatory like D-glyceraldehyde but D-fructose is laevo-rotatory. Nevertheless, since all the common aldoses and hexoses belong to the D-glyceraldehyde family, the configuration of the groups attached to the final asymmetric carbon atom of each molecule, as the formulae above indicate, is the same as that of D-glyceraldehyde.

Both aldoses and ketoses contain a carbonyl group and have reducing properties similar to those of aliphatic aldehydes.[#] The apparently anomalous position of the ketoses should be noted, for aliphatic ketones are not reducing agents; it would appear that reducing properties are conferred upon the ketoses by the presence in the molecule not only of a carbonyl group but also of primary and secondary alcohol groups. All monosaccharides consequently reduce Fehling's solution and Barfoed's solution (both of which are changed from a blue cupric compound to red cuprous oxide), and form a silver mirror with ammoniacal silver nitrate.

In general, free monosaccharides play only a minor role in biochemistry. Nearly always it is as derivatives such as phosphate esters that they enter into the metabolic pathways of the cell.

Disaccharides are sugars formed by the elimination of a molecule of water from between two molecules of monosaccharide:

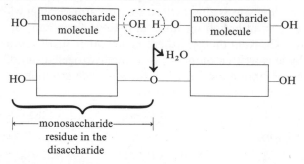

Polysaccharides can similarly be regarded as more complex derivatives of monosaccharides comprising large numbers of monosaccharide residues linked to form straight or branched chains. In the next section some of the structural features of a typical monosaccharide are illustrated, using glucose as an example.

3.2 Glucose—an aldohexose sugar

Glucose is a monosaccharide of considerable biochemical importance and for this reason has been chosen to illustrate and extend some of the considerations outlined in the previous section. Although it does not very often occur in large quantities in living organisms, glucose forms the unit structure of cellulose, which occurs abundantly in unlignified plant cell walls, as well as of starch and glycogen, which are storage carbo-hydrates in plants and animals; it is also one of the two units which unite to form the molecule of the disaccharide, sucrose, which is 'household' sugar.

The simplest structure that can be given to glucose is shown in Fig. 3.1a, which shows that it has five hydroxyl groups on carbon atoms 2 to 6, the aldehyde carbon atom being conventionally designated as carbon atom 1. Since the valency bonds of carbon are tetrahedrally distributed and four of these carbon atoms have four different monovalent groups attached to them, it follows that 2^4 different isomers of aldohexose exist. (If the reader is unfamiliar with the phenomenon of stereoisomerism,[#] he is strongly advised to construct models of sugars, remembering that the test of identity is superimposability and that pairs of isomers are related as object to mirror image.)

It was explained in the previous section that almost all naturally-occurring monosaccharides, irrespective of their optical rotation, are members of the D-series of sugars, the configuration around the penulti-mate carbon atom (number 5 in this case) being the same as that in D-glyceraldehyde (p. 61). The configuration around the other asym-metric carbon atoms can likewise be represented in two dimensions by writing the hydrogen atom either on the left or on the right; the arrange-ment in the case of glucose (left, right, left, left) is shown in the figure.

Like other aldehydes, glucose is a reducing agent, although in many reactions it is weaker in this respect than might have been anticipated. This diminished reducing power is probably a consequence of a complica-tion which occurs when solid glucose and other hexoses are dissolved in water—namely, that a tautomeric change occurs in which a large pro-portion of the molecules take on a cyclic form. This tautomeric[#] change can be regarded as an internal addition reaction, in which the hydrogen atom of the hydroxyl group attached to carbon atom 5 attaches itself to the oxygen atom of the carbonyl group on carbon atom 1, so as to form a new hydroxyl group (Fig. 3.1). As a result of the opening of the car-bonyl double bond in this manner, the free valency of the carbonyl group completes a heterocyclic, six-membered *pyranose* ring by union with the oxygen atom attached to carbon atom 5. This ostensibly somewhat unlikely event can be readily appreciated by making a model, when the valency angle between carbon atoms will be found quite

Carbohydrates and nucleotides

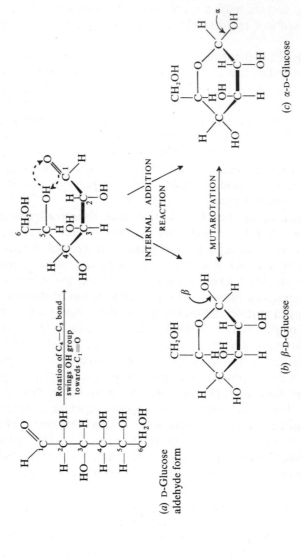

Fig. 3.1 The structure of D-glucose
((a) The straight chain form. The pyranose ring forms (b) and (c) result from this by an internal addition reaction)

(a) D-Glucose aldehyde form

(b) β-D-Glucose

(c) α-D-Glucose

Rotation of C_4—C_5 bond swings OH group towards C_1=O

INTERNAL ADDITION REACTION

MUTAROTATION

naturally to result in a close juxtaposition of the carbonyl group with the hydroxyl group of carbon atom 5.

This cyclization leads in turn to a further complication, for two tautomeric ring forms are possible. The reason for this is exactly the same as that which leads to, say, the formation of two forms of aldehyde cyanhydrin when hydrogen cyanide adds on to the double bond of acetaldehyde; in both addition processes, the addition reaction leads to the production of asymmetry where, previously, a symmetrical carbon atom existed. This happens because, after the addition, four different groups are attached to the carbon atom of what was originally the carbonyl group. In the case of the sugar, there are, at the moment of ring formation, theoretically equal chances of either of the two carbonyl double bonds rupturing, and according to which of the two breaks, two different sequences of the four monovalent group around this carbon atom are possible. The two forms can be represented, as in Fig. 3.1b,c, by directing the newly-formed hydroxyl groups either upwards or downwards, the plane of the ring itself being imagined as *projecting at right angles* out of the paper. The form for which the hydroxyl group is shown directed downwards is known as α-D-glucopyranose, the one in which it is directed upwards being the β-form.

The β-form of D-glucopyranose is somewhat more stable in solution than its α-counterpart, probably because, in the β-form, the hydroxyl groups on carbon atoms 1 and 2 are farther apart. Consequently, even though, theoretically, the carbonyl double bonds may have equal chances of opening, a 50:50 mixture of α- and β-forms is not achieved in practice owing to the intervention of a final complication—mutarotation. This process leads to the interconversion of α- and β-forms, until an equilibrium mixture is established. Thus, commencing with the pure α-isomer, upon contact with water, more and more of the β-form is produced until equilibrium is reached. For glucose, the equilibrium mixture contains 64% of the β-form, about 36% of the α-form, and a trace of the non-cyclic aldehydic reducing form; starting with the pure β-form, exactly the same equilibrium mixture is achieved. The process of mutarotation is accelerated by the enzyme *aldose mutarotase.*

The difference between α- and β-D-glucose has been stressed because, when this and other monosaccharides unite to give disaccharides and polysaccharides, the linkage is sometimes of one type and sometimes the other and, very often, the physical properties of the product are intimately bound up with the nature of the linkage. Cellulose, for example, is constructed from β-glucose groups, and comprises zig-zag chains of mono-saccharide units, whereas starch and glycogen, which contain α-glucose units, have molecules where the chains tend to wind round one another.

3.3   Other aldoses

In addition to glucose, there are two other aldo-hexoses of considerable biological importance.   These are D-galactose and D-mannose, and their structures are shown in Fig. 3.2.   They resemble glucose except that the configuration of the groups around certain of the asymmetric carbon atoms is different—galactose differs at carbon atom 4 and mannose at carbon atom 2.   Each has an α- and β-form analogous to the two forms of glucose—a comparison of models will show that α- and β-forms of a sugar are technically quite different substances, for they neither super-impose nor are mirror images of one another.

D-Glyceraldehyde is an aldo-triose formed *in vivo* when glycerol is oxidized.   Its phosphorylated derivatives are of great importance in the glycolytic pathway, a reaction sequence which results in the breakdown of hexose sugars.   D-Erythrose (Fig. 3.2) is an example of a tetrose; its phosphate is involved in the photosynthetic carbon cycle and in a route by which sugars are oxidized known as the pentose phosphate pathway.   Aromatic amino acids cannot be made by all organisms but, in those which can, D-erythrose phosphate frequently participates in the biosynthesis.

There are two important aldoses with five carbon atoms.   These are D-ribose and D-xylose.   In addition, there are two other materials, 2-deoxy-D-ribose and the ribose reduction product, ribitol, which have functional (as distinct from energetic) uses of great significance.   Though related to the pentoses, these are not, strictly speaking, monosaccharides. Deoxyribose, a component of DNA (deoxyribonucleic acid), has two hydrogen atoms attached to carbon atom 2, instead of having one hydrogen atom and one hydroxyl group.   Ribitol is a component of riboflavin, which is a cofactor for a number of enzymes which catalyse biological oxidations (p. 140).   The configurations for all these compounds are shown in Fig. 3.2.

D-Ribose, deoxy-D-ribose, and D-xylose exist largely in the *pyranose* 6-membered ring form when free in solution.   But when present as part of larger molecules they usually occur in the form of 5-membered rings (Fig. 3.3).   This *furanose* ring structure arises from the straight

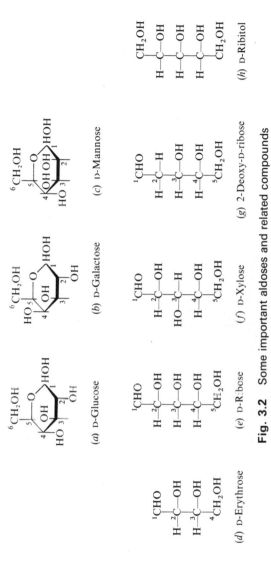

Fig. 3.2 Some important aldoses and related compounds

(a) D-Glucose

(b) D-Galactose

(c) D-Mannose

(d) D-Erythrose

(e) D-Ribose

(f) D-Xylose

(g) 2-Deoxy-D-ribose

(h) D-Ribitol

Carbohydrates and nucleotides

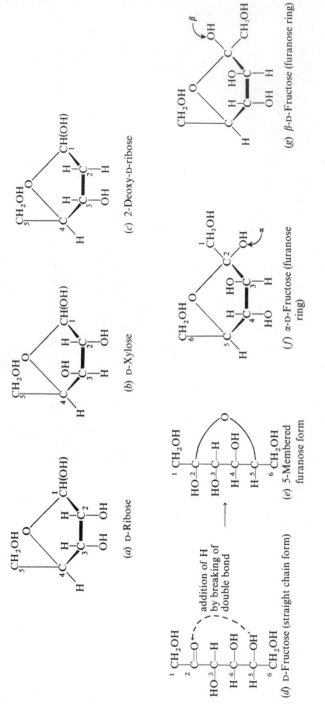

Fig. 3.3 Examples of furanose (5-membered) ring structures for mono-saccharides

chain form in a comparable way to the pyranose ring formation described for glucose.

3.4 Ketoses

Fructose is the commonest of the keto-hexoses. Its configuration resembles glucose—except, of course, at carbon atom 2, which in ketoses carries a double bond and in consequence is not asymmetric. In solution, fructose exists largely in the pyranose form but it resembles the pentoses by assuming a furanose ring form in many of the biologically important compounds of which it is a component part. This resemblance is not accidental, but is related to the fact that the carbonyl group, being on carbon atom number 2, effectively isolates one of the hexose carbon atoms so far as ring formation is concerned, leaving a molecular pattern which is akin to that in pentoses (Fig. 3.3).

Some other ketoses of biological interest are depicted in Fig. 3.4. The simplest, dihydroxyacetone, is a triose and is related to glyceraldehyde; indeed, the phosphates of these two trioses are enzymatically interconvertible in the cell. D-Ribulose contains five carbon atoms, and is

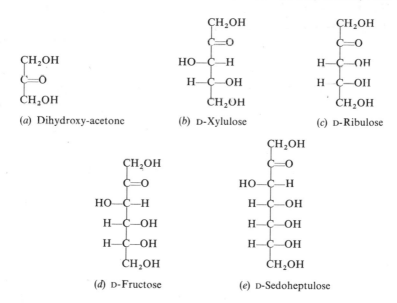

(a) Dihydroxy-acetone   (b) D-Xylulose   (c) D-Ribulose

(d) D-Fructose   (e) D-Sedoheptulose

Fig. 3.4   Ketoses (straight chain forms)

configurationally related to D-ribose. Note that ketonic sugars are often distinguished from their aldehydic counterparts by the ending *-ulose*. D-Sedoheptulose is a sugar containing seven carbon atoms; its phosphate, like those of ribulose and xylulose, plays an important role in the photosynthetic carbon cycle. Readers are, however, advised that it is unnecessary at this stage to commit all these names and formulae to memory. Such a process is tedious and only an overall impression of nomenclature is necessary until individual compounds are encountered in relation to particular aspects of cell metabolism.

3.5 Derivatives of the monosaccharides

Simple monosaccharides are seldom directly involved in biochemical reactions, but are first transformed to esters, to ether-like compounds, or to oxidation products. Occasionally, reduction products or amino sugars are of importance as well. Taking glucose as the principal example, the structure and nomenclature of some of these derivatives of sugars will now be illustrated.

The hydroxyl group on carbon atom 1 (number 2 in ketoses) is manifestly rather different from those on the other carbon atoms, for it originates from the carbonyl group of the straight chain form. It is not unusual to find that it is, in certain circumstances, more reactive than the other hydroxyl groups. Phosphoric acid, for example, readily esterifies glucose, *in vitro*, the product being D-glucose-1-P (in formulae and equations, phosphate is often abbreviated to P; inorganic phosphate is similarly represented as P_i). In the formulae for sugars which follow,

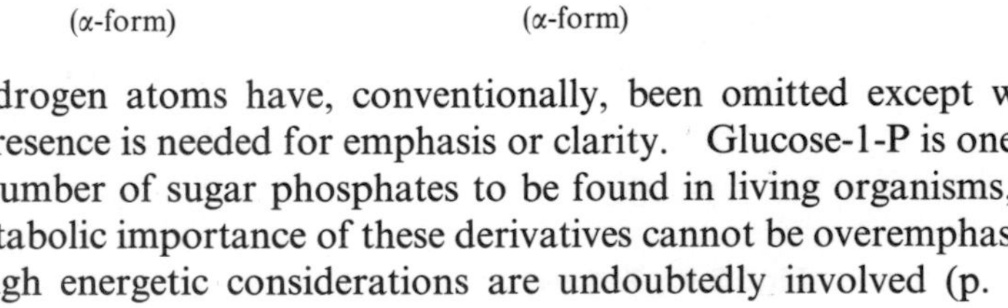

D-Glucose
(α-form)

D-Glucose-1-P
(α-form)

the hydrogen atoms have, conventionally, been omitted except where their presence is needed for emphasis or clarity. Glucose-1-P is one of a large number of sugar phosphates to be found in living organisms, and the metabolic importance of these derivatives cannot be overemphasized. Although energetic considerations are undoubtedly involved (p. 123), the almost universal occurrence in biochemical reactions of sugar phosphates rather than simple sugars is a fact not always readily explained. Another glucose phosphate, glucose-6-P, is also well known, as is a diphosphate in which the hydroxyl groups on carbon atom 1 and carbon atom 6 are both esterified. This diphosphate and its very important fructose counterpart have the following formulae:

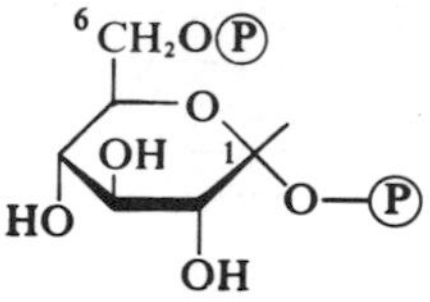

D-Glucose-1,6-di-P
(a pyranose compound)

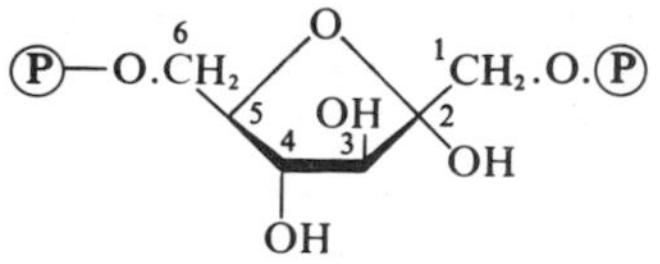

D-Fructose-1,6-di-P
(a furanose compound)

Under suitable conditions, alcohols react with the 'carbonyl-derived' hydroxyl group of monosaccharides to yield ether-like substances termed *glycosides*. The term 'ether-like' is used, for sugars are not typical alcohols and the resulting compounds, although structurally like ethers, are far less stable to hydrolysis than ordinary ethers, being in many respects more like esters:

$$R{-}O{-}H + H{-}O{-}R \longrightarrow R{-}O{-}R + H_2O$$

2 molecules of an alcohol   an ether

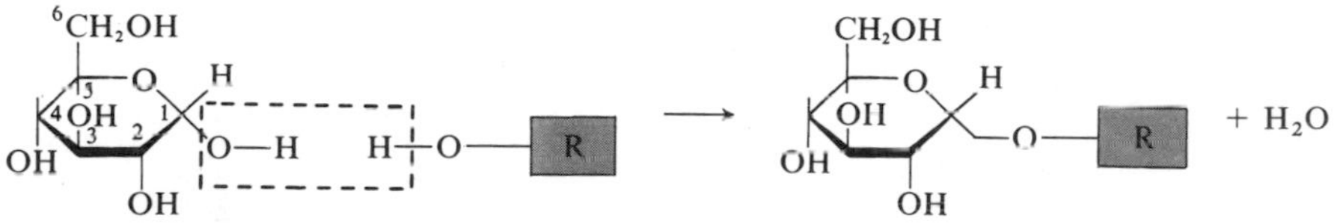

1 molecule glucose
(α-form)

R-α-glucoside, or,
1-R-α-D-glucose

Glycoside is a general name which can be used whenever it is unnecessary to specify the participation of any one particular sugar—it can be compared to such general names as alkyl, aryl, and acyl. Glycosides exist, of course, in α- and β-forms since the 'carbonyl-derived' hydroxyl group is involved in their formation.

For the alcohol which helps to form the glycoside linkage a second sugar molecule may be substituted—sugars contain alcoholic hydroxyl groups—and when this occurs a disaccharide is formed. Thus two α-D-glucose residues unite together in the following manner to give maltose, a sugar which is, essentially, a glucose-α-glycoside. That is to

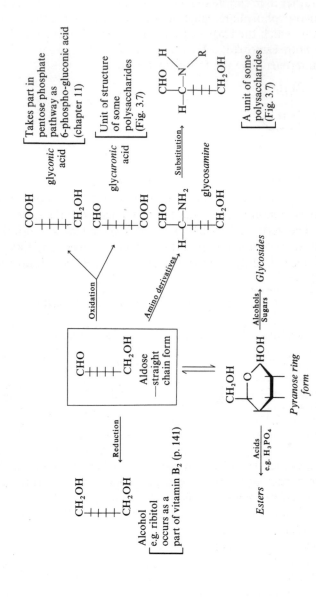

Fig. 3.5   Some derivatives of monosaccharides of biological importance

say, in the molecule of maltose, the sugar of the 'glycoside' part of the molecule is also glucose:

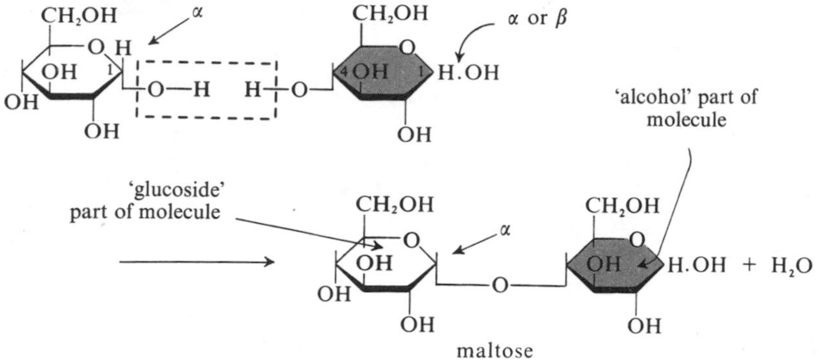

maltose

The repetition of this process in such a way that many monosaccharide units are held together by α- or β-glycoside linkages leads to the formation of polysaccharides; conversely, the hydrolysis of appropriate polysaccharides results in the liberation of the monosaccharide units from which they are constructed.

Several other monosaccharide derivatives play important but more restricted roles in cells or in the tissues of particular groups of organisms. They include oxidation and reduction products of monosaccharides as well as 2-amino and substituted 2-amino sugars. Some of these, with page or figure references, are included in Fig. 3.5.

3.6 Disaccharides

Maltose is a disaccharide in which *two* glucose units are joined through an α-1,4-glycoside linkage (Fig. 3.6). It is readily attacked by dilute acids and by *maltase*, an enzyme which specifically hydrolyses α-glycoside linkages and which has no action upon β-glycosides. Consequently, *cellobiose*, which is a disaccharide comprising glucose units joined by a β-glycoside linkage is not attacked by maltase, although it is hydrolysed by β-glycosidases and by dilute acids. *In vivo*, maltose is formed by the partial hydrolysis of starch; cellobiose is a breakdown product of cellulose.

Both maltose and cellobiose are reducing sugars, reducing cupric copper to cuprous oxide in the Fehling's test and the silver ion to metallic silver in the silver mirror reaction. However, they are not such powerful reducing agents as the monosaccharides glucose, galactose, and fructose,

Carbohydrates and nucleotides

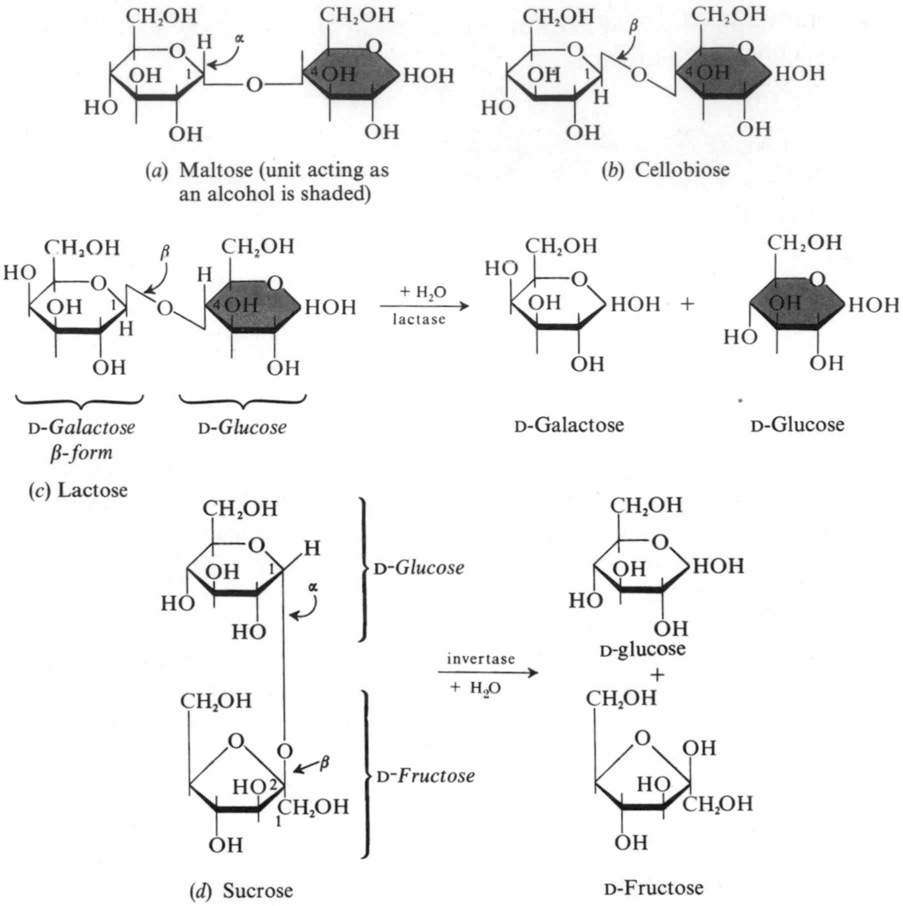
(a) Maltose (unit acting as an alcohol is shaded)
(b) Cellobiose
D-Galactose β-form
D-Glucose
+ H₂O
lactase
D-Galactose
D-Glucose
(c) Lactose
D-Glucose
D-Fructose
invertase
+ H₂O
D-glucose
+
(d) Sucrose
D-Fructose

Fig. 3.6 Some important disaccharides

and do not readily reduce Barfoed's reagent to cuprous copper. The reason for this difference is that, while carbon atom 1 of one of the two glucose residues is free to revert back to the carbonyl (reducing) form, that on the second residue is stabilized in the cyclic form by the existence of the glycoside linkage:

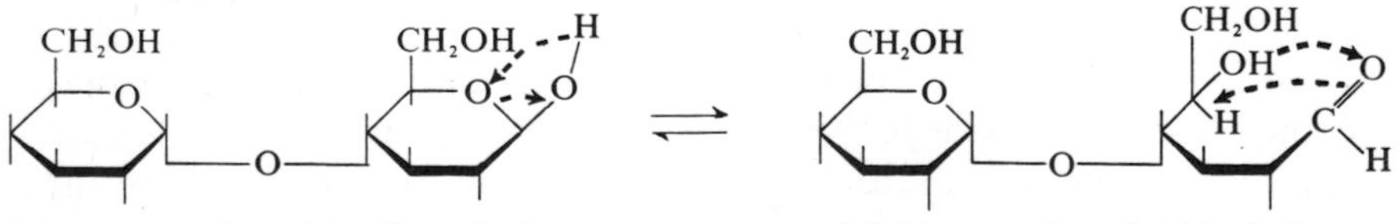

Maltose—outline of double cyclic form | Maltose—outline of aldehyde form

Some of the disaccharides which occur naturally are constructed from two different monosaccharide units, important examples being D-lactose, which is the main carbohydrate in milk, and sucrose, which is especially abundant in the sugar cane and sugar beet. D-Lactose consists of D-galactose in its β-form, linked to carbon atom 4 of D-glucose (Fig. 3.6c). It is hydrolysed to these two monosaccharides by *lactase*, a mammalian digestive enzyme.

Sucrose comprises a molecule of D-glucose linked through its 1-carbon atom to carbon atom 2 of fructose (Fig. 3.6d). Here, however, not only is the linkage α with respect to the glucose and β with respect to the fructose, but the glucose is present in its 6-membered pyranose form while the fructose is in the 5-membered furanose form. Since the two carbon atoms participating in the glycoside linkage are the two carbonyl-carbon atoms of the individual monosaccharides, neither half of the molecule can revert back to the straight chain form unless the molecule is destroyed by hydrolysis. Since there is no potential carbonyl group anywhere in the molecule, sucrose is a *non-reducing disaccharide*. An enzyme, *invertase*, capable of hydrolysing sucrose, is present in plant cells and in the absorptive cells of the small intestine of mammals. The enzyme acquires its name from the fact that, whereas sucrose is dextro-rotatory, the equimolar mixture of glucose and fructose obtained from it is laevo-rotatory; the 'inversion' of rotation occurs because fructose ('laevulose') is more laevo-rotatory than an equimolar quantity of glucose ('dextrose') is dextro-rotatory.

3.7 Polysaccharides

Polysaccharides are constructed from monosaccharide units by elimination of water from a number of such units. *Cellulose* is the most abundant polysaccharide in nature, since it is a major component of unlignified plant cell walls (chapter 1). It has great tensile strength and is resistant to chemical attack. It can only be hydrolysed by the most vigorous acid treatment, and few, if any, animals have enzymes capable of destroying it. Its breakdown by soil bacteria is, however, of great practical importance, as is its metabolism by bacteria in the rumen of such herbivores as cattle.

Starch is not a single substance but a mixture of at least two components. The simplest, amylose, is a straight chain 'condensate' of glucose residues and thus resembles cellulose except that, whereas the glycoside linkages in cellulose are β-linkages, those in starch are of the

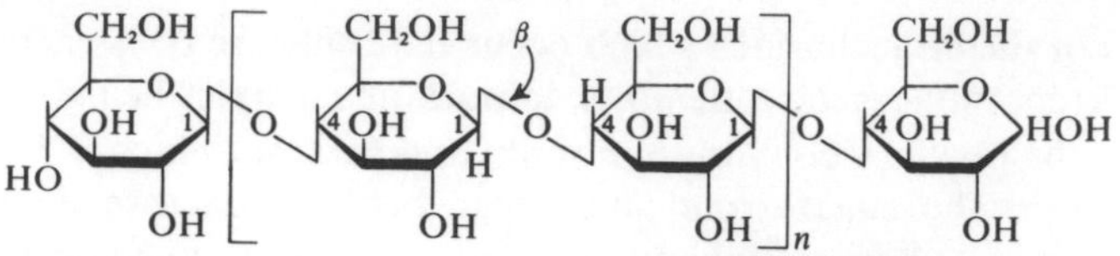

(a) Cellulose

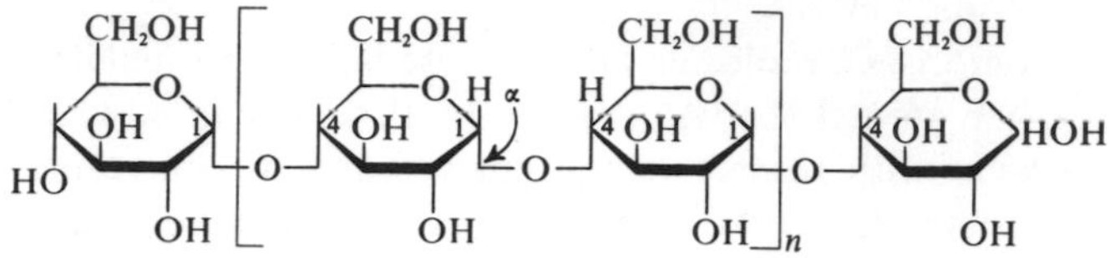

(b) Amylose

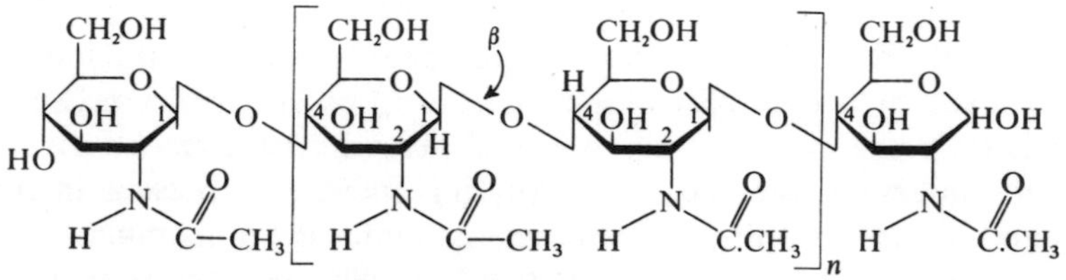

(c) Chitin (residues of N-acetyl glucosamine)

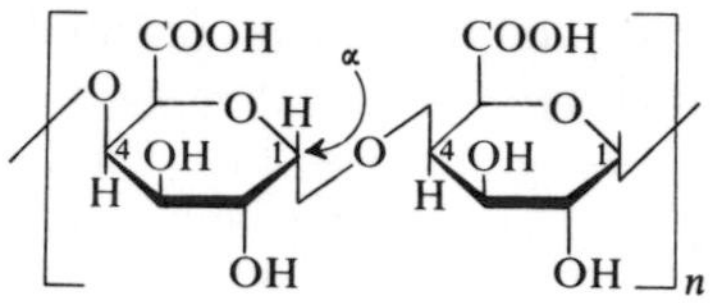

(d) Repeating unit of Pectin (residues of galacturonic acid)

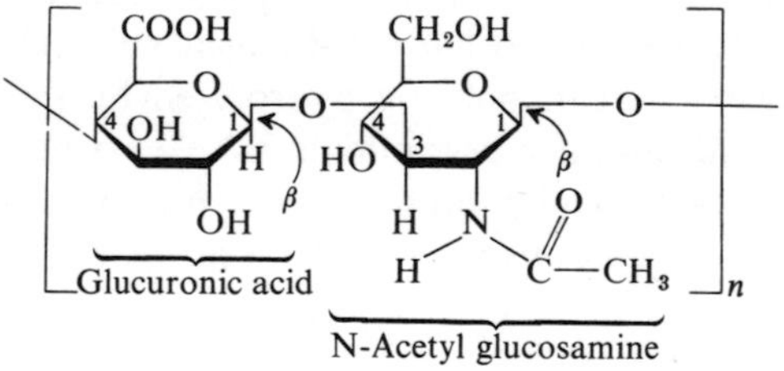

(e) Repeating unit of Hyaluronic acid

Fig. 3.7 Structure of some polysaccharides

α-form (Fig. 3.7b). Starch is much more readily hydrolysed than cellu-lose, hydrolysis being possible either by means of dilute acid or by the action of enzymes termed *amylases*. The amylases hydrolyse amylose with the formation of maltose; the disaccharide is then normally further degraded by maltase. Amylose reacts with iodine to give the well-known dark blue colour which disappears on warming and reappears on cooling. In common with other polysaccharides, starch is a non-reducing carbohydrate since carbonyl groups of all units (except one of the two terminal ones) participate in the glycoside linkages. The second component of starch is considered on p. 78.

The bulk of the carbohydrate in cells occurs as polysaccharide. This may be of structural importance as in the case of cellulose and the poly-saccharide-like substances chitin and pectin (Fig. 3.7c,d) or it may be a storage material such as starch in plants and glycogen in animals. The polysaccharides have molecular weights ranging from 10^3 to about 10^7, and, unlike the mono- and disaccharides, they are not readily soluble in water (some, such as starch, do however form colloidal solutions). In the cell, therefore, monosaccharide units, once incorporated into poly-saccharides, are effectively prevented from exerting any substantial osmotic pressure. Mono- and disaccharides tend to be the forms of carbohydrate which are transported between cells, e.g., glucose is carried about the mammal in the blood, and sucrose is transferred in plants from one organ to another in the phloem. Once inside the cell, however, monosaccharides are usually converted back into polysaccharides except for the small amount which is required immediately for metabolic pur-poses; the latter is instead converted to monosaccharide phosphates.

The two polysaccharides considered so far, amylose and cellulose, are linear polymers of glucose residues linked by 1,4-glycoside linkages. But some polysaccharides have a branched structure. For example, besides amylose, starch contains a compound known as amylopectin. On digestion with pancreatic amylase, this yields not only maltose but also a second disaccharide, iso-maltose, in which two glucose residues are united by a 1,6-α-glycoside linkage:

Iso-maltose
(glucose 1,6-α-glucoside)

Carbohydrates and nucleotides

This suggests that amylopectin consists of chains of glucose residues resembling amylose (α-1,4-glycoside linkages) but having secondary branches which are united to the main chains by 1,6-glycoside linkages. The structure of amylopectin is indicated in Fig. 3.8. The average chain

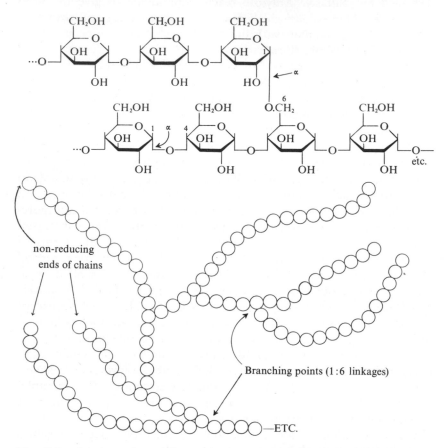

Fig. 3.8 Amylopectin. (Above, the interchain link. Below, a small fragment of amylopectin showing branching. Each circle represents a glucose unit)

contains 20 to 25 glucose units and the number of glucose residues between branching points within a chain is 5 to 8. Glycogen, the major carbohydrate reserve of animals, resembles amylopectin in structure but is more highly branched, the average chain length being 10–14 glucose residues, and the number of glucose units between branching points only 3 or 4.

Plant cell walls usually contain considerable quantities of cellulose, although a variety of other compounds is often present in lesser amounts. The cellulose chains frequently aggregate together forming stiff threads (microfibrils) readily visible under the electron microscope. Together the fibres form a tough network or basket around the cell. During growth of plant cells, the cellulose wall develops gradually, and very young plant cell walls contain several materials more flexible than cellulose. One of these, *pectin*, consists of a long chain of sugar-like units derived from galactose. The unit of structure is *galacturonic acid*, i.e., galactose in which the primary alcohol group of carbon atom 6 has been oxidized to a carboxylic acid group. The galacturonic acid units are joined by α-1,4 linkages (Fig. 3.7d). Pectin has jelly-like properties allowing easy expansion of the growing cell. The carboxyl groups are often in the form of their calcium salts. The jellifying properties of pectin are frequently employed in making jams 'set'.

In animals, too, polysaccharides play an important structural role. The cuticle of insects, for example, contains *chitin*, a polymer of a derivative of glucose in which the hydroxyl group attached to carbon atom 2 is replaced by amino groups which are in turn acetylated. Chitin consists of residues of *N-acetyl glucosamine* joined by β-1,4 links (Fig. 3.7c). The same structural unit is present in *hyaluronic acid*. This polysaccharide-like substance occurs as a component of several gelatinous tissues of vertebrates including the vitreous humor of the eye, and the synovial fluid which lubricates the joints between bones. Hyaluronic acid is composed of alternating units of N-acetyl glucosamine and glucuronic acid (the glucose equivalent of galacturonic acid). Hyaluronic acid is of interest in that its molecule contains, in addition to the familiar 1,4 type of glycoside linkage, the much less common 1,3-glycoside bond (Fig. 3.7e).

3.8 Nucleotides and nucleosides

These compounds are of very great biochemical importance. Since they contain ribose, or a near relative of ribose, they can be introduced conveniently as carbohydrate derivatives at this point. Their very varied functions include their roles as carriers of *energy*, as carriers of *electrons* and as carriers of *information*. These functions are considered in later

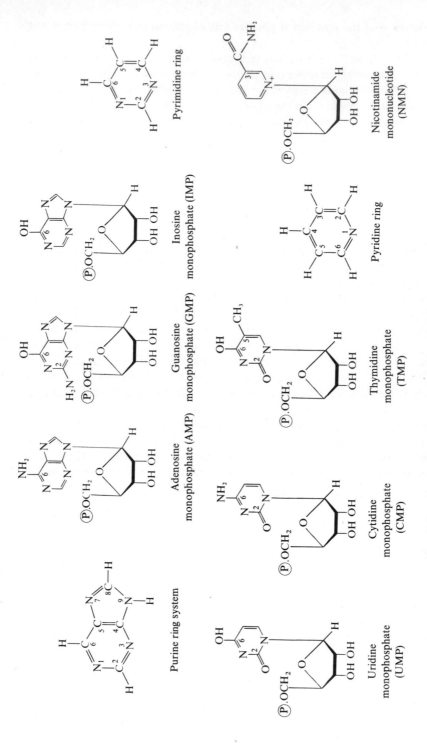

Fig. 3.9 Seven important nucleotides and their parent ring systems

chapters, the intention of the present section being to provide an overall impression of their structure and importance.

Nucleotides consist of three essential parts: a *heterocyclic base* linked to carbon atom number one of a *ribose* (or deoxyribose) molecule (section 3.3) which, in turn, has a *phosphate group* attached to it. Almost all the nucleotides discussed in this book have the phosphate group linked to carbon atom number five of the pentose and they will be represented thus:

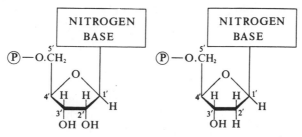

A ribose nucleotide   A deoxyribose nucleotide

The name of the heterocyclic nucleotide base (Table 3.1) will be inserted each time in the rectangle. In most cases the base is a derivative of *purine* or *pyrimidine*, though one important base, nicotinamide, is a *pyridine* derivative (Fig. 3.9). The commonest purine nucleotides are adenosine monophosphate (AMP) and guanosine monophosphate

Table 3.1   The major purine and pyrimidine bases and their respective nucleosides and nucleotides

Base	Nucleoside	Nucleotide
1. Purines:		
Adenine (6-amino purine)	Adenosine	Adenosine monophosphate (AMP), or adenylic acid
Guanine (2-amino-6-oxypurine)	Guanosine	Guanosine monophosphate (GMP), or guanylic acid
Hpoxanthine (6-oxypurine)	Inosine	Inosine monophosphate (IMP), or inosinic acid
2. Pyrimidines:		
Uracil (2,6-dioxypyrimidine)	Uridine	Uridine monophosphate (UMP), or uridylic acid
Cytosine (2-oxy-6-amino pyrimidine)	Cytidine	Cytidine monophosphate (CMP), or cytidylic acid
Thymine (5-methyl uracil)	Thymidine	Thymidine monophosphate (TMP), or thymidylic acid

Note: Since tautomeric forms exist, 2-oxy, 6-oxy can equally well be called 2-hydroxy, 6-hydroxy.

Carbohydrates and nucleotides

(GMP).   Two methods of numbering the atoms of the pyrimidine ring system are currently employed; in this book the scheme adopted is the one which resembles that used for the six-membered ring of purine.   To differentiate the carbon atoms of the pentose from those in the base, it is often convenient to refer to them as 1′, 2′, 3′, etc.   Thus the base-sugar linkage in AMP involves N-9 and C-1′.

Figure 3.9 shows three frequently-encountered pyrimidine nucleotides, namely, uridine monophosphate (UMP), cytidine monophosphate (CMP), and thymidine monophosphate (TMP).   Note that in these, as in the purine nucleotides, the base is attached directly to the ribose carbon atom by a β-N-glycosidic bond.   For the present, only the nucleo-

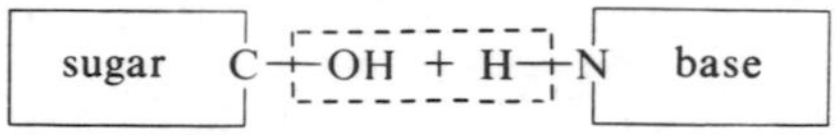

$$\boxed{\text{sugar} \quad \text{C}\!-\!\!\text{OH} + \text{H}\!-\!\!\text{N} \quad \text{base}}$$

tides containing D-ribose will be considered, for generally, the deoxyribose nucleotides occur only in a polynucleotide, deoxyribonucleic acid (DNA), the structure of which is discussed in chapter 14.

Nucleosides are similar to the nucleotides except that they *lack the phosphate group*.   For example, adenosine is a nucleoside corresponding to adenosine monophosphate (AMP, Fig. 3.9) but the hydroxyl group attached to carbon atom 5′ of the ribose is not esterified by attachment to phosphate.   Nucleosides, as such, only very rarely take part in cell metabolism whereas nucleotides participate directly in biochemical reactions.   Table 3.1 lists the names of the principal purine and pyrimidine bases and their corresponding nucleosides and nucleotides.   The reader is recommended first to get to know the names of the nucleotides, and then, at a later stage, the names and formulae of the bases.

The formulae of the nucleotide bases in Fig. 3.9 are the simplest structures which can be ascribed to them.   In reality, they are resonance hybrids of several structures and, moreover, the hydrogen atoms of hydroxyl groups undergo tautomeric changes.   For example, the base guanine can be represented by either of the two formulae:

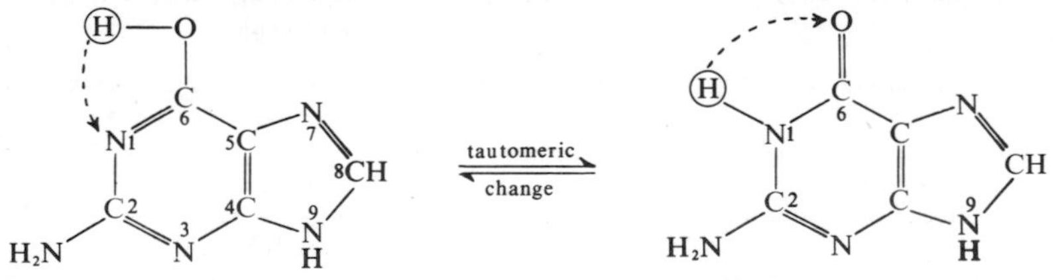

The hydrogen atom of nitrogen atom 9 is the one which is lost when the base unites with ribose or deoxyribose.   It is also worthy of note that the hydrogen atoms attached to the nitrogen and oxygen atoms of the nucleotide bases readily form hydrogen bonds which are vital in maintaining the structure of DNA.

Adenosine monophosphate (AMP) is a very important nucleotide for not only is it one of the components of DNA and RNA, substances which can be regarded as *carriers of information*, but it is also part of an *electron carrier* called NAD (nicotinamide adenine dinucleotide). Moreover, it plays a vital role as the organic component of the ubiquitous *energy carrier*, ATP, adenosine triphosphate. When energy of an appropriate form is available, inorganic phosphate (P_i) will combine with AMP to form, first, adenosine *di*phosphate (ADP) and then adenosine *tri*phosphate (ATP). The phosphate linkages of these last two compounds resemble those of inorganic pyrophosphate. Rather a large amount of energy is associated with these linkages—or, more precisely, is made available when they are broken.

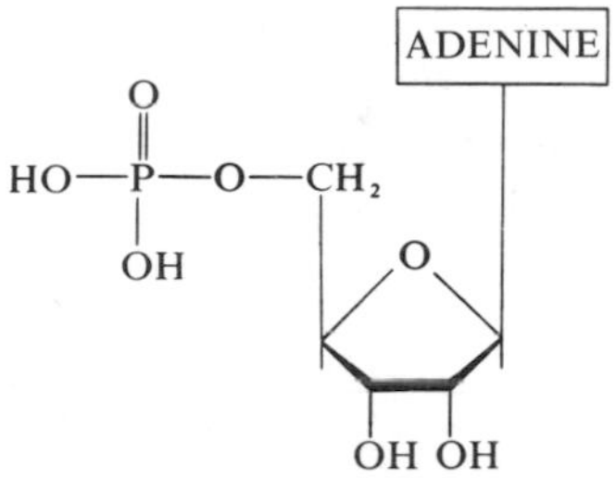

Adenosine monophosphate
(AMP)

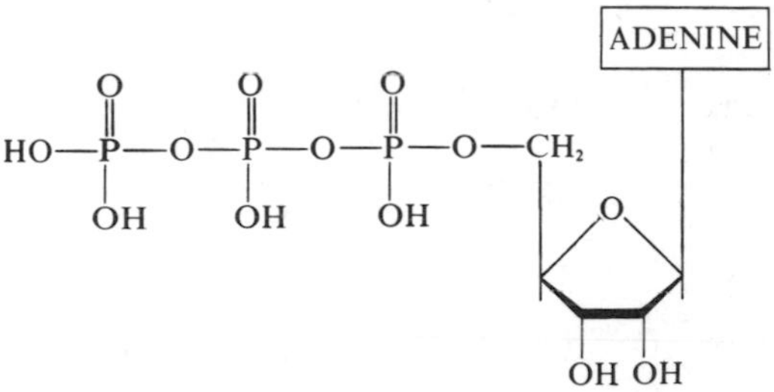

(unionized forms)

Adenosine triphosphate
(ATP)

The function of ATP in energy transfer was mentioned in chapter 1 and will be discussed in relation to bioenergetics (chapter 5). Briefly, however, when the terminal phosphate group of ATP is removed, energy is released which, in the presence of appropriate machinery, is not squandered as heat (as it is when hydrolysis takes place), but can be utilized to cause a muscle to contract or to drive reactions which only occur if a supply of energy is available.

The other nucleotides shown in Fig. 3.9 also form di- and triphosphates but these are rather less frequently concerned in energy transfer than are their adenosine counterparts. Nevertheless, they have specialist uses in this connexion; in particular, they will be encountered in relation to the biosynthesis of di- and polysaccharides (chapter 10) and of the nucleic acids (chapter 14).

3.9 Dinucleotides

Rather as two monosaccharides can be linked to form a disaccharide, two mononucleotides can be linked together by their phosphate groups to give biologically important compounds called dinucleotides:

$$\underbrace{\text{BASE-RIBOSE-}\textcircled{P}}_{\substack{\text{first}\\ \text{mononucleotide}}}\underbrace{\textcircled{P}\text{-RIBOSE-BASE}}_{\substack{\text{second}\\ \text{mononucleotide}}}$$

The most frequently encountered dinucleotide consists of the (mono) nucleotide, AMP, linked to the nucleotide which has nicotinamide as the heterocyclic base (Fig. 3.9). It is known as *nicotinamide adenine dinucleotide* (NAD) and it can be represented in the following way:

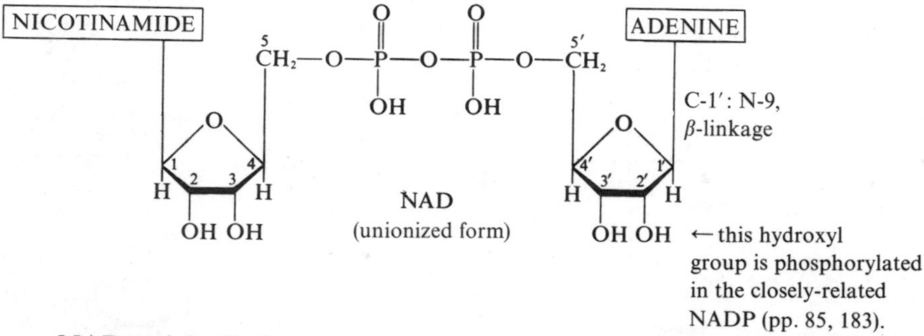

NICOTINAMIDE | ADENINE

NAD
(unionized form)

C-1′: N-9,
β-linkage

← this hydroxyl group is phosphorylated in the closely-related NADP (pp. 85, 183).

NAD, originally known as coenzyme I and later as diphosphopyridine nucleotide (DPN), is of great importance in cell biochemistry because it participates in numerous oxidation reactions. Only the *nicotinamide* part of the molecule is directly involved, although no doubt the shape of the molecule is important in relation to stereospecific attachment to enzymes. It can exist in oxidized and reduced forms with the following structures:

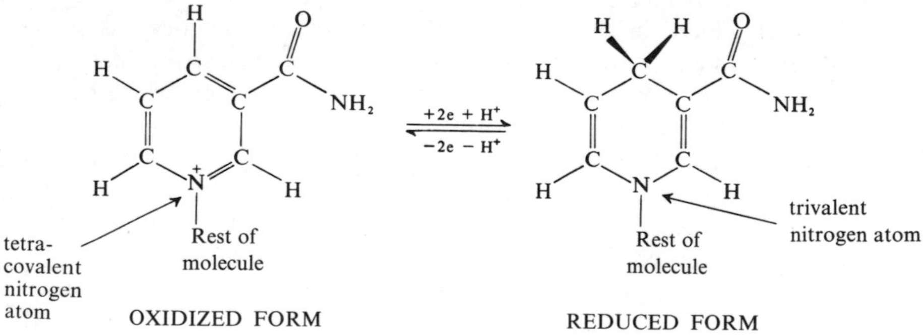

$$\xrightleftharpoons[-2e - H^+]{+2e + H^+}$$

tetra-covalent nitrogen atom

Rest of molecule

OXIDIZED FORM

Rest of molecule

trivalent nitrogen atom

REDUCED FORM

Oxidized NAD is often written as NAD^+ because it has a positive charge on its pyridine tetracovalent nitrogen atom. (A somewhat similar relationship is shown by that of NH_4^+ to NH_3.) In the reduced form, NAD possesses two *electrons* and one *proton* more than the oxidized form and is often written as NADH.

$$NAD^+ + 2\,\text{electrons} + H^+ \rightarrow NADH$$

Numerous oxidation reactions occur in the cell and many of these require the cooperation of NAD. The mechanism can be explained briefly by considering a hypothetical organic compound AH_2. Its oxidation involves the removal of the two hydrogen atoms (i.e., 2 protons together with 2 electrons) and can be represented thus:

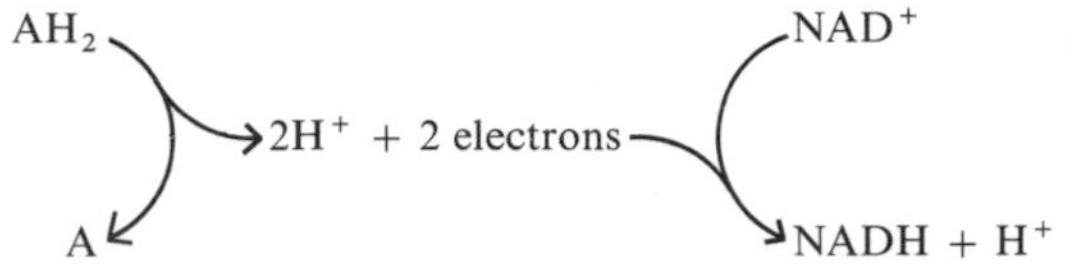

The two electrons are taken up by NAD^+ together with one proton to give reduced NAD (i.e., NADH). The reaction is normally written:

$$AH_2 \quad + \quad NAD^+ \quad \rightleftharpoons \quad NADH \quad + H^+ \quad + \quad A$$

Organic compound	Oxidized form of NAD	Reduced form of NAD	Oxidized form of organic compound

In subsequent chapters the role in the cell of this important *coenzyme* will become evident. It is noteworthy that there is a similar dinucleotide which differs from NAD only in having the hydroxyl group on carbon atom 2′ of the ribose in the adenosine part of the molecule esterified by an additional phosphate group. It is known as nicotinamide adenine dinucleotide phosphate or NADP. (Normally, in speaking of such compounds, initials are used because the full chemical names are cumbersome.) NADP was known originally as coenzyme II or triphosphopyridine nucleotide (TPN).

Three substances related to mono- and dinucleotides play vital roles as cofactors in various enzyme-catalysed reactions. Two of these three cofactors, known as FMN and FAD, are more conveniently considered elsewhere (p. 146), but the third, a compound known as *coenzyme A*, must be briefly mentioned now. This coenzyme is not strictly a dinucleotide but its resemblance to dinucleotides invites the title of 'pseudodinucleotide'. The functions of coenzyme A can be understood without a detailed consideration of its structure for almost invariably only one part of the molecule, a *thiol group* (H—S-) directly takes part in the biochemical reaction. As with other coenzymes, however, the rest of the molecule is undoubtedly important in relation to attachment to enzymes. Generally, coenzyme A is written as *HS—coenzyme A* (or coenzyme A—SH) to emphasize the part played by the thiol group.

Coenzyme A consists of the 'pseudo-mononucleotide', phosphopantetheine, attached to the nucleotide AMP by a pyrophosphate link. True dinucleotides are composed of mononucleotides united in exactly the same way. The AMP part of the coenzyme A molecule has an additional phosphate group on carbon atom 3':

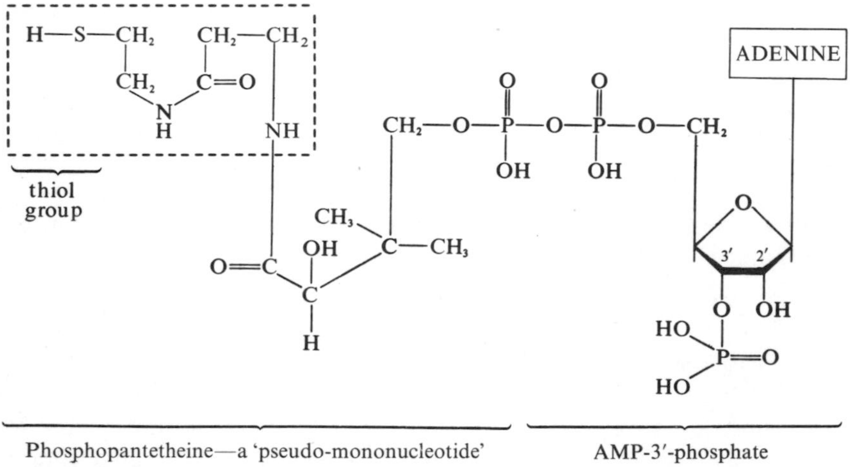

Phosphopantetheine—a 'pseudo-mononucleotide'   AMP-3'-phosphate

The major function of coenzyme A in the cell is to unite with carboxylic acids to form 'high-energy' acyl derivatives which are analogous in certain ways to acyl chlorides.

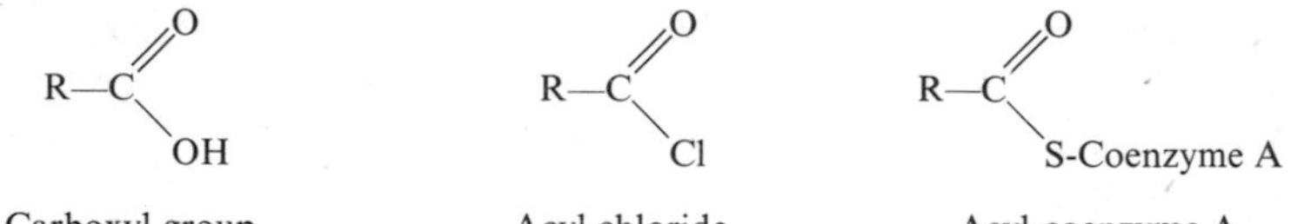

Carboxyl group   Acyl chloride   Acyl-coenzyme A

The formation in the cell of acylated derivatives of coenzyme A requires a supply of energy. Conversely, the high energy content of acyl coenzyme A derivatives renders them, like acyl chlorides, much more reactive in particular circumstances than are the uncombined carboxylic acids. It will be recalled that coenzyme A derivatives were mentioned in chapter 1 (p. 27). The part played by these high energy compounds in cellular energetics forms a fascinating subject which is considered in later chapters.

4

Enzymes

4.1 Enzymes as catalysts

Enzymes are catalysts formed by cells but capable of functioning *in vitro* if the conditions are appropriate. In the simplest case the reaction will occur if a solution of the enzyme is added to its *substrate* (the reactant), held at a suitable temperature and pH. Sometimes, however, as in other catalytic reactions, co-factors of various kinds may be essential, and, early in the history of biochemistry, it was discovered that the purification of an enzyme was frequently associated with diminution of enzymic activity owing to loss of the cofactors. For living cells enzymes are of vital importance for they accelerate a very large number of essential biochemical reactions. The cell must perform chemical reactions within a relatively narrow range of physical conditions—high temperature and extremes of pH are incompatible with the existence of the living cell. Enzymes thus enable reactions to occur under physiological conditions which would otherwise be unacceptably slow.

All enzymes so far isolated are proteins or polypeptides and are therefore of large molecular weight and do not dialyse. In common with other proteins, they are denatured by various agents and when denatured they lose their activity. They are often adversely affected by heat although a few can tolerate damp heat near the boiling point—heat lability is one important characteristic of enzymes which distinguishes them from most inorganic catalysts. A second difference is the *specificity* enzymes usually show for the *substrate* they act upon—by contrast, inorganic catalysts often accelerate several quite different reactions. Platinum, for example, catalyses numerous oxidation reactions including the Contact process, the catalytic oxidation of ammonia, and the combustion of hydrogen in oxygen. Enzymes are often named from the substrate they act upon, the ending -*ase* being added to the root of the substrate's name (e.g., RNA-ase, lipase). Sometimes the name also incorporates the *type* of reaction involved (e.g., succinate dehydrogenase, cytochrome oxidase).

4.2 The specificity of enzymes

It is possible to subdivide enzymes into groups according to the degree of specificity they exhibit. The most striking example of specificity is demonstrated by enzymes exhibiting an *absolute specificity* for one substrate. Urease, for example, catalyses the reaction:

$$H_2O + NH_2.CO.NH_2 \rightarrow CO_2 + 2NH_3$$
$$\text{urea}$$

Sometimes, and particularly when the reaction involves substrate oxidation, it is found that an absolute dependence also exists for a *second* substance, such as one which accepts electrons from the substrate. The acceptor is, in a sense, a secondary substrate, since its molecules are aligned on the enzyme surface, just as are the molecules of the substrate. Thus *lactate dehydrogenase* removes two electrons and two hydrogen ions from lactate, but only if the acceptor or carrier of electrons, nicotine adenine dinucleotide (NAD), is present:

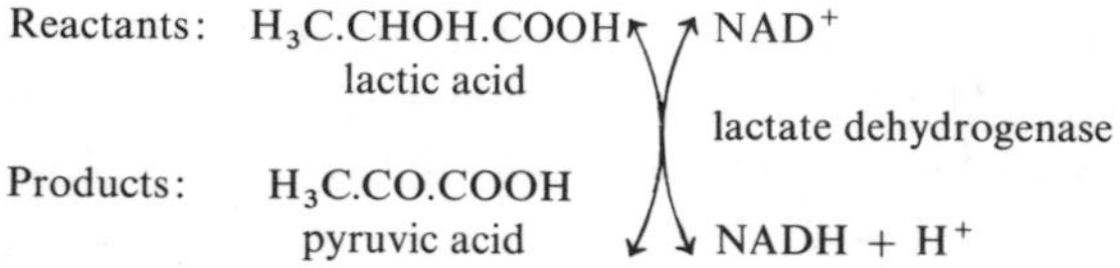

Reactants: $H_3C.CHOH.COOH$ NAD^+
             lactic acid

                           lactate dehydrogenase

Products: $H_3C.CO.COOH$
             pyruvic acid $NADH + H^+$

The manner in which NAD^+ is able to accept electrons was indicated in the last chapter (p. 84).

It is often convenient to write more complex biochemical changes in the way we have just indicated; in this example, the double arrows indicate that the reaction is *reversible*, i.e., if reduced carrier and pyruvate are put together in the presence of the enzyme, some lactate and NAD^+ are formed (p. 207). Such reversibility can lead to a semantic difficulty, for, if the enzymic reduction of pyruvate is encountered in isolation, it might appear as though the enzyme responsible, lactate dehydrogenase, had been named incorrectly by reference to the product rather than the reactant.

Lactate dehydrogenase of animal muscle is one of many enzymes which exhibit *stereochemical specificity*, for it only oxidizes L(+) lactic acid. More strikingly, perhaps, when acting in reverse, it reduces pyruvic acid only to L(+) lactic acid, whereas chemical synthesis by addition of hydrogen to the carbonyl double bond always gives the racemic mixture. Similarly, it is usual for only one of a pair of geometrical isomers to be formed by enzymic action. *Succinate dehydrogenase*, for example, converts fumarate to succinate, but has no action on the corresponding *cis*-isomer, maleic acid:

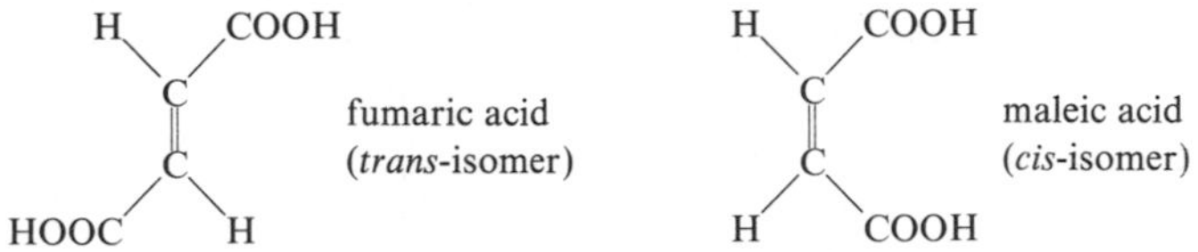

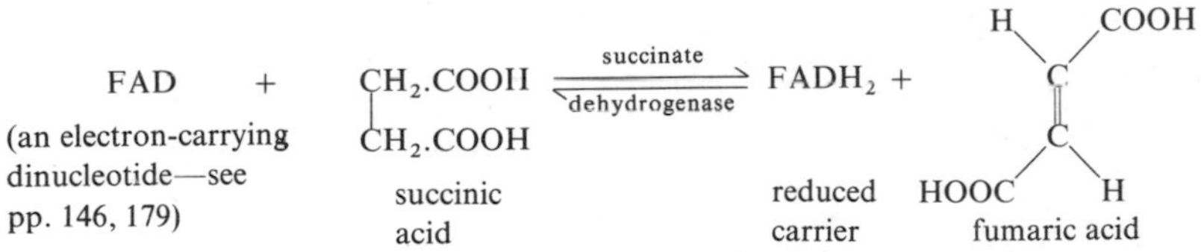

fumaric acid
(*trans*-isomer)

maleic acid
(*cis*-isomer)

FAD   + $CH_2.COOH$ $\xrightleftharpoons[\text{dehydrogenase}]{\text{succinate}}$ $FADH_2$ +

(an electron-carrying dinucleotide—see pp. 146, 179)

$CH_2.COOH$

succinic acid

reduced carrier

fumaric acid

A further type of specificity shown by some enzymes is known as *group specificity*.   As the name implies, the enzymes catalyse reactions involving a series of substrates which have in common one identical group but differ in some other way.   For example α-*glucosidase* hydrolyses several α-glucosides (Fig. 4.1).   In each case a hydroxyl compound (R—OH) such as methanol, ethanol, or glucose is removed from the α-glucoside,

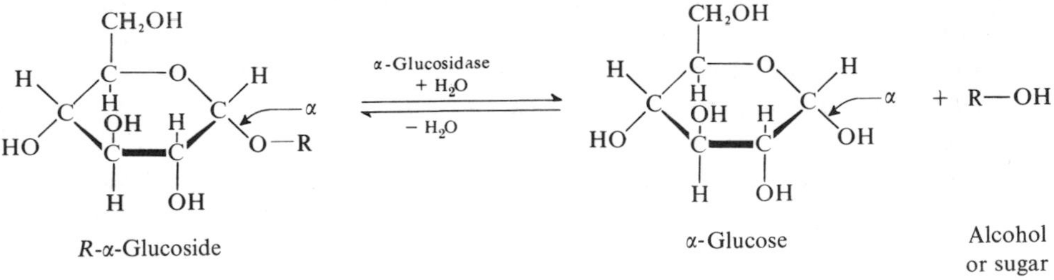

R-α-Glucoside

α-Glucose

Alcohol or sugar

Fig. 4.1   A group-specific enzyme, α-glucosidase.   (α-glucosidase hydrolyses α-glucosides but not β-glucosides)

the other product in each case being glucose. The enzyme maltase, which splits maltose into two molecules of glucose, is an example of an α-glucosidase, for maltose is an α-glucoside (p. 73). But in addition to hydrolysing maltose, maltase will attack a range of α-glucosides, though it attacks each at a different rate, indicating that the affinity of the enzyme with each substrate is different. On the other hand, maltase does not hydrolyse cellobiose because this sugar is a β-glucoside.

A final group of enzymes comprises members of *low specificity*; here, the only requirement from the substrate is that it should possess a particular linkage. Thus certain esterases will attack a wide range of compounds of the type R—COO—R′, where both R and R′ can be varied considerably. Lipase, for instance, will hydrolyse (at different rates) most of the natural fats—if this were not so, there would have to be a specific lipase for each of the several dozen different glycerol–carboxylic acid linkages which exist in dietary fats and oils.

$$3H_2O + \text{glyceryl trioleate} \xrightarrow{\text{lipase}} \text{glycerol} + 3 \text{ oleic acid}$$

$$3H_2O + \text{glyceryl tristearate} \xrightarrow{\text{lipase}} \text{glycerol} + 3 \text{ stearic acid}$$

$$3H_2O + \text{glyceryl tripalmitate} \xrightarrow{\text{lipase}} \text{glycerol} + 3 \text{ palmitic acid}$$

4.3 Classification of enzymes

Over the years, the naming of enzymes has developed somewhat haphazardly, but usually in accordance with the principle mentioned above—that the substrate for the reaction provides, directly or indirectly, the stem of the name of the enzyme. Enzymes can often be grouped into families, united by the common factor of the type of substrate they attack; there are, for example, numerous proteinases (e.g., pepsin), numerous esterases, and numerous carbohydrases (e.g., invertase and maltase). Rather similarly, the dehydrogenases all remove electrons and protons (often apparently in the form of hydrogen atoms) from an oxidizable substrate and pass them on to NAD or some other carrier, while phosphatases remove phosphate groups from phosphate esters. Yet others, such as the transaminases and transacetylases, form families named according to the nature of the groups they transfer; thus cysteine transaminase removes the elements of ammonia ($H.NH_2$) from the amino acid, cysteine, and inserts an oxygen atom instead, while tyrosine transaminase does the same for tyrosine.   The following is a typical transamination reaction:

$$\underset{\substack{\text{L-Amino}\\\text{acid}}}{R-\overset{\displaystyle H}{\underset{\displaystyle NH_2}{C}}-COOH} \;+\; \underset{\substack{\text{Keto acid}-\\\text{often }\alpha\text{-ketoglutaric}\\\text{acid}}}{X-\overset{\displaystyle }{\underset{\displaystyle O}{C}}-COOH} \quad\xrightarrow[\text{transaminase}]{\text{R-specific}}\quad \underset{\substack{\text{Keto acid}-\\\text{corresponding to}\\\text{original amino acid}}}{R-\overset{\displaystyle }{\underset{\displaystyle O}{C}}-COOH} \;+\; \underset{\substack{\text{New amino}\\\text{acid}}}{X-\overset{\displaystyle H}{\underset{\displaystyle NH_2}{C}}-COOH}$$

In the light of the gradual isolation of a wider and wider range of enzymes, and the development of an increasing insight into the mechanism of enzyme action, it has become apparent that many of the names given to the earlier-discovered enzymes are in various ways inadequate.   Nevertheless, well-established names die hard—especially when they have the advantage of being reasonably short; amylase, decarboxylase, hexokinase, even succinate dehydrogenase, are names which are fairly easily memorized.   Short names are, however, sometimes misleading, especially when, in addition to enzyme and substrate, some obligatory third substance actively participates in the reaction.   For this reason, the Enzyme Commission of the International Union of Biochemistry, in 1964, devised a scheme of classification which brought considerable order to the rather chaotic situation then existing.

In effect, the Enzyme Commission allocated a unique four-part 'quantum number' and a systematic name to each known enzyme.   The principal 'quantum' number (as it may perhaps be called) is a number from 1 to 6, placing the enzyme in one of the six major groups listed in Table 4.1.   Each of these six is then subdivided into subquanta, and each of these again subdivided.   Individual enzymes are then designated by a fourth number.   For example, all enzymes of group 3 are *hydrolases*, but those in sub-group 3.1 act only on *ester* linkages.   Ester linkages are themselves of several types, the *carboxylic* ester hydrolases being designated 3.1.1.   Within this last group are numerous enzymes, each with its separate number; 3.1.1.7, for example, is the enzyme which splits acetylcholine into the base, choline, and acetic acid.

In this book, a rigid choice between the older, shorter nomenclature and the newer, more systematic one, has not been made; instead, the method appearing most satisfactory in a particular context has been employed.   Nevertheless, a cross-reference system relating the earlier

Table 4.1   Classification of enzymes into six main groups (See also Appendix 1)

1. *Oxidoreductases:*	These add or subtract electrons, oxygen, or hydrogen; e.g., oxidases, dehydrogenases.
2. *Transferases:*	These transfer a group from one organic molecule to another; e.g., methyl group transferases, acyl group transferases, phosphotransferases, aminotransferases.
3. *Hydrolases:*	These split a molecule in two by the action of water; e.g., those acting on ester bonds, on polysaccharide linkages, on peptide linkages.
4. *Lyases:*	These remove groups non-hydrolytically, leaving a double bond, or, in reverse, add groups to double bonds; e.g., decarboxylases, citrate condensing enzyme, aldolase.
5. *Isomerases:*	These bring about a redistribution of atoms, or groups of atoms, within a molecule; e.g., epimerases, mutases.
6. *Ligases:*	These link together two molecules, always at the expense of a high-energy compound, usually ATP; e.g., the enzymes which unite amino acids to their specific s-RNAs, the enzymes leading to acyl CoA by uniting the free acid and coenzyme A.

names with the newer ones has been provided in Appendix 1 for many enzymes mentioned in the text.

4.4   The chemical and physical characteristics of enzymes

Sumner, in 1926, prepared a pure crystalline specimen of urease, and, since then, a large number of other enzymes has been produced in crystalline form.   Some enzymes are simple proteins but others are conjugate, having a non-protein group more or less closely associated with the protein *apoenzyme*; the total complex is termed the *holoenzyme*. When the non-protein part of the complex is firmly bound, it can be regarded as an integral part of the structure and is then called a *prosthetic* group, whereas, if it is very loosely bound, it can more properly be considered as a separate entity, termed a *coenzyme*.   In practice, however, there are intermediate examples, and no sharp dividing line exists.   Many coenzymes are of relatively low molecular weight and can be dialysed away from the protein apoenzyme, whereas true prosthetic groups remain attached to the apoenzyme.

Enzyme-catalysed reactions, like all others, proceed more rapidly as the temperature is raised, but since enzymes are denatured by heat, a point is eventually reached where the rate of destruction of the enzyme is so great that it off-sets the increase in the rate of the reaction catalysed by enzyme which is still functioning. Consequently, an *optimal temperature* exists for enzyme reactions (Fig. 4.2). Nevertheless, even for one enzyme operating under standard conditions, this optimum is not uniquely fixed, for the factor of time must also be taken into account. In reactions lasting only a few seconds, for example, the optimum might be 20–30°C higher than when the same enzyme-catalysed reaction is studied over a longer period of time.

The catalytic efficiency of any enzyme is markedly affected by the pH of the surroundings, the activity rising rapidly to a maximal value at the

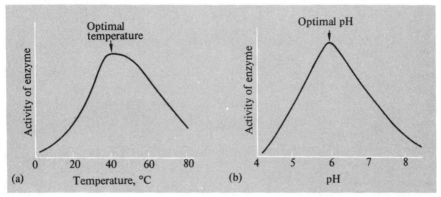

Fig. 4.2 The effect (a) of temperature, and (b) of pH on the activity of an enzyme. The actual optimal values vary with the enzyme concerned

optimal pH and then falling off again as the pH is increased further (Fig. 4.2). This occurs because only one of the many possible ionic forms the protein can assume (p. 36) possesses catalytic activity, and at the optimal pH it is this ionic form which predominates. When the pH is altered a *little* from the optimum, the loss of activity is reversible, thus differing from the (irreversible) loss of activity which occurs when the pH change is too great, or when the enzyme is denatured by heating or in some other way.

Enzymes, being proteins, are also inactivated by numerous chemical reagents which denature proteins (p. 58). Examples of chemical denaturing agents are heavy metal ions, trichloroacetic acid, and tannic acid. However, these materials very often markedly inhibit enzyme activity before visible precipitation of protein occurs and it is therefore likely that relatively small changes in the secondary or tertiary structure of the protein can result in loss of enzyme activity.

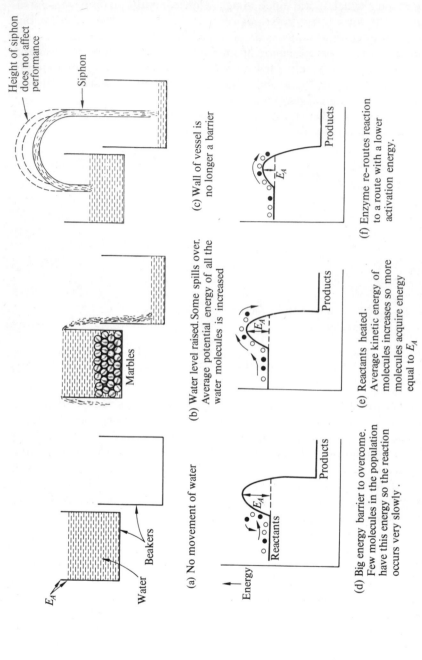

(a) No movement of water

(b) Water level raised. Some spills over. Average potential energy of all the water molecules is increased

(c) Wall of vessel is no longer a barrier

(d) Big energy barrier to overcome. Few molecules in the population have this energy so the reaction occurs very slowly .

(e) Reactants heated. Average kinetic energy of molecules increases so more molecules acquire energy equal to E_A

(f) Enzyme re-routes reaction to a route with a lower activation energy.

Fig. 4.3 An analogy drawn between physical and chemical systems to illustrate the meaning of potential energy and 'activation'.

4.5 How enzymes accelerate reactions

The function of enzymes is to hasten attainment of the equilibrium state. Without them, many cellular reactions would occur too slowly to support life. For a reaction to be possible, it is necessary for reacting molecules to possess a certain minimal energy, E_A, the activation energy,[#] which is an attribute of a particular chemical reaction occurring by a particular reaction mechanism. When molecules of potential reactants which possess less than this minimal energy are brought together, they fail to react; indeed, for most reactions, only a very small proportion of the total number of collisions between molecules of reactants results in a chemical change.

Possession of energy in excess of E_A is a fundamental requirement if molecules are to have a *chance* of reacting, but, even then, reaction may not occur, for it is also necessary that the molecules of each reacting species should be appropriately orientated with respect to the other when they come together. Enzymes probably affect the rate of biochemical reactions both by enabling more molecules to overcome the energy barrier for the reaction, and by increasing the probability of correct orientation at the moment of 'collision'.

Enzymes enable reactions to by-pass high activation energy barriers by *re-routing* reactions; re-routed reactions are essentially different reactions, with their own activation energies, even though both the substances disappearing and the substances being formed are the same as when no enzyme is present. This is usually achieved by reactant molecules forming an intermediate complex with the enzyme so that they are suitably placed and in an appropriate state of electron activation for the reaction to occur.

An analogy which illustrates some aspects of activation may be helpful so long as it is recognized that no analogy is perfect. Water will not flow spontaneously from one beaker to another placed beneath it, because the walls of the upper vessel act as a barrier. However, the barrier can be overcome by several means; an obvious one would be to put marbles into the beaker so as to raise the water level (Fig. 4.3). Such a method affects all the water just as *heating* a reaction mixture raises the average kinetic energy of the molecules in the system. A more efficient way to move the water is to use a siphon; this causes water to flow without the necessity of raising the level (i.e., the potential energy) of the whole mass of water, i.e., only the water in the siphon is 'activated'.

An enzyme no more affects the energy change occurring when reactants form products (Fig. 4.3f) than the siphon affects the relative height of the two beakers—*the enzyme, like the siphon, merely organizes a transitional stage.* Water stops flowing through the siphon as soon as the level of water is the same in the two beakers. Similarly, a reaction catalysed by an enzyme effectively ceases when chemical equilibrium is reached. The analogy can be taken a step further, for if water is constantly removed from the lower beaker to ensure that the water levels do not equalize, water will continue to flow through the siphon. So too, an enzyme continues to catalyse a reaction if the products of the reaction are constantly removed by the intervention of some additional biochemical system.

It should be noted that the magnitude of the activation energy has nothing to do with the difference in energy between reactants and products (a subject discussed in the next chapter). Moreover, energy put into molecules of reactants to activate them is released again as they surmount the energy barrier. Similarly, in the analogy above, the height of the bend of the siphon tube above the wall of the beaker does not affect the way it enables water to escape from the higher to the lower reservoir.

Catalysts re-route reactions by forming intermediate complexes with one or more reactants, and often with cofactors as well. The classic work of Keilin still offers one of the most elegant and simple demonstrations of enzyme-substrate complex formation. The enzyme peroxidase, which catalyses the reaction

$$H_2O_2 + X \rightarrow H_2O + XO$$

is a conjugate protein containing a haem group. This haem group possesses a very distinctive absorption spectrum with four prominent lines, but when hydrogen peroxide is added to the enzyme, these four absorption bands disappear and two new ones appear. Such a change in pattern of light absorption is an indication that a change has occurred in the molecular structure of the haem prosthetic group. If the enzyme-substrate (i.e., peroxidase—H_2O_2) complex so formed is separated and a substance is added capable of accepting oxygen from that complex, it is found that the four-banded spectrum of the peroxidase reappears and simultaneously the added substance, X, is oxidized:

$$H_2O_2\text{—haem} + X \rightarrow H_2O + XO + \text{haem}$$
$$\underset{\text{apoprotein}}{\vert} \qquad \qquad \qquad \underset{\text{apoprotein}}{\vert}$$

The highly specific nature of many enzyme-substrate interactions and the loss of enzyme activity which accompanies denaturation leave little doubt about the fundamental importance of the three-dimensional structure of proteins in relation to enzyme action. Furthermore, three-dimensional structure is itself an intricate function of the attractive and repulsive forces inherent in the hydrophilic and hydrophobic qualities of the side chains of constituent amino acids (p. 57). An interesting theory of enzyme action is based upon these facts. If a groove or crevice in the enzyme topography known as the 'active centre' allows the substrate molecule to fit neatly into it, new forces will inevitably come into operation whenever enzyme molecule and substrate molecule come into close proximity. This, in turn, implies that the balance of forces which originally led to the enzyme's tertiary structure becomes 'upset' as *the substrate molecule fits into position*, with the result that the enzyme molecule will *change its shape*. It can be imagined, in particular, that, in the immediate vicinity of the substrate molecule, a profound redistribution of forces is likely, resulting in both a stretching and a contraction of parts of the enzyme structure, distorting the substrate molecule as it does so (Fig. 4.4).

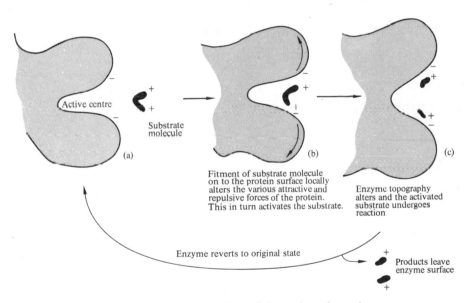

Fig. 4.4 Schematic representation of the action of certain enzymes

This probably causes certain linkages in the substrate molecule to become more unstable than they were before complex formation. If this indeed happens, relatively little extra energy is necessary for the bonds concerned

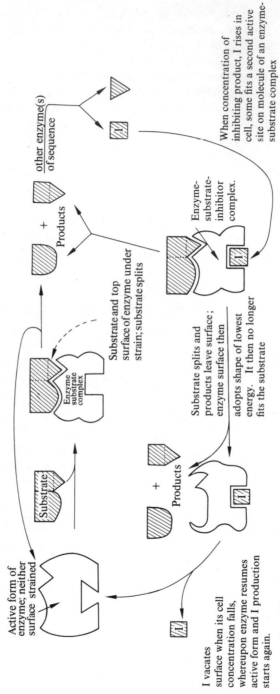

Fig. 4.5 One possible interpretation of allosteric inhibition of an enzyme by a product of the reaction sequence

to break or to undergo reaction with a second reactant—i.e., the activation energy is low. The products eventually leave the surface of the enzyme which then reverts to its original shape.

In some instances it has been found that the rate of an enzyme-catalysed reaction is slowed down reversibly by a compound quite unrelated structurally to the substrate of the enzyme. For example, when a series of reactions is catalysed by a number of enzymes working in sequence, an accumulation of the final product may produce a reduction in the first enzyme's activity. It probably does this by reacting with a second site on the enzyme; its presence there causes a *change in shape* of the protein molecule which inactivates the active centre. This is an elegant device whereby a final product of a reaction sequence can act as a controlling device preventing excess of it being formed. Enzymes which show this property are known as *allosteric* ('other shape') *enzymes* and the phenomenon is called an allosteric effect (Fig. 4.5).

4.6 Enzyme-substrate interaction and the Michaelis constant

Evidence such as that provided by Keilin's work leads to the conclusion that the substrate and enzyme combine together—and therefore that the enzyme can be regarded as a *temporary* reactant which, unlike the substrate, S, is not used up during the course of the reaction. Since the rate of every reaction, whether catalysed or not, is governed, *inter alia*, by the Law of Mass Action, let us consider a situation in which the concentration of enzyme is fixed, and only the concentration of the substrate varies. In such a case, the initial reaction velocity will be expected to increase with increasing levels of substrate concentration until a limiting initial reaction velocity is reached (Fig. 4.6). This will happen when the substrate occupies all the active sites of all the enzyme molecules. Similarly, if the reaction is repeated with half as much enzyme, the limiting value of the initial reaction velocity will be half its former value. Note that the term 'initial reaction velocity' is used, for we are *not* discussing here the progress of a reaction with passage of time.

Such observations are explained by the application of classical equilibrium concepts and provide a useful constant, the Michaelis constant. This constant gives a measure of the relative activity of an enzyme in different circumstances. In particular, it is of interest in relation to the activity of enzymes of rather low specificity—i.e., of enzymes capable of catalysing several related reactions but each to an extent determined by the exact nature of the substrate.

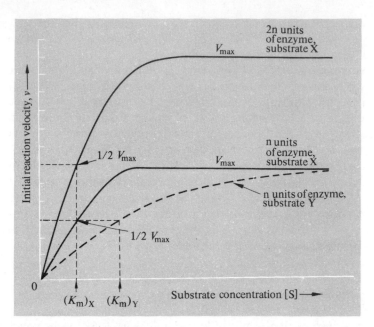

Fig. 4.6 Relationship between reaction velocity and level of substrate. (The continuous lines show V_{max} when n and 2n units of enzyme are present. The dotted and lower continuous lines show the effect on K_m of equal amounts (n units) of enzyme on two related substrates, X and Y; the affinity of enzyme for substrate X is greater than for substrate Y.

The derivation of the Michaelis constant in the simplest possible case assumes that one molecule of substrate S forms an intermediate complex, ES, with one molecule of the enzyme E, and that the products A and B do not readily form a complex with the enzyme:

$$E + S \underset{k_2}{\overset{k_1}{\rightleftharpoons}} ES \overset{k_3}{\to} A + B \qquad (4.1)$$

In order that these reactions can be considered from a kinetic viewpoint, the concentrations of the substances participating can be represented by square brackets, thus: Free Enzyme [Free E], Enzyme-substrate complex [ES], and substrate [S]. k_1, k_2, and k_3 are the velocity constants of the reactions in the directions indicated by the arrows. Using these symbols and applying the ideas of mass action:

(a) At fixed temperature, and at any moment of time,

rate of formation of ES = k_1[Free E][S]

The *total* enzyme concentration [E] consists of two fractions, the enzyme which is free at any instant and that which is bound to substrate.

Therefore,

$$[\text{Free E}] = [\text{E}] - [\text{ES}]$$

and rate of *formation* of ES $= k_1([\text{E}] - [\text{ES}])[\text{S}]$

Note that the rate of formation of ES is dependent upon time, since the concentration of the substrate [S] will decrease as time passes— for isolated enzyme systems of the sort now being considered.

(b) The disappearance of ES can be considered in similar terms, except that equation (4.1) shows that two different reactions lead to its disappearance.

Rate of *destruction* of ES $= k_2[\text{ES}] + k_3[\text{ES}] = (k_2 + k_3)[\text{ES}]$

(c) At equilibrium, the rate of destruction and the rate of formation of ES are equal, and therefore

$$k_1([\text{E}] - [\text{ES}])[\text{S}] = (k_2 + k_3)[\text{ES}]$$

The Michaelis constant K_m is defined in terms of the three rate constants in the following manner:

$$K_m = \frac{k_2 + k_3}{k_1} = \frac{([\text{E}] - [\text{ES}])[\text{S}]}{[\text{ES}]} \qquad (4.2)$$

Equation (4.2) is rarely suitable for estimating K_m since the concentrations of the enzyme and the enzyme-substrate complex are usually difficult to measure. However,

(d) A simple and rather ingenious modification of the equation leads to quantities readily determined by experiment. The rate of formation of the products A and B from E is, by definition, the overall velocity, v, of the multiple reaction represented by equation (4.1). Therefore,

$$v = k_3[\text{ES}] \qquad (4.3)$$

In the presence of a *large excess of substrate*, almost all the enzyme becomes bound to the substrate to form ES. Therefore, when the system is 'overwhelmed' with substrate there is hardly any difference between [ES] and [E], and inevitably, the enzyme is working at maximal velocity, V_{max}. In other words, the maximal value of the concentration of enzyme-substrate complex, $[ES_{max}]$, approaches, but cannot exceed, the concentration of enzyme E originally present.

$$V_{max} = k_3[ES_{max}] \simeq k_3[E] \qquad (4.4)$$

Therefore, from equations (4.4) and (4.3):

$$\frac{V_{max}}{v} = \frac{k_3[E]}{k_3[ES]} = \frac{[E]}{[ES]} \qquad (4.5)$$

From (4.2),

$$K_m = \left(\frac{[E]}{[ES]} - 1\right)[S]$$

Substituting equation (4.5) into this expression:

$$K_m = \left(\frac{V_{max}}{v} - 1\right)[S] \qquad (4.6)$$

If the observed initial velocity of the reaction, v, is measured under otherwise standard conditions but at varying levels of substrate concentration $[S]$, right up to complete saturation with substrate, K_m can be calculated. In fact, as substitution of $2v = V_{max}$ in equation (4.6) demonstrates, K_m is numerically equal to the substrate concentration when the reaction is working at half its maximal rate.

An important consequence of the Michaelis relationship is shown in Fig. 4.6. Cholinesterases, for example, show varying affinity for different choline esters. One of these enzymes, in addition to hydrolysing acetylcholine is able to attack propionylcholine and butyrylcholine, but the K_m value is different in all three cases. As the figure illustrates, the more active an enzyme is in yielding products from a particular substrate, the smaller is the value of the Michaelis constant.

Lineweaver and Burk have re-arranged equation (4.6) to provide an alternative and often more convenient graphical method of expressing experimental data for determining K_m:

$$\frac{K_m}{[S]} = \frac{V_{max}}{v} - 1; \qquad \frac{V_{max}}{v} = \frac{K_m}{[S]} + 1$$

Therefore,

$$\frac{1}{v} = \frac{K_m}{V_{max}} \cdot \frac{1}{[S]} + \frac{1}{V_{max}} \qquad (4.7)$$

Upon comparison of this equation with that for a straight line

$$y = mx + c \qquad (4.7a)$$

the following relations are apparent:

1. When the reciprocal of initial reaction velocity is plotted against reciprocal of the substrate concentration which resulted in that velocity, a *straight line* is to be expected. This happens because, in the Lineweaver–Burk equation, $y = 1/v$ and $x = 1/[S]$.
2. The *slope* of this line is K_m/V_{max} since this is the term in equation (4.7) corresponding to m in equation (4.7a).
3. In the general equation (4.7a), when $x = 0$, $y = c$, i.e., the value of y when $x = 0$ is numerically equal to the 'constant' term of the equation. For the Lineweaver–Burk equation, when $1/[S] = 0$, the intercept on the $1/v$-axis is numerically equal to $1/V_{max}$.
4. Similarly, in the general equation, when $y = 0$, $x = -c/m$. In the present instance, when $1/v = 0$, in equation (4.7):

$$\frac{1}{[S]} \cdot \frac{K_m}{V_{max}} = \frac{-1}{V_{max}} \quad \text{or,} \quad \frac{1}{[S]} = \frac{-1}{K_m}$$

It is apparent from these relations that K_m and V_{max} can be readily estimated in simple cases by measuring the velocity of the reaction at several levels of substrate concentration (Fig. 4.7).

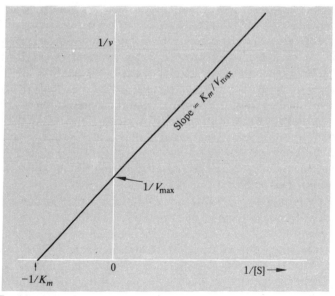

Fig. 4.7   Lineweaver–Burk plot of the reciprocal of the initial velocity against the reciprocal of the substrate concentration

Two aspects of the Michaelis constant and its applications are particularly worthy of note: (a) the Michaelis constant has the *same units as the substrate concentration*, and K_m is consequently expressed as moles per litre (actual values are often within the range 10^{-2} to 10^5 moles per litre). (b) Since the value of K_m is estimated by measuring initial reaction velocity at various substrate concentrations, but without reference to the quantity of *enzyme* present, it is possible to obtain estimates of K_m using very impure preparations of enzymes. Most other measures of enzyme activity require a knowledge of the exact amount of enzyme present and sometimes even of its molecular weight—for example, the *turnover number* of an enzyme is the number of moles of reactant converted to product per minute by one mole of enzyme.

4.7 Inhibition of enzymes

Several different ways exist in which enzymes may be inhibited, that is, partially or completely put out of action, by addition of other substances. Chemical reagents which coagulate, precipitate, and denature proteins necessarily destroy enzyme activity, but, in addition, other chemical reagents form more or less stable complexes with important centres in many enzymes. Thus, iodoacetate ($ICH_2.COO^-$) unites with sulphydryl centres of certain dehydrogenases, while hydrogen cyanide and hydrogen sulphide attack some haem prosthetic groups, and especially the one present in an important enzyme called cytochrome oxidase.

There is, however, yet another way in which the efficiency of many enzymes can be impaired. When a compound exists which is sufficiently like the natural substrate in structure for it to be able to take the place of that substrate on the enzyme surface, forming a *loose* complex with the enzyme, its presence there prevents natural substrate molecules from attaching themselves. The type of enzyme inhibition which results is called *competitive inhibition*, for mass action considerations ensure that it can be partially reversed by addition of more of the natural substrate.

The most famous example of competitive inhibition is the effect of malonate on the activity of succinate dehydrogenase, the enzyme which catalyses the formation of fumarate from succinate (p. 89). Malonic acid ($HOOC.CH_2.COOH$) like succinic acid, is a dicarboxylic acid, but is one methylene group smaller; it becomes attached to the enzyme, blocking sites normally occupied by succinate. Its effect is illustrated by the following equations, where S stands for succinate and M for malonate; the other terms have the same meaning as in section 4.6:

$$E + S \rightleftharpoons ES \rightarrow A + B + E$$
$$E + M \rightleftharpoons EM \text{ (inactive; no further action)}$$

Since S and M are clearly competing for a limited amount of enzyme, the concentration of ES and hence the rate of formation of A and B, depends on the dissociation constants of ES and EM and upon the relative concentrations of S and M. A and B in this instance are $FADH_2$ and fumaric acid.

4.8 Enzymes acting in sequence

So far we have considered single enzymes acting in isolation but, in practice, most biochemical processes involve the participation of several enzymes working together in sequence:

$$A \xrightarrow{\text{enzyme } E_1} B \xrightarrow{\text{enzyme } E_2} C \xrightarrow{\text{enzyme } E_3} D$$

The enzymes of such a sequence are sometimes located in a subcellular membrane. The intermediates B and C must then move from one enzyme to the next rather like an object passed along a factory production line. In a production line it is the slowest of the processes which determines the overall rate of production and exactly the same is true of sequential biochemical reactions. Thus, if one enzyme, e.g., enzyme (E_2) is either slower acting or in shorter supply than neighbouring enzymes it controls the rate at which the whole sequence works. In this situation enzyme E_1 would supply so much B to enzyme E_2 that enzyme E_2 would be converted almost entirely to E_2-B complex. Enzyme E_2 would therefore work at maximum rate (p. 101). No matter how fast enzyme E_3 is able to function *in vitro*, it cannot, when part of such a sequence, act faster than enzyme E_2 provides it with substrate C. Consequently, enzyme E_2 will act as a brake, or *pacemaker*, determining the rate at which D is formed from A.

One of the chief characteristics of living cells is their ability to regulate their activities according to the circumstances. Suppose, for example, the cell has need of a certain amount of a particular metabolite D. Clearly, for optimal efficiency, it must have neither an excess nor a deficiency of D and, in practice, there are several ways in which the level of D can be controlled.   One way is for an excess of D to inhibit the activity of one of the enzymes involved in its biosynthesis.   For example, if enzyme E_1 were an allosteric enzyme (p. 99), the product D might have an allosteric action on it.   Alternatively if enzyme E_3 is dependent on a cofactor or secondary substrate also required by enzyme E_1, the formation of D could reduce the activity of enzyme E_1.   Such effects of metabolites on the rate at which enzymes act are examples of *enzyme suppression*. There is a third mechanism whereby the quantity of D can influence the rate at which the whole sequence of reactions operates.   This is termed *enzyme repression*, and might occur where excess of D reduces the rate of *synthesis* of one of the enzymes in the sequence.   On the other hand, cases are known where a deficiency of D in the cell leads indirectly to the synthesis of more D by initiating in some way the formation *de novo* of the enzyme(s) involved in its synthesis.   When this occurs the process is known as *enzyme induction* or *adaptative enzyme synthesis*.

In the previous chapter it was mentioned that amylase converts starch to maltose and that this may be hydrolysed in turn to glucose by the action of maltase:

$$\text{Starch} \xrightarrow{\text{amylase}} \text{Maltose} \xrightarrow{\text{maltase}} \text{Glucose}$$

Such a sequence occurs, for example, in germinating wheat and barley seeds, the glucose being metabolized by the developing embryo.   The rate at which glucose is made available has to be controlled to provide the appropriate level of glucose.   It has been found that, when present in excess, maltose *suppresses* the rate at which amylase acts; indirectly, therefore, it regulates the amount of glucose produced from starch.   The amylase is suppressed some 60% by 0·1 molar maltose but is not affected by other sugars.

4.9   Coenzymes and activators

Cofactors of various kinds play an indispensable role in numerous enzyme-catalysed reactions.   Their existence was first recognized in 1905 when Harden and Young showed that dialysed yeast juice did not ferment sugars, but that undialysed juice, or dialysed juice to which some of the dialysate was subsequently added, fermented sugars actively.   These cofactors, removed from the enzymes by dialysis, were found to be of relatively small molecular weight; in the case of yeast juice, NAD later proved to be one of them and magnesium ions another.   It is now known that many enzyme-catalysed reactions require the participation of inorganic ions or of organic molecules which, compared to enzymes, are of relatively small molecular weight.

Numerous enzymes have metal cations as cofactors; magnesium, iron, calcium, manganese, copper, and zinc are all important examples. When some of these ions are absent, deficiency diseases occur in all types of organisms but can often be remedied by addition of the deficient cation to the diet or growth medium. *Trace elements* are elements which, for the reason above, are only needed in minute quantities.

Many coenzymes contain organic sub-units which animals cannot synthesize for themselves, and so must be provided in the diet. Such sub-units are called *vitamins*. Amongst them are nicotinamide (a part of NAD), riboflavin, pyridoxal, thiamine, and folic acid. Since they or certain of their derivatives are parts of essential cofactors, a dietary deficiency of any one of them may cause specific deficiency diseases in animals, and may even lead to death. Plants must, of necessity, be able to synthesize all the organic cofactors that they require, but some micro-organisms do need vitamins for growth.

Clearly, cofactors are of several kinds; some are small organic molecules and are, in the strictest sense, coenzymes, while other cofactors are inorganic in nature. Generally speaking, coenzymes participate in enzymic changes at or *after* the formation of the intermediate compound with the substrate, but other accessory factors, termed *activators*, may be necessary *before* the intermediate compound comes into being. Some remove blocking devices which prevent enzymes from functioning; others are necessary to stabilize a particular protein conformation essential to enzyme action, and yet others probably form an integral part of the intermediate complex, ES.

A few examples of enzymes which require cofactors of various kinds will help the reader to appreciate the roles of various kinds of cofactors in the total enzyme-catalysed reaction system.

Aldolase is an example of the simplest kind of enzyme; it requires no cofactor other than water. The remarkable qualities of this solvent being frequently—and wrongly—taken for granted, an enzyme of this sort is usually said to require no supplementary substance to stabilize it or aid it in its catalytic effect. Aldolase catalyses (reversibly) the addition of dihydroxyacetone phosphate to the double bond of glyceralde-hyde phosphate, the product being hexose diphosphate (Fig. 4.8). Operating in reverse, the enzyme catalyses the splitting of fructose 1,6-diphosphate to triose phosphate in an important pathway of carbohydrate

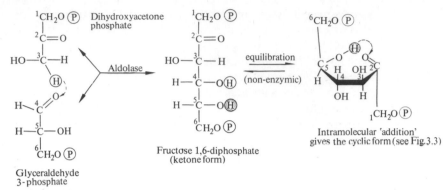

Fig. 4.8 The action of aldolase and subsequent ring formation. (The ringed hydrogen atom on C_4 is important in relation to the left-hand reaction; that on C_5 in relation to cyclization on right). See also p. 200

catabolism. Urease (p. 88) and α-glucosidase (p. 89) are two other examples of enzymes which require no cofactor other than water.

Phosphoglucomutase is an enzyme which requires a small soluble organic molecule, glucose 1,6 di-phosphate, as cofactor. This cofactor can be made by animals, so it and its precursors are not vitamins. The enzyme catalyses the interconversion of glucose-1-phosphate and glucose-6-phosphate. The phosphate group does not, in fact, migrate from one end of the molecule to the other, but the change is brought about by the intervention of the cofactor.

Glucose-1-P (substrate) ⟶ Glucose-1,6-di-P (coenzyme)

Glucose-1,6-di-P (coenzyme) ⟶ Glucose-6-P (product)

This allows the phosphate group to change sides with the result that, in one cycle, it is the *coenzyme* molecule, not the substrate, which becomes the product of the reaction; at the same time, however, a new coenzyme molecule is formed by phosphorylation of the substrate molecule. In other reactions the contribution of the coenzyme is often much more complex than this. Especially is this so for systems dependent on thiamine and pyridoxal, both of which are well-known vitamins; the biochemical roles of these coenzymes are considered in later chapters.

Alcohol dehydrogenase catalyses the oxidation of primary alcohols to aldehydes, NAD^+ and zinc ions being essential cofactors. The NAD^+ acts in its typical role of electron acceptor (p. 84) and it must therefore be orientated on the surface of the enzyme in such a way that electrons can pass to it from the enzyme-substrate complex. The zinc ions are

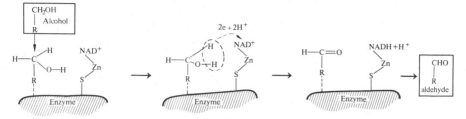

Fig. 4.9   The action of alcohol dehydrogenase

apparently involved in the attachment of the NAD^+ to the enzyme surface, probably by bonding which involves the sulphydryl groups of cysteine residues (Fig. 4.9).

Amylase, as well as having intracellular functions in plants and animals, is secreted by the salivary glands and by the pancreas, its digestive role being, of course, to hydrolyse insoluble starches and glycogen to maltose and glucose. The soluble sugars are later absorbed by the intestinal absorptive cells. Salivary amylase is a metallo-protein, each molecule of the enzyme containing one ion of calcium. The role of this calcium is different from that of the zinc in the example above, for if the calcium ions are removed by chelating agents, the enzyme's activity is not necessarily impaired. The calcium ions apparently play some part in stabilizing the three-dimensional structure of the protein, for removal of calcium ions renders amylase more readily inactivated by proteolytic enzymes. On the other hand, the enzymic activity of amylase is directly dependent on chloride ions, for if these ions are removed by dialysis, amylase no longer hydrolyses complex α-glucosides.

Proteolytic enzymes, secreted from various sources into the alimentary canal, have been extensively used in the study of the primary structure of proteins (p. 49) and a brief description of these enzymes which possess an extracellular action is therefore appropriate at this point. *Pepsinogen* (mol. wt. 42,000) is a zymogen secreted by cells called chief cells, located in the wall of the stomach. It is activated by the hydrochloric acid secreted by other stomach cells, a pair of small peptides breaking off from one end of the protein chain to produce *pepsin*, an enzyme with a molecular weight of about 35,000. Pepsin catalyses the hydrolysis of peptide linkages mainly on the amino-side of phenylalanine and tyrosine residues (Fig. 4.10). The pancreas secretes three major pro-enzymes (or zymogens), which, after being activated by removal of small lengths of peptide chain, give rise to trypsin, chymotrypsin, and carboxypeptidase.

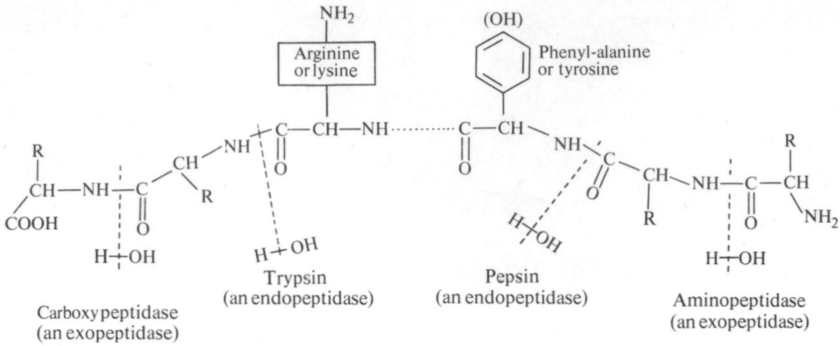

Fig. 4.10   How various peptidases attack proteins

Chymotrypsin is not very specific, but *trypsin* acts on the carboxyl side of the basic amino acids arginine and lysine. Pepsin, trypsin, and chymotrypsin are all *endopeptidases*, in that they hydrolyse peptide bonds within protein chains, so splitting such chains into smaller polypeptide units. On the other hand, carboxypeptidase acts only on terminal amino acid residues situated at that end of the chain which bears the free carboxyl group. Another *exopeptidase*, aminopeptidase, does the same thing at the other end of the chain (Fig. 4.10); the intestinal content of this enzyme probably arises principally from the breakdown of intestinal absorptive cells, within which it is abundant. The specificity of these various proteolytic enzymes provided the necessary means by which Sanger in 1955 was able to work out, for the first time, the structure of a protein (insulin).

5

Bioenergetics

5.1   Energy for the living cell

An athlete, on completing a sprint, may say, rather colloquially, that he has 'used up a lot of energy'.   In fact, he has done nothing of the sort; his muscles have converted certain chemical compounds into a variety of products by reactions which liberate energy.   This energy is then employed to do work; it makes his muscle fibres contract and expand and so propels him along.   During the exercise, he becomes 'hot'; he also stirs air as he moves along, and his feet produce frictional heat where they press against the ground.   No energy has actually been used up in the sense that it has been 'destroyed', but energy which was an integral part of certain molecules has been converted into a variety of other forms of energy.

Before exercise, the athlete needs to supply his muscles with energy-supplying materials, and his appetite is likely to be good.   An energy source is also needed to enable the tissues of his body to carry out numerous other processes which include growth, tissue repair, and the pumping of certain substances across cell membranes.

Food, very indirectly, is the fuel which drives a complex and efficient machine.   Subsequent chapters describe how useful energy is derived by living cells from chemical substances such as sugars and fats and how this energy is harnessed to do work of many kinds.   Together, these many forms of work occurring within the cell (or organism) constitute what is known as 'living' or 'being alive'.   The discipline of *thermodynamics* deals with such things as energy, heat, and work, and consequently provides explanations for many of the energetic processes occurring in cells.   In this book, the approach is deliberately *non-mathematical*, even at the expense of some loss of precision.   Having acquired a 'feeling' for the qualitative aspects of energetics, readers requiring a quantitative, mathematical approach will be in a position to proceed to more rigorous treatments. [#]

5.2 Interconversion of energy

The sprinter was described as converting chemical energy to the mechanical energy of muscular contraction. The conversion of energy from one form to another occurs in the inanimate world and in all living systems. Examples can be found involving several types of energy. Chemical energy can be changed to heat, electrical, or light energy; it can also be converted to osmotic energy. Light energy may be converted into electrical or chemical energy. Similarly, electrical energy may be transformed into chemical energy. Whenever the interconversion of two forms of energy occurs, it takes place according to the requirement of the first law of thermodynamics—that *energy can neither be created nor destroyed*. The law implies that, when a decrease in the quantity of one form of energy occurs this is *exactly compensated* by an equivalent increase in the quantity of other forms. The principle of conservation of energy was formulated mainly by reference to energy transformation in the inanimate world but it is equally applicable to living systems.

One of the most vital of all energy transformations occurs when light energy reaches the leaf of a plant. Some of the energy is absorbed and some reflected. A part of the absorbed light is converted to heat, but some excites or *energizes* certain electrons in chlorophyll molecules, and these electrons eventually provide reducing power to convert carbon dioxide to sugars. The latter possess high energy compared to the carbon dioxide and water from which they were made. In due course, and in contrast to the process of photosynthesis, the carbohydrate is catabolized by the plant—or by a foraging animal—with the loss of chemical (potential) energy from the carbohydrate molecule. This *loss* of potential energy by sugar as it is oxidized can result in an *increase of potential energy of some other compound* as that compound is changed in some way. In addition, a certain amount of *heat* is produced, as is energy associated with the random movement of carbon dioxide and water molecules resulting from the oxidation (the energy of *entropy*).

Since energy transformations occur with no *overall* gain or loss in energy, there is an exact equivalence between different forms of energy and hence the same unit for the measurement of energy can be applied to the various forms. This may be illustrated with a simple example. When electrical energy is passed through the coil of a water immersion heater, the heat energy gained by the water is exactly equivalent to the electrical energy lost by the power generator. Since the two forms of energy are equivalent, it follows that we can express electrical energy in terms of calories if we so desire, one calorie being the amount of heat required to raise the temperature of 1 gram of water at 14·5°C to a temperature of 15·5°C. There is, as it were, a precise rate of exchange from one currency of energy to another. If the current flowing through the coil is I amperes and the voltage V volts, the heat energy produced is IV joules or $IV/4·18$ cal (since 4·18 joules = 1 cal).

When several types of energy are encountered in a particular context it is often found convenient to express the various types in terms of the same energy unit. In biochemistry, the unit usually chosen is the calorie (cal), or the kilocalorie (kcal) which is equal to 1000 cal. However, this practice can be a *serious pitfall* to the unwary. Thus, if it is stated that a change in chemical energy of x cal occurs during a chemical reaction, this does *not* mean that heat—a form of energy which is *directly* capable of causing change in termperature—is necessarily involved, but only that energy *equivalent* to x cal of heat has been gained or lost.

5.3 Potential energy in physical and chemical systems

A factor of considerable importance in relation to the energetics of cellular processes is the direction in which chemical reactions tend to proceed, and a familiar situation from the inanimate world—the water-fall—will provide a clue to the principles involved. Static water cannot drive a water wheel but, if the water is allowed to flow downhill, the energy of the moving water can be harnessed so that it does the work of rotating a wheel at the foot of the fall. The greater the distance through which the water falls the more the energy harnessed by the water wheel. Thus the energy that unit quantity of water can provide to do work depends on the initial height of the water, relative to a wheel situated at the foot of the water fall. Clearly, there is a spontaneous movement of water downhill; it does not move *spontaneously* in the opposite direction, though suitable machinery, driven by an external energy supply, can be used to pump water uphill. The fall of the water is an energy-yielding process, for the water loses *potential energy* as it moves from a higher to a lower level. Any process, whether it is physical or chemical, which results in the surroundings gaining energy is termed an *exergonic* process, i.e., the system or substance becomes poorer in energy. Energy provided by the exergonic process can be harnessed to drive an energy-consuming or *endergonic* process, so long as appropriate machinery is available,

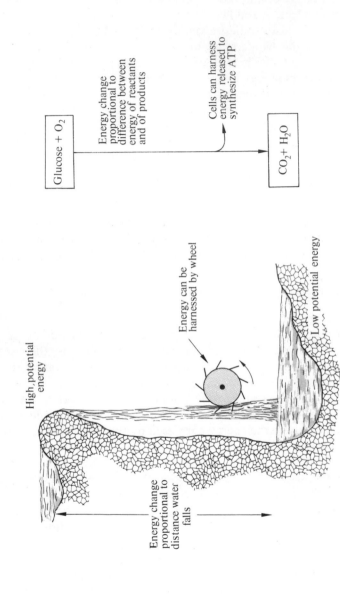

High potential energy

Energy can be harnessed by wheel

Low potential energy

Energy change proportional to distance water falls

Glucose + O_2

Energy change proportional to difference between energy of reactants and of products

Cells can harness energy released to synthesize ATP

$CO_2 + H_2O$

Fig. 5.1 The potential energy lost as water descends a waterfall (left) can be harnessed to do work just as energy released in a chemical reaction can be used by cells to drive endergonic processes although the harnessing is never 100% efficient. In the absence of this machinery the energy is not harnessed, but is just squandered, when it usually manifests itself as heat.

Precisely the same argument can be applied to chemical processes as to physical ones. When any *chemical reaction* occurs, the potential energy of reactants is replaced by the potential energy of products, and the difference between these two energy levels represents the energy liberated (or absorbed) in the reaction. Thus when glucose is oxidized to carbon dioxide and water, energy is released from the glucose molecules. The chemical energy inherent in the glucose molecule is a *potential* energy, and it is greater than the potential energy of the oxidation products. Glucose oxidation is thus analogous to the fall of water from a higher to a lower level—energy is released which can be harnessed to do work if appropriate machinery is available (Fig. 5.1).

When water flows downhill it does not undergo change in its physical state or its chemical structure. In chemical reactions, on the other hand, chemical bonds are broken or forged, molecules change in size, and, as in the case above, a highly organized crystalline solid may be replaced by small chaotically-moving gas molecules. For this reason, some part of the change in potential energy in a chemical reaction is associated with this change of organization or orderliness, so introducing an additional factor into the system (section 5.5).

The change in chemical potential energy occurring when reactants give rise to products in an exergonic reaction is of great importance from a bioenergetic viewpoint, because it is a measure of the maximal quantity of energy made available by the reaction for driving endergonic processes in the cell. Similarly, when the change is endergonic, the energy difference measures the energy that must be provided to force the reaction to proceed. The *free energy change* (ΔG), is thus defined as the difference between the (summed) free energy of the various reactants, ΣG_R, and the (summed) free energy of the various products, ΣG_P:

$$\Delta G = \sum G_P - \sum G_R \qquad (5.1)$$

Some books use the symbol F instead of G. For most practical purposes in biochemistry ΔF and ΔG can be regarded as identical.

In equation (5.1), the symbol Σ simply means 'the sum of', so ΣG_R for a reaction involving the combustion of sugar in oxygen would just mean 'the free energy of the sugar undergoing reaction plus the free energy of the oxygen undergoing reaction'. If there is only one reactant, ΣG_R and G_R have, of course, the same meaning. In practice the absolute values of $\Sigma\,G_R$ and $\Sigma\,G_P$ are of less importance than ΔG, the free energy released or absorbed as the reaction occurs. Just as a reaction can only proceed spontaneously if it is exergonic (ΔG is negative, since $\Sigma\,G_P$ is smaller than $\Sigma\,G_R$), so, if it is endergonic (ΔG positive) it will only occur if external energy is provided.

In considering the energy change accompanying a chemical reaction it is necessary to take account of the *mass of the reactants* in addition to the *change in potential energy*. This is analogous to the waterfall, where the energy change is proportional both to the distance the water falls and to the quantity of water flowing over the top. The capacity factor of chemical reactions is taken into consideration by the use of a quantity known as the *standard free energy change*, ΔG^0. This corresponds to the free energy change for a *molar* reaction. Thus, for the complete oxidation of one mole of glucose (180 g) to carbon dioxide and water, the standard free energy change (ΔG^0) is -691 kcal, and therefore if 360 g glucose were to be oxidized the *actual* free energy change, ΔG, would be -1382 kcal.

The overall equation for photosynthesis is the reverse of the one for glucose oxidation:

$$C_6H_{12}O_6 + 6O_2 \xrightleftharpoons[\text{photosynthesis}]{\text{oxidation}} 6CO_2 + 6H_2O + \text{Energy}$$

In the forward direction, free energy is released by the oxidation, whereas, in photosynthesis, light energy indirectly raises the energy level of carbon dioxide and water to the higher free energy of glucose plus oxygen. To compare the free energy change of the two reactions it is *not necessary to know details of the individual pathways* (which are, in fact, quite different) because the law of conservation of energy requires that the free energy intake in the one direction is inevitably the same as the free energy released in the other. This is shown in the following scheme:

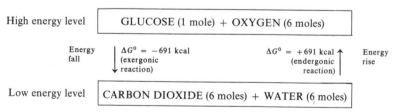

It should be noted that although the oxidation of glucose is spontaneous—the direction of the chemical change is always towards the formation of carbon dioxide and water—nevertheless, as is well known, glucose remains stable almost indefinitely in the presence of oxygen at room temperature. This is because oxidation only occurs at finite speed if either (a) energy is supplied in sufficient amount to ensure that enough molecules overcome the barrier presented by the activation energy, E_A, or (b) if the value of E_A is effectively reduced by the presence of a catalyst or enzyme (see p. 95). Note that the magnitude of E_A in no way affects the value of ΔG, since energy supplied to drive reacting molecules over the 'energy hump' is regained in full as the barrier is surmounted.

5.4 The driving force of the living machine (Re-routing of endergonic reactions)

Present knowledge of living processes and of the way that classical thermodynamic considerations can be applied to them, has shown that there is no difference in principle between energetic aspects of life and the energetic processes which drive inanimate machines. In particular, no 'vital force' need be invoked to account for how endergonic processes occur in living systems.

Energy changes are often conspicuous in certain reactions because *heat* is evolved. Early on in our scientific training most of us became aware of the existence of 'exo*thermic*' and 'endo*thermic*' reactions, perhaps without fully realizing that gain or loss of energy by chemical systems does not necessarily take the form of gain or loss of *heat*. In biochemistry, however, 'energy change' and 'heat change' seldom even approximately correspond, for heat, it must be remembered, is like the water of a waterfall, in that it can only do work when it changes potential, i.e., when it flows from a hotter to a colder body.

Clearly, since the living machine operates at an almost constant temperature, it is not driven by the movement of *heat* energy down a temperature gradient. Instead, it depends on *gradients of chemical potential energy*. Since changes in chemical (potential) energy require that a change in molecular structure should occur, and this in turn implies a redistribution of valency electrons, it follows that living organisms must depend on a redistribution of energy associated with a redistribution of electrons. Often, electron redistribution takes the form of flow down an oxidation-reduction gradient. In such cases, the energy of oxidation, instead of being released in the form of heat—as it is when sugar, wood or coal are burnt as fuels for a *heat*-engine—is effectively 'harnessed' or '*coupled' to endergonic processes in such a way that these energy-requiring processes are driven uphill.* This sort of 'coupling' is in no way unique to biochemistry: physical energy is similarly harnessed whenever a car-jack is employed, and a decrease in chemical potential energy is translated into an increase in physical potential energy whenever burning petrol drives a car uphill.

One of the reasons why energy transactions are handled efficiently by living cells is that potential energy changes occur in small discrete steps, and the cell has an elaborate organization both at the molecular level and above it, which ensures effective energy coupling. For example, the oxidation of glucose in a calorimeter results in a one-step heat change, whereas the cell splits glucose up, step by step, so that *at no one stage is there a massive release of energy.* Moreover, when the cell degrades glucose in steps, only a fraction of the energy so made available is manifest as heat, for the quantity of energy released at some of those steps is such that it can be passed on at once to molecules of some other substances nearby, thereby raising *their* potential. This 'trapped' energy is frequently used to drive *formally endergonic* reactions.

The phrase "formally endergonic" has been used because at this point a difficulty is encountered which relates more to the use of words than to any biochemical problem. Suppose we have a reaction $A \rightarrow B$ in which B is richer in energy than A. The formation of B from A is undoubtedly *endergonic*, and requires an input of external energy if it is to occur. But suppose that an exergonic reaction $C \rightarrow D$, instead of occurring in a simple or 'direct' manner, takes place in such a way that it involves the participation of substance A of the endergonic reaction $A \rightarrow B$. So long as the energy change $C \rightarrow D$ is greater than that of $A \rightarrow B$, a new spontaneous exergonic reaction becomes feasible:

$$A + C \rightarrow B + D \ (\Delta G, \text{ negative}) \tag{5.2}$$

In such a 'double' reaction, A does not change to B in quite the sense that the simple reaction $A \rightarrow B$ implies, yet B is still formed from A. The resolution of this paradox is quite simple; the reaction $A \rightarrow B$ has, in fact, been *re-routed*. Since we often need to take such 'coupled' reactions apart (it should be remembered that the reaction $C \rightarrow D$ can occur without A being present, just as falling water need not drive a water wheel) it is convenient to think of the *biologically-impossible* 'direct' reaction $A \rightarrow B$ as 'formally' endergonic.

The formation of the important substance adenosine triphosphate, or ATP, from the energy-poorer diphosphate ADP, is an apparently endergonic process which proceeds as a result of 'coupling' with an exergonic change. Figure 5.2 shows that ADP combines with inorganic phosphate

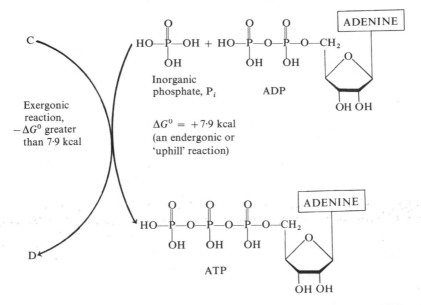

Fig. 5.2 The exergonic reaction, C → D, 'drives' the change ADP + P_i → ATP

to form ATP in a reaction requiring an input of about 7·9 kcal per mole. Thus ADP and phosphate together correspond to 'A' in the coupled example above, and ATP corresponds to 'B'. Except when the energy for the reaction originates from sunlight, as it does in photosynthesis, it is usually derived from one of several exergonic reactions C → D which are steps in the process of respiration.

5.5 Free energy, enthalpy, and entropy[#]

It was stated earlier that the capacity of a chemical system to do useful work often depends on the downhill flow of electrons; ΔG, the *free energy* available for such work, can be measured in potential energy units just as can the ability to do work of falling water. And just as the potential difference in the case of the aqueous system depends on the height the water falls, so in the chemical system the free energy difference ΔG is the difference between the sum of the free energies of all the reactants,

$\Sigma\ G_R$, and those for all the products, $\Sigma\ G_P$. Hence, we have equation (5.1):

$$\Delta G = \sum G_P - \sum G_R$$

Free energy G is a property of a system just as are volume and mass; once the size and certain environmental conditions have been fixed, G is accurately defined. Neither the previous history of the system nor the route whereby a chemical system of energy $\Sigma\ G_R$ approaches another one, of energy $\Sigma\ G_P$, plays any part in determining the magnitude of ΔG.

The change $\Sigma\ G_R \rightarrow \Sigma\ G_P$ with a consequent change ΔG in free energy, is the thermodynamic quantity of most significance to the biochemist for, as has already been seen,

(a) Its sign indicates whether a reaction is theoretically able to proceed spontaneously, though it does not indicate how *fast* the change will occur (sometimes, in fact, it may be almost infinitely slow).

(b) If the reaction is exergonic the magnitude of $-\Delta G$ indicates the maximum energy available for useful work.

In addition, it will be shown later that

(c) the 'driving force' $-\Delta G$ of a reversible reaction, declines as the mixture of reactants and products moves towards equilibrium;

A second thermodynamic quantity is the internal energy of a system. When the pressure is kept constant, this is known as the *enthalpy* and rather unfortunately, perhaps, is given the symbol H. Like volume and free energy (G), H is a state function; that is to say, its magnitude depends only on the condition and mass of the system before and after change. The change in enthalpy when a reaction takes place is therefore independent of the route, and is called ΔH.

The absolute enthalpy H is the sum total of numerous atomic and sub-atomic energy factors.* On the other hand, ΔH is a relatively simple and straightforward experimental quantity; it is the *heat of reaction as measured in a constant pressure calorimeter*. In other words, it is measurement of ΔH which leads to statements such as 'the heat of reaction when glucose is burnt in oxygen is 673 kcal'. A word of warning is, however, necessary, for ΔH is only the *heat* of reaction if the reaction remains uncoupled to other changes. If coupling occurs, the heat measured in the calorimeter is, of course, the heat evolved in the overall coupled reaction. Thus when the breakdown of glucose *in vivo* is coupled with

---

* Absolute values of H, G, and S are unknown.

endergonic reactions, only part of the enthalpy change is manifest as a heat change, and it is in this sense that the use of the symbol ΔH is somewhat misleading.

Most exothermic reactions (ΔH, negative) are also exergonic, and the oxidation of glucose is a case in point ($\Delta G^0 = -691$ kcal; $\Delta H^0 = -673$ kcal). Nevertheless, the heat of reaction is not always a good guide as to whether a reaction is exergonic and therefore theoretically spontaneous. It may seem strange that the complete oxidation of glucose to carbon dioxide and water yields more energy capable of doing useful work (ΔG) than the difference of the internal energy (ΔH) between the system oxygen/glucose and the oxidation products carbon dioxide/water. Additional energy to the extent of $\Delta G^0 - \Delta H^0$, i.e., 691–673 kcal, becomes available as a result of the oxidation. This additional energy is known as energy associated with *entropy* and is given the symbol ΔS. The entropy of a system, S, is a state property, just like G and H, and the quantitative relation between changes in these three forms of energy is given by the equation

$$\Delta G^0 = \Delta H^0 - T \Delta S^0 \qquad (5.3)$$

where T is the absolute temperature. Applying this equation at 20°C to the total oxidation of one mole of glucose, we get:

$$-691 = -673 - (293 \Delta S^0)$$
$$-18 = -293 \Delta S^0$$

Thus the entropy change in this reaction, ΔS^0, is $+18/293$ kcal per degree.

The quantities S and ΔS have unnecessarily worried many biologists. The entropy of a substance or system represents the freedom of movement of molecules within that substance or system, and of the atoms which constitute each molecular particle. The more the freedom of movement, the more the chaos and the higher the energy associated with entropy. Thus, as the temperature is raised, the energy of entropy increases; the increased chaos of gaseous molecules at higher temperatures is well known. Similarly, if a regular crystal lattice is destroyed by melting, or an ordered large covalent molecule is split up by oxidation, ΔS will be a positive quantity.

Glucose has a highly organized molecule with atoms arranged in an ordered structure and the molecules of solid glucose are arranged in a crystal lattice, whereas the products are less complex molecules moving about haphazardly. Thus, when solid glucose is oxidized, the total entropy of the products is greater than the total entropy of reactants and

ΔS is consequently positive. The molar figure is, in fact, 18 kcal per degree, at 20°C, as was seen above, when measured at 20°C.

5.6   Priming and coupling reactions

Reactions which are driven by physical energy such as heat, electricity, or light may be represented simply as

$$A + \text{energy} \rightarrow B$$

and are by definition endergonic reactions. Heat and electricity are excluded as major *sources* of energy for biological reactions. Light, on the other hand, drives the photosynthetic reduction process by 'exciting' certain electrons in chlorophyll molecules, so raising their reduction potential. This important biophysical process is discussed in chapter 11.

The immediate source of energy for all other energy-requiring reactions in biochemistry is a concomitant *decrease in chemical potential energy* of linked systems (p. 118).   Just as it is possible to couple loss of potential energy of one physical system with gain in potential energy in another (e.g., the descent of a mallet causes a weight to rise and ring a bell in the funfair), so, in effect, chemical potential energy can be transferred from one system to another. Special enzymic and structural machinery is necessary to enable this energy-coupling to occur just as physical machinery is clearly necessary to couple energy changes in the examples of the car-jack and of the mallet and weight.

When biochemical reactions are coupled, it is important to recognize that physical energy is not an intermediate—heat, for example, is *not* liberated by one system and absorbed by the other. We can illustrate the principle of coupling of potential energies in the following two ways by reference to the exergonic reaction $C \rightarrow D$ ($-\Delta G^0 = X$ cal) which drives a hypothetical formally endergonic reaction $A \rightarrow B$ ($+\Delta G^0 = x$ cal):

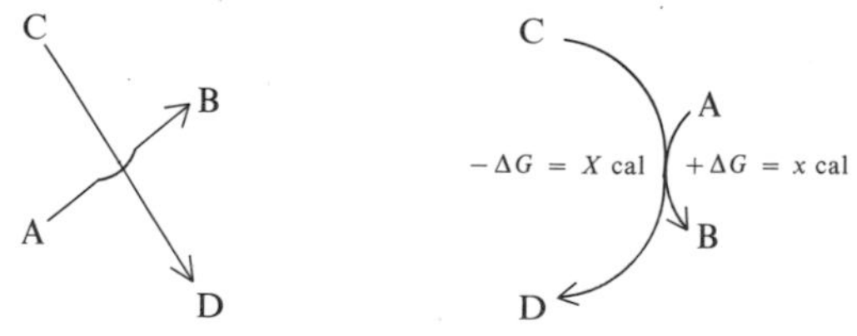

In the left-hand diagram, vertical distance represents the level of potential energy, so that A is driven up the energy scale to form B, a change which occurs at the expense of the larger energy drop from C to D. The diagram on the right is the form more usually adopted to represent coupled reactions. The standard free energy changes (ΔG^0) for the individual reactions which are coupled are, by definition, those occurring when one mole of each reactant is converted to one mole of each product in a molar solution of reactants and products. In this present example, if 1 mole of C and 1 mole of A are converted to D and B respectively, the *overall energy change* is $(X - x)$ cal; the coupled reaction is *exergonic* since X must exceed x if the formally endergonic reaction A → B is to be driven 'uphill'.

Precisely how the enzymic machinery functions in coupled reactions is not always understood, but it is very frequently found that coupling involves the formation of organic phosphate compounds. One important example of such coupling has already been mentioned (p. 119) namely, the formation of adenosine triphosphate (ATP).

The formation of ATP is, however, only a means to an end, for this substance does not accumulate in cells. If one mole ATP is hydrolysed *in vitro* to ADP and inorganic phosphate, 7·9 kcal are liberated as heat, but it is also possible for the terminal phosphate to be transferred to organic molecules if the appropriate enzymes are present. When *covalent phosphate linkages* are formed in this way, some —and sometimes much—of the energy which would have been liberated as heat if phosphate *ions* had been formed, is *transferred with the phosphate group* to an acceptor molecule. In other words, the potential energy of the acceptor (substance A below) is raised in a coupled reaction:

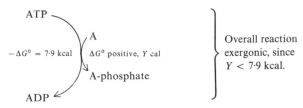

A *priming reaction* occurs when a compound is converted to a derivative of higher potential energy by the attachment of some additional group to it. Priming is by no means peculiar to biochemistry; if acetic acid is converted to acetyl chloride by the action of phosphorus pentachloride, the product, acetyl chloride, is far more reactive than the original acetic acid. So too is its biological counterpart, acetyl coenzyme A (p. 86).

A consequence of this is that reactions which are endergonic when acetic acid is the reactant may be exergonic when the (higher potential) acetyl chloride or acetyl coenzyme A is used instead.

Biochemical priming reactions frequently, but not always, involve the formation of covalent phosphate linkages. This often occurs, as in the example above, at the expense of ATP. In such cases energy from the ATP is transferred *with* the phosphate group (represented in equations as $-\text{\textcircled{P}}$) to substance A, but since the transfer of energy from the ATP $\rightarrow$ ADP system is never 100% efficient, the coupled reaction is *exergonic*. The 'primed' substance may then react in a second *exergonic* reaction with another substance F in such a way that B, the product of the simple endergonic reaction A $\rightarrow$ B, is formed:

$$\text{A} + \text{ATP} \rightarrow \text{A}-\text{\textcircled{P}} + \text{ADP} \ (\Delta G \text{ negative})$$

followed by

$$\text{A}-\text{\textcircled{P}} + \text{F} \rightarrow \text{B} + \text{P}_i \ (\Delta G \text{ negative})$$

The energy for the change A $\rightarrow$ B is thus provided by ATP, making it possible for both the priming reaction and the reaction coupled to F to be *exergonic*. This is illustrated in Fig. 5.3, where vertical distance corresponds to potential energy level.

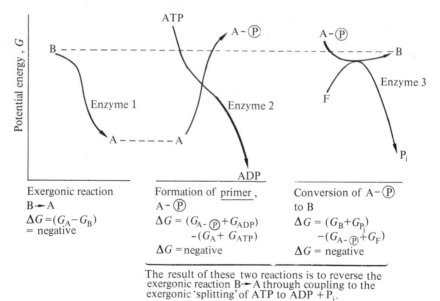

The result of these two reactions is to reverse the exergonic reaction B→A through coupling to the exergonic 'splitting' of ATP to ADP + P$_i$.

Fig. 5.3 Priming and coupling reactions

The enzyme responsible for the exergonic reaction B → A will not normally be concerned in the reactions which convert A back to B through such intermediates as A − (P) and F. Conversely, the enzymes involved in the formally endergonic (though actually exergonic) reverse pathway A → B will normally function uni-directionally, for the property of enzyme specificity (p. 88) makes it unlikely that they will act on substrates other than the intermediate reactants A—(P) and F of the re-routed reaction.

5.7 Free energy, equilibria, and the equilibrium constant

Most readers will be familiar with the basic ideas of chemical equilibrium[#], having already encountered such systems as the esterification of acetic acid. Commencing with several vessels containing varying amounts of acetic acid and ethyl alcohol, the (mathematical) product of the concentrations of the products of the reaction, divided by the (mathematical) product of the concentrations of *residual* reactants is eventually the same for all the systems, at one particular temperature, so long as sufficient time has elapsed for *equilibrium* to be reached. Using square brackets to represent molar concentrations at equilibrium, the equilibrium constant K is thus given by the expression

$$K_{(T)} = \frac{[\text{ethyl acetate}]\,[\text{water}]}{[\text{ethanol}]\,[\text{acetic acid}]} \tag{5.4}$$

Values of K have been determined for numerous biochemical reactions and are of considerable value to the biochemist. A reaction is said to be *reversible* if it does not proceed so near to completion, under the conditions of the experiment, that the denominator of equations such as that above tends to zero due to near-total depletion of one or both reactants.

Let us first illustrate the meaning of K by consideration of a very simple (exergonic) reaction B → A, for which K is found experimentally to be 3. This means that, at equilibrium, the concentration of A is three times that of B:

$$K = \frac{[\text{A}]}{[\text{B}]} = \frac{3}{1}$$

So, if four moles of B are allowed to react, eventually only one mole of B will remain while the rest will have been converted to 3 moles of A.

Thereafter, no further net change occurs.   Similarly, if pure A is left
to equilibrate under the same conditions, one quarter of it will eventually
have been converted to B.   It should be noted that nothing has been said
above about *how long* a reaction takes to reach equilibrium; it might be
reached in a fraction of a second or only after many years.   This is where
catalysts and enzymes play their part, for while their presence does not
in any way influence the *value of the equilibrium constant*, they do influence
those kinetic factors (such as activation energy and molecular orientation)
which determine the *rate of approach* to equilibrium (p. 95).

All spontaneous reactions are exergonic, so when pure A is converted
to an equilibrium mixture of A + B, free energy is liberated.   Similarly,
the approach to the equilibrium mixture, starting from pure B, is spontaneous and exergonic.   At equilibrium, no further change in concentrations occurs; the system is static.   A 'static' chemical system, like
water in a motionless reservoir, is incapable of doing any useful work
and, from the moment of equilibrium onwards, $\Delta G = 0$.   This is summarized in the following scheme:

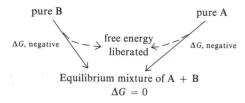

Note carefully that when a freely reversible reaction is being considered
(i.e., K is not too far from unity under the experimental conditions), it
is only formally true to write

$$A \underset{-\Delta G\ =\ x\ \mathrm{cal}}{\overset{+\Delta G\ =\ x\ \mathrm{cal}}{\rightleftharpoons}} B$$

for the 'uphill' part of the reaction is from the *point of minimal potential
energy*, namely, from the equilibrium mixture of A and B.

This is illustrated in Fig. 5.4 where the humps represent activation
energies.   These do not concern the present discussion because the value
of the activation energy does not affect the overall energy change in a
chemical reaction.   A calculation of the free energy change which would
occur if pure A were to be converted to pure B is not directly relevant
because, whatever the initial composition of a mixture of A and B, the
composition after equilibrium is reached is always the same, viz. three
times more A **than** B.   The *theoretical* energy change when pure A is
converted to pure B is $G_B - G_A$ and, when pure B is converted to A, it

is $G_A - G_B$, but to move from the *equilibrium mixture* towards *either* pure A or pure B requires the input of energy. What tends to happen in real situations in the cell is that, for substances A, B, *linked by a readily reversible reaction*, the lower-energy compound A can, in practice, be converted continuously to B provided that B is constantly *removed* from the equilibrium mixture. In other words, although the proportion of

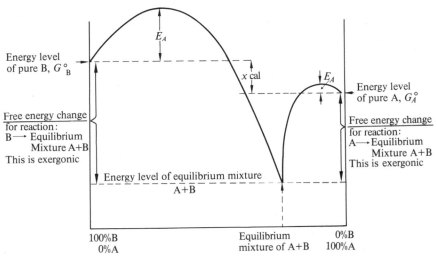

Fig. 5.4 Energy relationship for two substances, A and B for which $K = [A]/[B] = 3$. (Standard free energy change from pure A to pure B (x cal per mole) $= +\Delta G°$, or $(G_B° - G_A°)$ and the reaction is endergonic

B in the equilibrium mixture may be exceedingly small, the reaction is pulled over towards the higher-energy form B either by transfer of B to another part of the cell or by the conversion of B to other compounds.

When the equilibrium constant K for a reaction is measured under standard conditions, an approximate estimate of the free energy change in terms of calories *per mole* of reactant can be obtained from the expression

$$\Delta G^0 = -RT \ln K \qquad (5.5)$$

The logarithm used in this equation is a natural logarithm (ln), not a logarithm to base 10. Arithmetically, it is often more convenient to use $\log_{10} K$; this is related to $\ln K$ by a factor of $2·303$:

$$2·303 \log_{10} K = \ln K \quad \text{or} \quad \log_{10} K = 0·4343 \ln K$$

As an example, let us consider a reaction for which the equilibrium constant is 3 at 27°C (300° Absolute). Then

$$\Delta G^0 = -R \times 300 \times 2{\cdot}303 \log_{10} 3$$

Since

$$R = 1{\cdot}987 \text{ cal per deg per mole,}$$
$$\Delta G^0 = -1{\cdot}987 \times 300 \times 2{\cdot}303 \times 0{\cdot}477$$
$$= -655 \text{ cal.}$$

5.8 Electron flow: the relation of chemical to electrical potentials

Oxidation is best regarded as a loss of electrons, rather than as a loss of hydrogen atoms or a gain of oxygen atoms. Since this is as true for simple inorganic systems as it is for reactions involving the complex materials present in the living cell, let us first consider some reactions familiar to all who have encountered the electrochemical series.[#]

When magnesium is burnt in oxygen or in chlorine, the following equations can be written:

$$Mg + \tfrac{1}{2}O_2 \rightarrow Mg^{2+}O^{2-}; \quad x \text{ cal liberated per g atom Mg}$$
$$Mg + Cl_2 \rightarrow Mg^{2+}2Cl^-; \quad y \text{ cal liberated per g atom Mg}$$

The fundamental similarity of these two changes is obvious; in both, each atom of magnesium *loses two electrons* and an atom of a non-metal receives them. Since the electrons originate in both reactions from magnesium, they start at the same potential, and hence the difference in the energies of reaction (x and y cal) must be attributed to the fact that the electron acceptors, oxygen and chlorine, lie at different levels 'downhill'.

The energy change in an oxidation-reduction, or *redox*, process is as inevitably tied up with the height of the 'energy-fall' as is the change of potential of water with the height of the 'waterfall' (Fig. 5.1). The capacity factor, needless to say, is the mass of the reactant (p. 116). 1 g atom or 1 mole always consists of about 6×10^{23} particles (Avogadro's number), and therefore, in the equations above, the standard energy changes x, y cal represent the energy changes associated with the loss of 12×10^{23} electrons, for magnesium is divalent.

In reactions such as the burning of magnesium in oxygen, the energy change relates to the ionization energy of the electron donor and electron affinity of the electron acceptor. It is usually measured spectroscopically and expressed in kcal. The figures so obtained cannot be directly related to redox reactions involving the same ions in *solution* because in this second case there are additional complications such as the energy associated with the hydration of ions. Nevertheless, the potential energy changes which occur when electrons are transferred during a redox reaction in aqueous solution can be readily measured electrically, the energy change being initially expressed in electron volts. Since energy units are interconvertible, such voltage differences can be re-expressed as calories or joules, as the following example illustrates.

The standard electrode potentials,[#] or redox potentials, of reactions occurring at 25°C in aqueous solution are to be found in textbooks of physical chemistry. That for the change $Fe^{2+} \rightarrow Fe^{3+}$ is $+0.76$ V, while that for the change $\frac{1}{2}Cl_2 \rightarrow Cl^-$ is $+1.36$ V. From these data it is possible to calculate the energy change which occurs when 1 mole of ferrous sulphate in 1 litre of solution is oxidized by chlorine.

The reaction is exergonic, the energy change being $(1.36-0.76)$ or 0.60 V. For one equivalent of ferrous sulphate, i.e., for 6×10^{23} electrons flowing 'downhill', the energy change is $96,500 \times 0.60$ joules, for 96,500 coulombs of electricity (the Faraday) correspond to the charges on 6×10^{23} electrons. Or, expressed as calories:

$$\Delta G^0 = \frac{96,500 \times 0.60}{4.18} \text{ cal} = 13.8 \text{ kcal}$$

In biological systems, the substances which gain or lose electrons when oxidation-reduction occurs are more complex, but the energetic principles involved are the same. If the difference ΔG^0 between the redox potentials of the reducing and oxidizing systems is known, together with the number of electrons, n, transferred per *molecule* of reactant ($n = 1$ or 2 for biological systems), the standard free energy change ΔG^0 can be calculated:

$$\Delta G^0 \text{ (in kcal)} = \frac{n \times 96,500}{4.18 \times 1000} \times \Delta E^0 \qquad (5.6)$$

This equation is of great importance in relation to the functioning of the mitochondrial electron transport system, and its use will be illustrated in the next chapter.

5.9 Steady states

Reversible reactions catalysed by isolated enzymes, like other reversible chemical reactions, tend always to move to a state of equilibrium.

However, in the living cell, unlike most reactions *in vitro*, the overall set of chemical reactions tends to run at a nearly constant rate, since it is necessary for the cell to keep a number of integrated processes ticking over throughout its life. Performance of work is, however, inconsistent with the establishment of a general state of equilibrium, for at equilibrium no useful work is done ($\Delta G = 0$) and there is no net formation of products from reactants.

What actually occurs *in vivo* is that many (but not all) enzymic reactions are associated in chains in such a way that one or more of the products of the first reaction become the reactants of the next (p. 105). Clearly, the removal of the products of the first reaction prevents the establishment of equilibrium even though the concentrations of reactants and products in the mixture often depart little from what the equilibrium concentrations would have been under the same conditions. The same is true for all other reactions in the chain. In other words, $-\Delta G$ for any one reaction of the series tends to be minimal, but it never actually reaches zero because of the effect of the reactions which follow it. Such a *steady state* system resembles certain industrial assembly lines; each step operates at a rate governed by the provision of the appropriate reactants and by the removal of the products through the intervention of the next reaction in the sequence. This interdependence of the reactions ensures that the whole chain or cycle settles down to a constant level of performance unless additional or external factors disturb it in some way.

The velocity with which reactants are converted to products will be different at each step of the sequence and will depend not only upon the concentrations of reactants but also upon the *quantity and effectiveness* of the participating enzymes. Thus the steady state concentrations [A], [B], [C], [D] of compounds involved in a sequence $A \rightarrow B \rightarrow C \rightarrow D$ will be influenced by the activity of the enzymes, a situation which contrasts with the equilibrium condition in which, it will be recalled, the presence of enzymes or catalysts under constant conditions in no way influences the *position* of equilibrium but *only the rate* at which equilibrium is attained. In the industrial production line, the overall rate of production is determined by the rate of the slowest process; so also in a biochemical pathway the slowest reaction limits the rate of formation of intermediates and products. A reaction which limits the rate of a sequence or cycle of reactions is termed the *pacemaker*. In higher animals, the control of metabolism often operates by hormones increasing or decreasing the activity of the pacemaker.

5.10 Adenosine triphosphate and other 'energy-rich' compounds

Since energy is indestructible, it follows that when the energy-rich phosphorus bond of ATP is ruptured, 7·9 kcal of energy must become available per mole ATP.   If the rupturing is by hydrolysis, the energy appears as heat.   Under biological conditions, however, and in the presence of enzymes, much of the energy is converted to forms other than heat.   In particular, it can, as we have seen, be harnessed so that it can increase the potential energy of another system (p. 124).   In later chapters it will be shown that the energy can also be made to do mechanical, osmotic, or other forms of work.

In Fig. 5.5, the ADP/ATP cycle is presented as a coupling device, linking energy released from a catabolic change with absorption of energy by endergonic processes.   Often, the terminal phosphate of ATP is not released as inorganic phosphate ions, but instead becomes combined *covalently* with a compound whose chemical potential energy is thereby raised.   For example, the initial reaction in the metabolism of glucose is the formation of glucose-6-P, the catalyst being *hexokinase*:

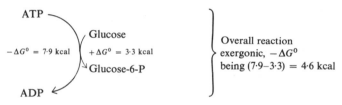

Glucose-6-phosphate, unlike unphosphorylated glucose, is of a suitable energy level for subsequent metabolic conversions.   Clearly, this is a 'priming' reaction (p. 122); the utilization of some of the energy of a molecule of ATP in the formation of glucose-6-P is a biochemically most worthwhile transaction, for eventually more than 30 ATP molecules are synthesized from 30 of ADP as the primed glucose molecule is oxidized to carbon dioxide and water.

No discussion of ATP or of bioenergetics is complete without reference to what is meant by the somewhat misleading term *'energy-rich'* or *'high' energy'* bond.   When glucose-6-P is hydrolysed to glucose and inorganic phosphate, the energy change takes the form of a release of heat of the order of 3·3 kcal.   Comparable energy changes occur when other sugar phosphates are hydrolysed (e.g., fructose-6-P and glyceraldehyde-3-P). It will be noted that this energy change is small compared with that

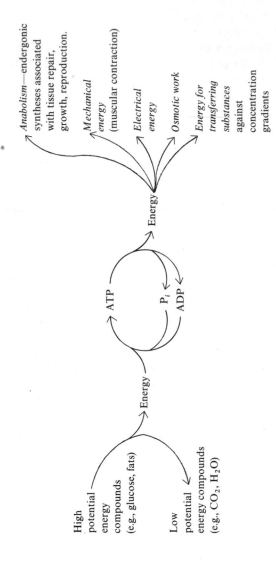

High
potential
energy
compounds
(e.g., glucose, fats)

Low
potential
energy compounds
(e.g, CO_2, H_2O)

Energy

ATP

P_i
ADP

Energy

Anabolism—endergonic
syntheses associated
with tissue repair,
growth, reproduction.

Mechanical
energy
(muscular contraction)

Electrical
energy

Osmotic work

Energy for
transferring
substances
against
concentration
gradients

Fig. 5.5 Central role of ATP in cellular energetics

occurring when the terminal phosphate group of ATP is removed by hydrolysis to form ADP. Compounds such as ATP which undergo reactions yielding a large change in free energy are often called *high energy compounds*, and the particular bond cleaved in the highly exergonic reaction is often described as a *high energy bond* and represented thus: $\sim$. In fact, ATP has two such bonds for, when the terminal and middle phosphate groups are removed by hydrolysis, in each case some 7·9 kcal per mole are released.   On the other hand, the removal of the phosphate group from AMP, adenosine *mono*phosphate, only results in the liberation of some 3 kcal.   To indicate the position of the high energy bonds ATP can be written as follows:

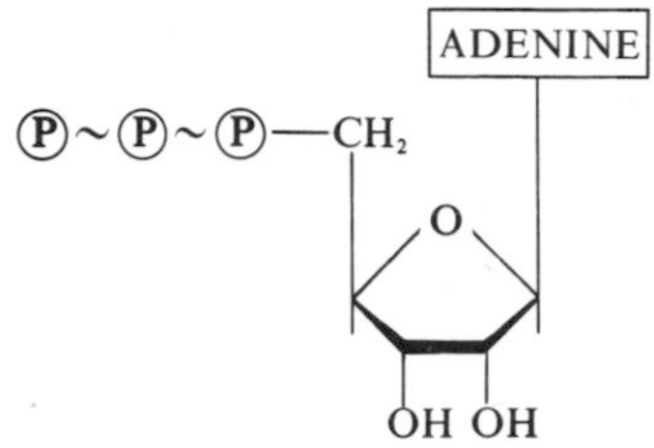

$$\textcircled{P} \sim \textcircled{P} \sim \textcircled{P} - CH_2 \quad \boxed{ADENINE}$$

$$OH \quad OH$$

Although there is no clear-cut distinction between a high and a low energy bond, arbitrarily an energy change upon hydrolysis of some 5 or 6 kcal is frequently taken as the upper limit for low energy bonds.

The total energy of ATP and of other high energy compounds is best regarded as being distributed *throughout* the molecule rather than being associated with a particular bond, though at the *moment of cleavage* the bond being cleaved is effectively associated with a disproportionately high level of energy.   Upon hydrolysis of ATP, the high energy change occurs because the pyrophosphate (anhydride type) linkage is replaced by more stable phosphate ions or covalent phosphates—and a 'stable' chemical substance is, of course, one which is relatively low in energy. Hence energy is released not because of any unique property of a particular bond, but because of the *overall chemical structure* of the ATP molecule. Nevertheless, the designation 'high energy bond' provides a *convenient shorthand way* of signifying at what locations in molecules particularly important energy changes occur when the compounds participate in biochemical reactions.

One important product of glucose metabolism is known as *phospho-enol pyruvate* (PEP). It is a high energy compound bearing a phosphate group.   When PEP undergoes simple hydrolysis pyruvate and inorganic orthophosphate are formed, the reaction being highly exergonic

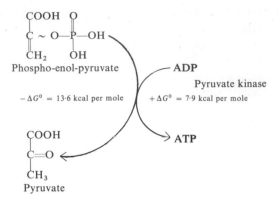

Phospho-enol-pyruvate

$-\Delta G^0 = 13.6$ kcal per mole

ADP

Pyruvate kinase

$+\Delta G^0 = 7.9$ kcal per mole

ATP

Pyruvate

(a) *Transfer of high energy phosphate from phospho-enol-pyruvate to ADP (ΔG^0 for coupled reaction = $-13.6 + 7.9$ or -5.7 kcal per mole)*

CREATINE

$-\Delta G^0 = 10.5$ kcal per mole

ATP

Creatine kinase

$+\Delta G^0 = 7.9$ kcal per mole

CREATINE ~ (P)

ADP

(b) *The 'Lohmann' reaction ($\Delta G^0 = -10.5 + 7.9 = -2.6$ kcal per mole).* For the formula of creatine, see p. 316

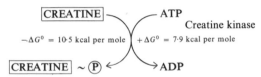

Compound with a suitable carbonyl group   Acetyl coenzyme A   Product of addition reaction

(c) *A typical reaction of high energy acetate (acetyl coenzyme A)*

Fig. 5.6   Three reactions involving high energy compounds

($-\Delta G^0 = 13.6$ kcal per mole).   However, within the cell, an enzyme, *pyruvate kinase*, normally catalyses the transfer of the phosphate from PEP to ADP, much of the energy of cleavage accompanying the phosphate, in the sense that it provides the energy needed to attach the phosphate to the ADP and so to raise the energy level of the latter to the higher level of ATP (Fig. 5.6a).   This is one way in which ATP is synthesized in the cell.   PEP is clearly an example of compound C in Fig. 5.2.   Pyruvate kinase is one of a family of *kinases* or phosphoryl transferases, each of which either transfers organic phosphate to ADP (so converting it to ATP) or, as in the case of hexokinase (p. 131), passes phosphate from ATP to an organic compound, thereby converting it to a phosphate ester.

A rather similar reaction is known as the Lohmann reaction but it differs from the preceding examples in that it is *reversible*. In vertebrate muscle, ATP reacts with creatine to form a high energy compound, creatine phosphate (Fig. 5.6b). The reaction, catalysed by creatine kinase, is of particular importance because ATP never accumulates in cells. When the muscle cell needs more ATP than it can synthesize in the time available, creatine phosphate donates its phosphate group to ADP and so regenerates ATP. For this reason, creatine phosphate acts as an energy store and is called a *phosphagen*. The creatine kinase reaction is reversible because the free energy change is relatively small. By contrast, the reaction catalysed by pyruvate kinase is not reversible for the coupled reaction involves too high an energy change for pyruvate to be directly phosphorylated by ATP.

Acetyl coenzyme A is yet another example of a high energy compound. Figure 5.6c shows one sort of reaction in which this compound can participate—it is an addition reaction which resembles the aldol condensation# of classical organic chemistry. The 'acetyl' part of acetyl coenzyme A has a higher energy than free acetate (p. 86) and, unlike the latter, is able to 'condense' with suitable carbonyl groups. Several examples of this special kind of addition reaction will be given in later chapters.

It is not possible to present or to acquire a very meaningful picture of what biochemistry is all about without giving consideration to the energy changes which accompany biochemical reactions, for all reactions which occur in cells must of necessity be both *mechanistically* and *energetically feasible*. This should be constantly kept in mind as, in later chapters, the various biochemical pathways are described. Interesting though the machinery of the living engine frequently proves to be, it is nevertheless the energetic function that the machinery fulfils which is worthy of special consideration, for this function is the *raison d'être* for its existence.

6

Mitochondrial electron transport and oxidative phosphorylation

6.1   The purpose of electron transport

In chapter 1, the mitochondrion was described as the 'power house of the cell' and it is proposed now to show why it merits this name.   To do this, it will be necessary to make use of certain energetic considerations outlined in the previous chapter, and especially the idea of electron loss as an essential factor in oxidation.   To show the subject matter of this chapter in perspective, the overall process of mitochondrial oxidation and electron transport will be briefly described first; then follows a more detailed consideration of the mechanisms of electron transfer and of an important process in which the energy lost by the electrons leads to the formation of ATP.

The oxidation of carbohydrates and of fats results in the production of *acetyl coenzyme A* (p. 27), the acetyl part of which undergoes controlled oxidation with the eventual production of ATP, carbon dioxide, and water.   The controlled oxidation is brought about by an elegant mechanism called the tricarboxylic acid cycle, which is a catalytic device involving a cyclic series of reactions.   The nature of these reactions is not our present concern; they represent the *machinery* of the oxidation whereas this chapter is concerned with the *energetic purpose* for which that machinery exists.   All that needs to be stressed now is that, by means of this machinery, the breakdown of the acetyl unit is subdivided into four oxidative stages, and that the electrons released at each of these four stages are passed down a redox gradient until in each case they reduce oxygen to water.   Thus, if the machinery of oxidation is represented as

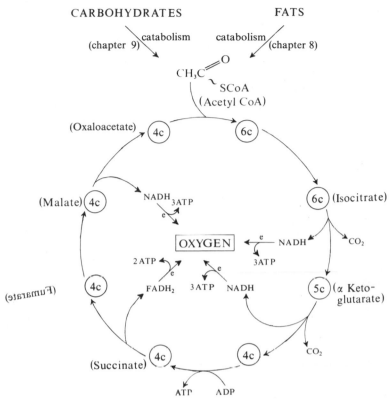

Fig. 6.1 A simple outline of the tricarboxylic acid cycle. (Acetyl coenzyme A, resulting from carbohydrate or fatty acid catabolism, enters the cycle by combining with a four carbon compound. One turn of the cycle produces 3NADH, 1FADH$_2$, and 2CO$_2$. The reduced carriers are the source of energy for ATP synthesis by oxidative phosphorylation. $\xrightarrow{\text{e}}$ indicates passage of electrons from carriers to oxygen) $\xrightarrow[\text{ATP}]{\text{e}}$

the circumference of a wheel (Fig. 6.1) the electron transport redox systems can be represented as four spokes directed to a central hub of oxygen.

The oxidizable substrates which are the electron donors for the four 'spokes' are *isocitrate*, *α-ketoglutarate*, *succinate*, and *malate*. Each of these, when oxidized, loses two electrons together with two hydrogen ions, the electrons passing down the energy gradient of the transport system. When electrons originate from succinate, the transport system is in certain small respects different from that which operates when the electrons come from the other three substrates, but for the moment any one of the four substrates can be represented by AH_2:

$$AH_2 \xrightarrow{\text{AH}_2 \text{ dehydrogenase}} A + 2H^+ + 2e^-$$

The electrons are eventually accepted in turn by each member of a series of electron carriers neatly arranged on the inner membranes (cristae) of the mitochondrion; they pass along the chain of carriers, each of which has a potential lower than the previous one, till the electrons finally reach oxygen; the process is, energetically, a 'downhill' process, even though it is 'along' a chain of carriers. The total system can be likened to a waterfall divided into several separate cascades by the intervention of water wheels situated at different heights. Just as the water wheels harness the energy from the falling water so the mitochondrion uses the energy of oxidation to drive a concomitant 'uphill' reaction in which inorganic phosphate and ADP react to give the high-energy compound ATP (Fig. 6.2). This latter

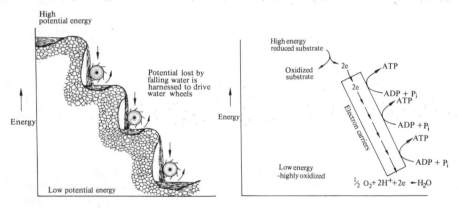

Fig. 6.2 The flow of electrons along a series of electron carriers from a high energy level to a lower level is used to drive the endergonic reaction ADP + P_i → ATP. The diagram on the right represents one 'spoke' of the previous figure. At the lowest energy level, the electrons finally reduce oxygen to water. The system is energetically analogous to the energy capture of cascading water by a series of water wheels (left)

process is termed *oxidative phosphorylation* and contrasts with reactions such as the one catalysed by pyruvate kinase (Fig. 5.6a) in which high energy phosphate is transferred to ADP without the intervention of oxygen. Reactions of this latter type are called *substrate level phosphorylations*.

6.2 NAD, the first electron acceptor

The coenzyme NAD is the initial electron acceptor in the oxidation of three of the substrates mentioned in section 6.1, namely, isocitrate, α-keto-glutarate, and malate. NAD is therefore the first of the series of electron carriers so important in oxidative phosphorylation.

The mechanism of the reversible oxidation-reduction of NAD was described in chapter 3: the reduction of NAD^+ to NADH involves the addition of *two electrons* (and one proton) to the nicotinamide part of the molecule. The nicotinamide moiety is thereby converted from a heterocyclic ring with *three* double bonds to a ring with *two* double bonds (section 3.9). This process is similar to numerous redox reactions involving organic compounds; reduction involves the gain of two electrons and the loss of a double bond, while conversely, the loss of two electrons and the formation of a double bond is frequently observed when organic compounds are oxidized (Fig. 6.3).

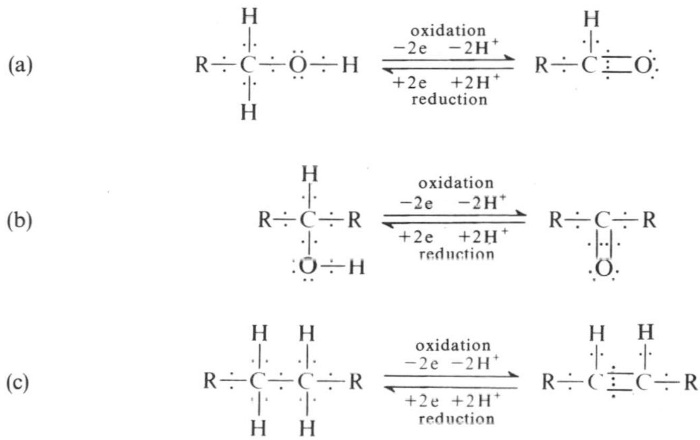

Fig. 6.3 Three common types of oxidation of organic compounds. Electrons indicated as (O). Each redox change involves addition or removal of two electrons and two protons

A specific example will illustrate how NAD^+ links with substrate oxidation. Malic acid is oxidized in the mitochondrion to oxaloacetic acid; the reaction is catalysed by the enzyme *malate dehydrogenase*, for which NAD^+ is cofactor. The oxidation and reduction phases of the reaction are closely coupled but can be described in two parts. Firstly, two electrons and two protons are removed from the malate, yielding

Mitochondrial electron transport and oxidative phosphorylation

oxaloacetate. Secondly, the two electrons and a proton combine with NAD^+ producing the reduced form, NADH.

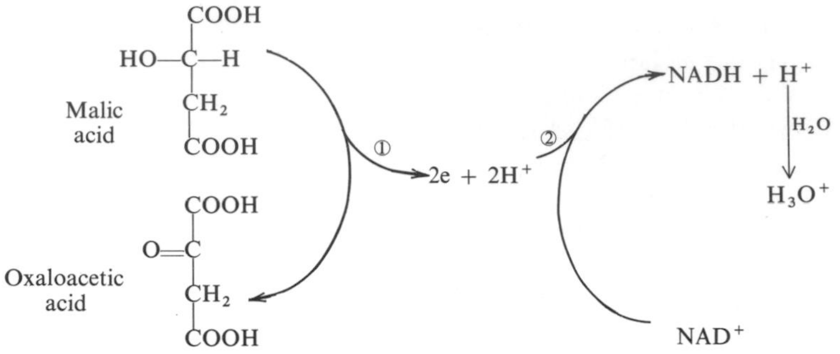

Only one of the two protons removed from the malate is transferred to NAD^+; the other combines with water, tending to make the (buffered) solution slightly acidic. Usually, this reaction is written in the simpler form:

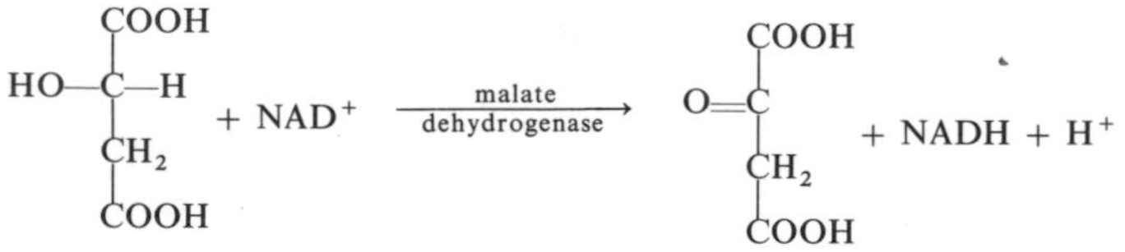

For notes on how to remember the formulae of malate and oxaloacetate, see pp. 147–150.

6.3   Flavin mononucleotide (FMN)

The NADH produced by the malate dehydrogenase reaction and by certain other mitochondrial enzymes is reoxidized to NAD^+ by passing the two electrons to the prosthetic group of a lipoprotein. This protein, like many other components of the electron transport chain, forms an integral part of the inner mitochondrial membrane. The prosthetic group, flavin mononucleotide (FMN), provides the mechanism for electron transport by this, the second electron transport carrier of the chain.

FMN is, in fact, *riboflavin phosphate*, the 'nucleotide' base being dimethyl isoalloxazine (Fig. 6.4a). In the strictest chemical sense it is not a nucleotide because it contains ribitol and not ribose, but nature is not so insistent upon strict adherence to definitions as chemists tend to be.

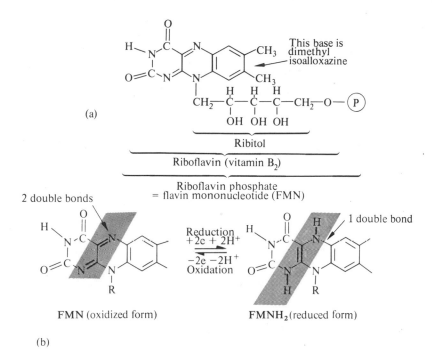

(a)

(b)

Fig. 6.4 Flavin mononucleotide (FMN)
(a) Complete structure
(b) Oxidized and reduced forms

Like NAD (and each of the remaining electron carriers) it exists in oxidized and reduced forms. The reduced flavin has two more electrons and two more protons than oxidized FMN and is usually written $FMNH_2$. The two forms are illustrated in Fig. 6.4b.

The oxidation of NADH by FMN can be indicated thus:

$$NADH + H^+ \qquad 2e + 2H^+ \qquad FMNH_2$$
$$NAD^+ \qquad\qquad\qquad\qquad FMN$$

The NAD^+ so formed is then ready to take part in further NAD-linked dehydrogenase reactions, while the $FMNH_2$ is reoxidized by passing two protons and two electrons to the next electron carrier. Note that here, too, the reduced form possesses one less double bond than does the oxidized form.

Sci Amer Jan '64 Mitochondria

See Green

6.4 Ubiquinone

Mitochondria contain considerable quantities of a quinone known as ubiquinone, or coenzyme Q. Its reduction involves the addition of two electrons together with two protons, resulting in a decrease in the number of double bonds from four to three (Fig. 6.5). $FMNH_2$ is oxidized by

See Lehninger

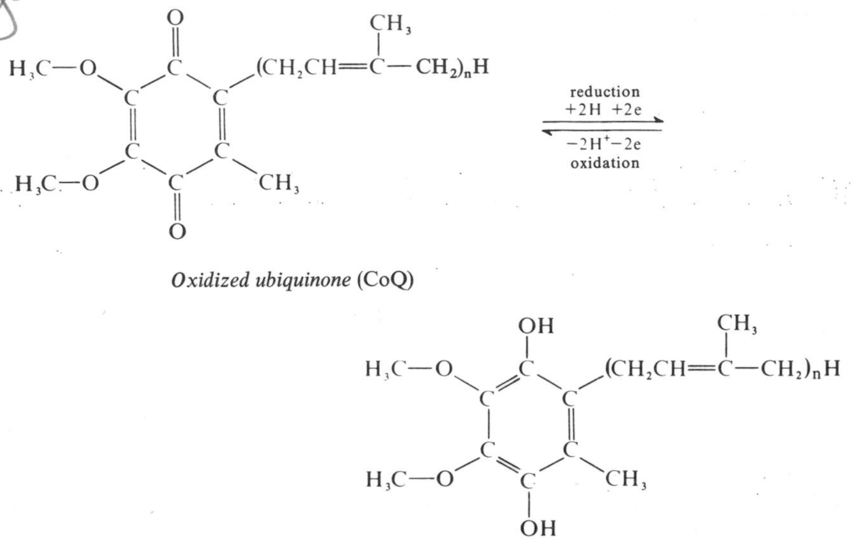

Oxidized ubiquinone (CoQ)

Reduced ubiquinone ($CoQH_2$)

Fig. 6.5 The structure of oxidized and reduced ubiquinone (coenzyme Q). (n = 9 and 10 in mammalian mitochondria; n = 6, 7, or 8 in microorganisms)

mitochondrial ubiquinone to FMN, the ubiquinone being thereby reduced to a diphenol.

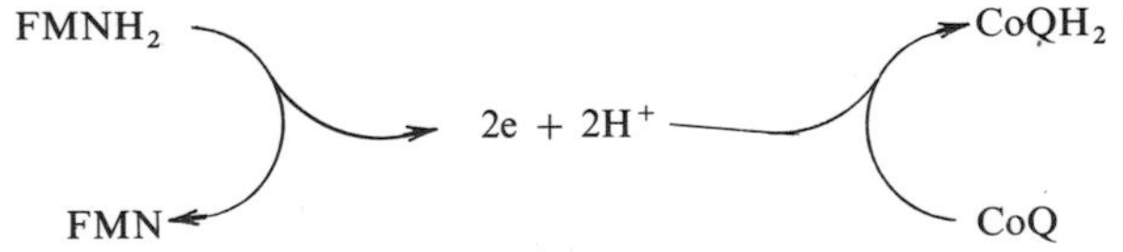

Ubiquinone thus represents the third electron carrier, NAD being the first and FMN the second. It subsequently passes on the electrons to the next carrier, cytochrome **b**.

6.5 Cytochrome b

Reduced mitochondrial ubiquinone (CoQH$_2$) must be reoxidized if it is to continue to function as an electron carrier capable of accepting electrons from FMNH$_2$. The two electrons are transferred to a lipo-protein, cytochrome **b**, which, like all other cytochromes, contains an iron ion which undergoes reversible oxidation and reduction between the ferrous and ferric states.

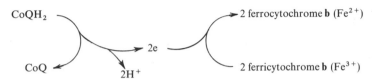

Since the reaction (Fe^{3+} + e $\rightleftharpoons$ Fe^{2+}) requires the transfer of only *one* electron, two molecules of cytochrome **b** are involved in the oxidation of one molecule of reduced ubiquinone.   The two protons released in the reaction are temporarily accommodated in the buffered solution nearby, being (statistically) accounted for at the cytochrome oxidase stage described later.

Cytochrome **b**, like several other cytochromes participating in the mitochondrial electron transport system, is of interest because it possesses an iron-containing haem prosthetic group.   The group present in cyto-chrome **b** is called *protohaem*; the same prosthetic group is an essential constituent of haemoglobin and myoglobin (chapter 15).   Protohaem belongs to a family of compounds known as *porphyrins*, many of which are important biological pigments.   Like other porphyrins, protohaem has, as its nucleus, four pyrrole rings (Fig. 6.6). This nucleus is highly 'aromatic', each bond having some single and some double bond charac-teristics, so that, as in the case of benzene or the purine and pyrimidine bases, it is not possible to draw a formula which accurately represents the position of the two 'different' types of bond.   In the various por-phyrins, different side groups are often attached to the nucleus at the eight outer carbon atoms.   For protohaem (Fig. 6.6) the side groups, numbered 1 to 8, are indicated in light type.   The porphyrin nucleus is planar, and frequently, as in the case of protohaem in haemoglobin or cytochrome **b**, there is a central iron ion which is held by four coordinate links to the four nitrogen atoms.   Although iron is often coordinated to porphyrins in this way, other cations may be present instead, a notable case being magnesium, which is present in the porphyrin-like group of

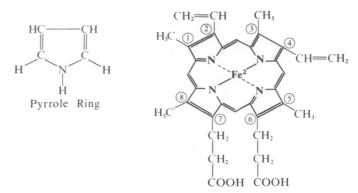

Fig. 6.6 Protohaem—the prosthetic group of cytochrome **b**, haemoglobin, and myoglobin. The porphyrin nucleus (heavy type) has eight positions for side groups; those for protohaem are indicated in light type. The iron ion has four coordinate links with nitrogen atoms.

chlorophyll. In different flavins and cytochromes, the prosthetic groups are also, of course, attached to different *proteins*.

6.6 The other cytochromes

Reduced cytochrome **b** is reoxidized by passing an electron on to a second porphyrin lipoprotein, cytochrome c_1 the iron of which is thereby converted from the ferric to the ferrous state:

reduced cytochrome **b** (Fe^{2+})   → reduced cytochrome c_1 (Fe^{2+})

e

oxidized cytochrome **b** (Fe^{3+}) ←   oxidized cytochrome c_1 (Fe^{3+})

Thus electrons derived from an enzyme reaction such as that catalysed by malate dehydrogenase reach cytochrome c_1 by the following route:

$$\text{malate} \rightarrow \text{NAD} \rightarrow \text{FMN} \rightarrow \text{ubiquinone} \rightarrow \text{cyt. } \mathbf{b} \rightarrow \text{cyt. } \mathbf{c}_1$$

Their final destination is oxygen and they reach it from cytochrome c_1 via at least two additional cytochromes, known by the letters **c** and **a**. Like cytochrome **b**, these contain a porphyrin ring with iron at the centre. The transfer of electrons is associated with a ferric to ferrous valency change on the part of the acceptor cytochrome, and of a valency change in the other direction on the part of the donor. Each change is comparable to the reduction of cytochrome c_1 by reduced cytochrome **b**.

The porphyrins of cytochrome c_1 and **c** are identical, being similar to the protohaem of cytochrome **b** except that the vinyl groups of side chains 2 and 4 (Fig. 6.6) are linked to two cysteine residues of the lipoprotein molecule. The porphyrin of cytochrome **a** resembles protohaem except that the methyl group at position 8 is replaced by —CHO, and there is a long branched aliphatic side chain at position 2 instead of the vinyl group. All but one of the mitochondrial cytochromes are insoluble in water and have proved very difficult to study. The exception is cytochrome **c**, which is soluble in water, and considerable progress has been made in investigating the amino acid sequence of the protein of this carrier.

A lipoprotein complex which contains both cytochrome **a** and copper forms the terminal link of the electron transfer chain. It is known as *cytochrome oxidase* since it passes on to molecular oxygen the electrons derived from cytochrome **c**. The latter becomes oxidized while the oxygen is reduced to water, protons from the surrounding solution being needed at the same time.

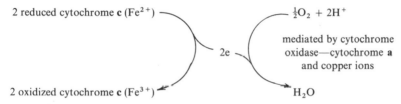

2 reduced cytochrome **c** (Fe^{2+}) — $\frac{1}{2}O_2 + 2H^+$

2e — mediated by cytochrome oxidase—cytochrome **a** and copper ions

2 oxidized cytochrome **c** (Fe^{3+}) — H_2O

Note that, because two electrons and two protons are required to convert $\frac{1}{2}O_2$ to H_2O, it is necessary for stoichiometric reasons to indicate the participation of two molecules of reduced cytochrome **c**. The oxidase is poorly understood: the electrons probably pass from a sub-component sometimes called cytochrome **a**, via copper (which may undergo a cuprous-cupric valency change) to oxygen.

There are probably other components in the electron transport chain about which little is known. In particular, in at least three places in the chain, a protein (or proteins) which contain iron *not* bound within a porphyrin ring, are known to be present. These are called non-haem iron proteins. Additionally, it should be remembered that, as in all cases where conjugate proteins play roles as enzymes or carriers, the lipoprotein *apoenzymes* probably contribute to the catalytic effect. For example, electrons accepted by a cytochrome may combine with the protein part of the enzyme before they are passed on, by an unknown mechanism, to the ferric ion in the centre of its prosthetic porphyrin group. When thinking of electron transport, it is easy to forget that the flavins and porphyrins are only very small parts of much larger molecules.

6.7 Succinate dehydrogenase

Of the four oxidizable substrates derived from acetyl coenzyme A, succinic acid is exceptional in not passing electrons to NAD. There is instead a lipo-flavoprotein which specifically catalyses the oxidation of succinate to fumarate and is known as succinate dehydrogenase.

Flavoproteins are conjugate proteins containing a derivative of riboflavin as prosthetic group. The protein which oxidizes NADH (section 6.3) is an example of one in which the prosthetic group is flavin *mononucleotide*, FMN. By contrast, succinate dehydrogenase contains *flavin adenine dinucleotide* (FAD). FAD consists of the nucleotides AMP and FMN linked together by a pyrophosphate bond to form a dinucleotide:

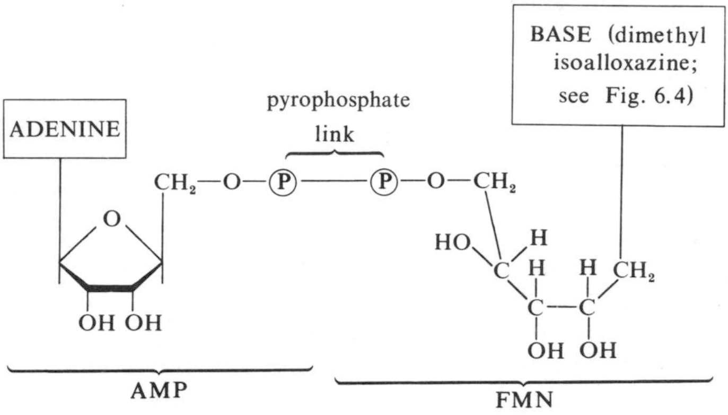

The structure of FAD should be compared with that of the dinucleotide NAD (p. 84). The riboflavin part of the FAD molecule undergoes a reversible redox reaction identical to that outlined for FMN (Fig. 6.4b).

When succinate dehydrogenase oxidizes succinate to fumarate, two protons and two electrons are transferred to FAD which is thereby reduced to $FADH_2$. Succinate dehydrogenase is *inhibited competitively* by malonate (chapter 4) which, owing to its structural similarity to succinate, is able, when present in excess, to bind to the active site of the enzyme.

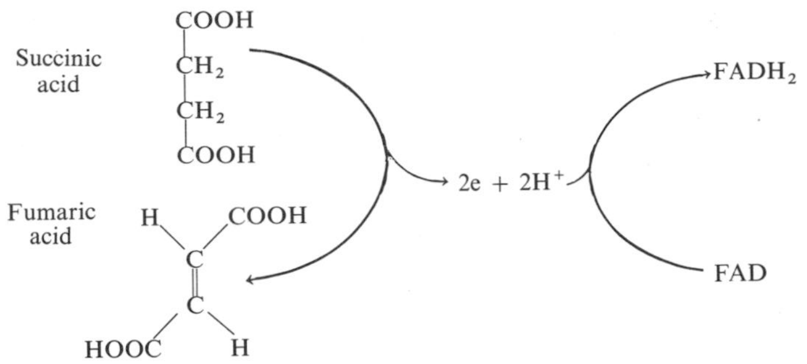

It is noteworthy that succinate is oxidized to the *trans* unsaturated dicarboxylic acid, fumarate, and never to the *cis* isomer.

The reduced FAD of succinate dehydrogenase is reoxidized by passing its electrons to the electron transport assembly previously outlined. The point of entry is not firmly established; cytochrome **b** may receive the electrons directly or, more probably, it receives them via ubiquinone and a non-haem iron protein.

It is evident therefore that the electron transport system has two major initial routes which merge at, or near, ubiquinone; the route commencing with succinate is truncated relative to the other, for NAD is not involved in electron transport from succinate. Nevertheless, after the passage of electrons along the transport chain, oxygen is the final electron acceptor for each of the four major oxidizable substrates derived from acetyl coenzyme A and hence, eventually, from fats and carbohydrates. The electron transport system is summarized in Fig. 6.7.

6.8 Formulae and names of some important intermediates

This section may be helpful to those whose background of chemistry is limited and who are not familiar with compounds such as oxaloacetate, α-ketoglutarate and citrate, for these and some other substances referred to below will be encountered again and again in biochemical systems.

(a) Fumarate, succinate, and malate

Readers may have encountered the unsaturated dicarboxylic acid, fumaric acid, in relation to geometrical (*cis–trans*) isomerism. Fumaric acid is the *trans* acid. If fumaric acid is saturated by addition of hydrogen, succinic acid is obtained, and if it is saturated by addition of the elements

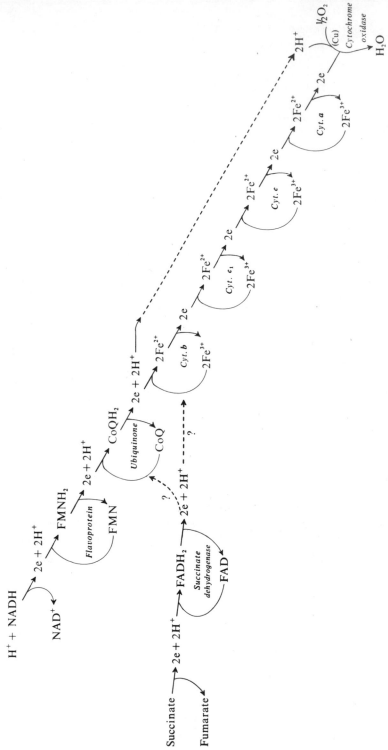

Fig. 6.7 Electron transport chain. (Arrows indicate direction of electron flow. Vertical distance, both in chain and individual cycles, shows relative energy levels of components. Note that initially electrons derived from succinate follow a different route from those of NADH)

of water, malic acid results.   The latter has an asymmetric carbon atom and therefore exists in L and D forms.   However, in the cell, only L-malic acid is involved in metabolism in most situations.

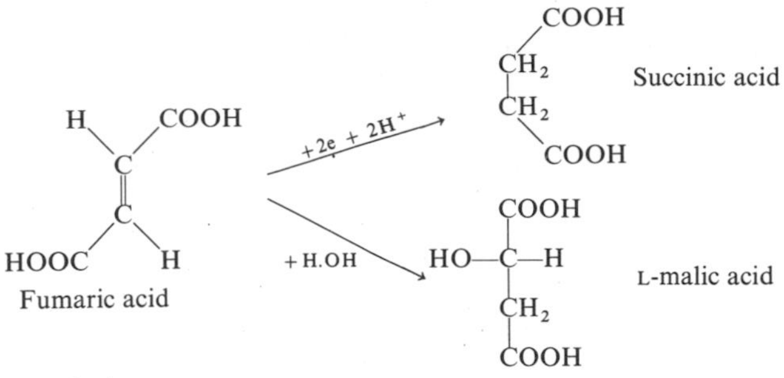

$$H_3C—CO\underline{OH}\ \underline{H}—CH_2—COOH \rightarrow CH_3.CO.CH_2.COOH$$

Fumaric acid, Succinic acid, L-malic acid structures

$$+2e + 2H^+$$

$$+ H.OH$$

Succinic acid

$$\begin{array}{c} COOH \\ | \\ CH_2 \\ | \\ CH_2 \\ | \\ COOH \end{array}$$

L-malic acid

$$\begin{array}{c} COOH \\ | \\ HO—C—H \\ | \\ CH_2 \\ | \\ COOH \end{array}$$

Fumaric acid

$$\begin{array}{c} H \qquad COOH \\ \ \ \ \diagup \\ C \\ \| \\ C \\ \diagup \qquad \diagdown \\ HOOC \qquad H \end{array}$$

(b) Acetoacetate, oxaloacetate, and oxalosuccinate

Each of these can be regarded as a composite structure resulting (theoretically) from the reaction of the hydroxyl group of the acid forming the initial part of the name with an '*alkyl*' *hydrogen atom* of the acid indicated by the last part of the name:

$$H_3C—CO\underline{OH}\ \underline{H}—CH_2—COOH \rightarrow CH_3.CO.CH_2.COOH$$
Acetic acid Acetic acid Acetoacetic acid

$$HOOC—CO\underline{OH}\ \underline{H}—CH_2—COOH \rightarrow HOOC.CO.CH_2.COOH$$
Oxalic acid Acetic acid Oxaloacetic acid

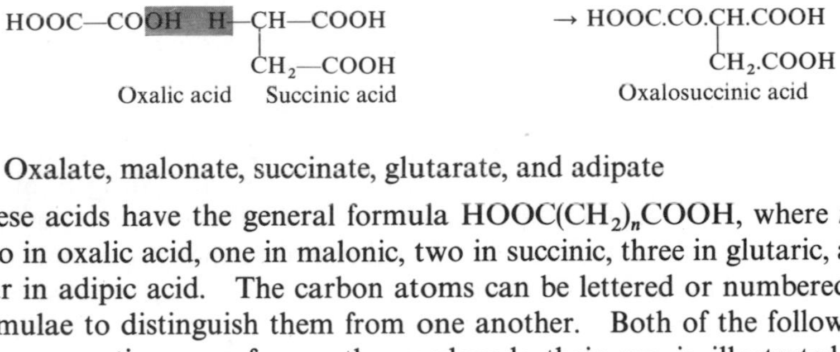

$$HOOC—CO\underline{OH}\ \underline{H}—\begin{array}{c}CH—COOH \\ | \\ CH_2—COOH\end{array} \qquad \rightarrow \begin{array}{c}HOOC.CO.CH.COOH \\ | \\ CH_2.COOH\end{array}$$
Oxalic acid Succinic acid   Oxalosuccinic acid

(c) Oxalate, malonate, succinate, glutarate, and adipate

These acids have the general formula $HOOC(CH_2)_nCOOH$, where n is zero in oxalic acid, one in malonic, two in succinic, three in glutaric, and four in adipic acid.   The carbon atoms can be lettered or numbered in formulae to distinguish them from one another.   Both of the following two conventions are frequently employed; their use is illustrated by reference to glutaric acid.

$$HOOC—CH_2—CH_2—CH_2—COOH$$

1	2	3	4	5
—	α	β	γ or α'	—

Thus α-hydroxyglutaric acid is:

$$\text{HOOC}-\overset{\alpha}{\underset{\underset{\displaystyle \text{OH}}{|}}{\text{CH}}}-\text{CH}_2-\text{CH}_2-\text{COOH}$$

The oxidation of this secondary alcohol yields α-ketoglutaric acid:

$$\text{HOOC}-\overset{\alpha}{\underset{\underset{\displaystyle \text{O}}{\|}}{\text{C}}}-\text{CH}_2-\text{CH}_2-\text{COOH}$$

(d) Citrate and isocitrate

Citric acid is a tricarboxylic acid with the carbon skeleton of glutaric acid but with the two hydrogen atoms on the β carbon atom replaced by a hydroxyl group and a carboxyl group. Note that the structure is symmetrical about the β carbon atom. Its isomer, isocitric acid, differs only in the position of the hydroxyl group, for it is on the α instead of the β carbon atom.

$$
\begin{array}{ll}
\begin{array}{l}
\text{COOH} \\
| \\
\text{CH}_2 \\
| \\
\text{HO}-\text{C}-\text{COOH} \\
| \\
\text{CH}_2 \\
| \\
\text{COOH} \\
\text{Citric acid}
\end{array}
&
\begin{array}{l}
\text{COOH} \\
| \\
\text{HO}-\text{C}-\text{H} \\
| \\
\text{H}-\text{C}-\text{COOH} \\
| \\
\text{CH}_2 \\
| \\
\text{COOH} \\
\text{Isocitric acid}
\end{array}
\end{array}
$$

$\leftarrow \alpha \rightarrow$ $\leftarrow \beta \rightarrow$

(e) The naming of weak acids and their salts

In previous sections, it will no doubt have been observed that the words 'fumarate' and 'fumaric acid' etc., have apparently been regarded as interchangeable, the most convenient form of the word-pair having been employed in headings and in text. The reason is that the organic acids considered in this section, while capable of losing one hydrogen ion for each carboxylic acid group present, are nevertheless only partially dissociated at physiological pH. In other words, they are examples of *weak acids*, in contrast to *strong* or nearly completely dissociated acids such as sulphuric acid. For an acid such as succinic acid, both the undissociated acid and its conjugate base, the succinate ion, are present in cells in equilibrium with one another. Therefore, although a given enzyme may conceivably act only upon the conjugate base or upon the acid, it is not necessary to indicate the specific form since the two are related by a self-adjusting equilibrium. The ratio of the two forms of any such substance depends only on the dissociation constant of the acid and on the pH of the environment (p. 38).

6.9 Redox potentials

Electrons derived from metabolites such as malate or succinate move from one carrier to the next along the electron transport assembly to oxygen which, as has been seen, is the final electron acceptor. Normally they move via electron carriers which tend, percentage-wise, to be more in the oxidized and less in the reduced state the nearer they are to oxygen in the chain. The flow of electrons is always towards oxygen because electrons in the water molecule can be regarded as being at a lower energy level than when they maintain compounds such as NAD or cytochrome **c** in the reduced state. As they are transferred in steps from malate to oxygen they are descending an energy gradient just as is water when it flows downhill (section 5.3).

The tendency for substances to undergo redox reactions differs considerably and depends on the relative ease with which they accept or

Table 6.1   Redox potentials of some important cellular reactions

Redox reaction	E_0' (25°C or 30°C; pH, 7·0) (V)
Oxidized form → reduced form:	
Succinate → α-ketoglutarate	$-0·67$
CO_2 + α-ketoglutarate → isocitrate	$-0·48$
Ferredoxin (ox.) → (red.)	$-0·43$
NAD → $NADH_2$	$-0·32$[a]
Riboflavin → reduced riboflavin	$-0·21$[b]
Acetaldehyde → ethanol	$-0·20$
Pyruvate → lactate	$-0·18$
Oxaloacetate → malate	$-0·17$
Fumarate → succinate	$+0·03$
Cyt. **b** Fe^{3+} → Fe^{2+}	$+0·04$
Ubiquinone (CoQ) → $CoQH_2$	$+0·10$
Cyt. **c** Fe^{3+} → Fe^{2+}	$+0·22$
Cyt. **a** Fe^{3+} → Fe^{2+}	$+0·29$
Cu^{2+} → Cu^+	$+0·35$
Oxygen, $\frac{1}{2}O_2$ → H_2O	$+0·82$

(a) NADP has the same redox potential as NAD
(b) Riboflavin as part of FAD or FMN in flavoproteins has a wide range of redox potentials.

give up electrons.   This tendency can be measured, and, for comparative purposes, can be referred to the readiness with which hydrogen gas is oxidized to hydrogen ions:

$$H_2 \rightleftharpoons 2H^+ + 2e$$

Compounds in the upper part of Table 6.1 lose electrons more readily than hydrogen gas, while those in the lower part have a lesser tendency to become oxidized.   For each substance the tendency of its reduced form, under standard conditions, to give up electrons and so become oxidized, is known as the standard redox potential E_0' and is measured in volts.   (Standard conditions need to be specified, for the actual redox potential varies with such things as concentration and pH; E_0' refers to a molar solution containing equal quantities of the oxidized and reduced forms of the compound at pH 7·0.)   When E_0' is *negative* the substance tends to become oxidized more readily than hydrogen gas, i.e., a high negative E_0' indicates a high *electron-donating*, or reducing, capacity. Conversely, substances with positive values of E_0' tend to become reduced because they readily accept electrons, and so act as oxidizing agents.

Table 6.1 shows the values of E_0' for a number of donors and carriers of electrons; *reduced forms* of compounds at the top of the table lose electrons to the *oxidized forms* of those lower down—i.e., to those with a more positive value of E_0'.   Thus, since NADH tends to lose electrons more readily than does reduced cytochrome **c**, the oxidized cytochrome is likely to accept them from NADH either directly, or indirectly via other carriers.   As electrons move down to the lower energy levels of carriers with more and more positive redox potentials, a little of the energy is dissipated as heat but much of it is transferred to other systems—in the case of the mitochondrial carriers, to ADP + P_i—which are thereby raised to a higher energy level to form ATP.

If we write the electron carriers in the order they appear in the mito-chondrial chain, it is found that the sequence almost corresponds to the magnitudes of their redox potentials:

$$E_0' = \begin{array}{cccc} \text{NAD} \rightarrow & \text{riboflavin in FMN} \rightarrow & \text{ubiquinone} \rightarrow & \text{cytochrome } \mathbf{b} \\ -0{\cdot}32 & -0{\cdot}21 & +0{\cdot}1 & +0{\cdot}04 \end{array}$$

$$\begin{array}{cccc} \text{O}_2 \leftarrow & \text{Cu} & \leftarrow \text{cytochrome } \mathbf{a} \leftarrow & \text{cytochrome } \mathbf{c} \\ +0{\cdot}82 & +0{\cdot}35 & +0{\cdot}29 & +0{\cdot}22 \end{array}$$

An apparent anomaly is the E_0' value of ubiquinone, but, as will be seen later, the *actual* value of the potential of a redox compound may differ significantly from its *standard* redox potential E_0', for it by no means follows, in the living cell or elsewhere, that the environmental conditions are such as to assure a 50 : 50 ratio of oxidized and reduced forms (contrast, p. 152). Furthermore, the biochemist is confronted with another problem, for the E_0' values in the table are those observed *in vitro*, often for the isolated cofactor, whereas those operative *in vivo* are those for the cofactor attached to one or more proteins—and the E_0' value of a prosthetic group is often considerably modified by the protein to which it is attached.

The energy change associated with the movement of electrons down the redox gradient towards oxygen can be calculated in the following manner. When electrons move from NADH to oxygen, the distance they move down the redox scale is from -0.32 to $+0.82$ V, so the potential difference $\Delta E_0'$ is 1.14 V. This energy change can be expressed in calories in the way described on p. 129. Applying equation (5.6), n is 2, since two electrons are transferred and therefore

$$-\Delta G^0 = \frac{2 \times 96{,}500 \times 1.14}{4.18 \times 1000}$$

$$= 52.6 \text{ kcal}$$

Since the reaction is exergonic, some 52 kcal of energy is made available when one mole of NADH transfers its electrons to oxygen.

A second important example is succinate oxidation, oxygen being again the final electron acceptor.

$$\Delta E_0' = 0.82 - 0.03, \quad \text{or} \quad 0.79 \text{ V}$$

The number of electrons involved in the equation is again 2, and consequently,

$$-\Delta G^0 = \frac{2 \times 96{,}500 \times 0.79}{4.18 \times 1000}$$

$$= 36.5 \text{ kcal}$$

The reaction is exergonic but less energy is released than when NADH is the electron donor. It might therefore be correctly anticipated that fewer molecules of ATP are produced when succinate is oxidized than when a substrate first donates electrons to NAD.

From what has been said so far it may seem surprising that malate can reduce NAD since its standard redox potential is lower than that of NAD on the redox scale (Table 6.1). The reason why compounds like malate, lactate, and ethanol are able to reduce NAD in the presence of suitable enzymes (dehydrogenases) is that the actual redox potentials operative *in vivo* are somewhat greater than the standard values E_0'. Thus, when there is more reduced form (e.g., malate) than oxidized form (oxalacetate), the redox potential is more negative than the standard value of E_0'. If the ratio

$$\frac{[\text{reduced form}]}{[\text{oxidized form}]}$$

is great enough, the effective redox potential is such that NAD^+ is readily reduced. In other words, the reducing power when a large proportion of the reduced form is present is greater than when a smaller proportion is present. In the case of the succinate/fumarate system, the redox potential is too low on the scale for the system to reduce a finite amount of NAD^+, so the intermediate electron acceptor for succinate dehydrogenase is FAD. This carrier has a lower redox potential (that is to say, one nearer to the potential of the oxygen/water system) than has the system $NAD^+/NADH$.

6.10 Oxidative phosphorylation

The major function of the mitochondrial electron transport system is the synthesis of ATP. The quantity of inorganic phosphate, P_i, which joins with ADP to form ATP can be measured, as can the quantity of oxygen converted to water. This has been achieved using both intact isolated mitochondria and inner membrane particles derived from mitochondria by their partial disintegration. The ratio of the number of atoms of inorganic phosphorus incorporated into ATP, to the number of atoms of oxygen consumed, is known as the P:O ratio. The measured P:O ratio for NADH oxidation is approximately 3·0, while for succinate oxidation it is 2·0. This is in agreement with the prediction in the previous section, for it means that less ATP is synthesized when succinate is oxidized than when the substrate is malate, isocitrate, or α-ketoglutarate.

In the previous chapter, it was explained that ATP synthesis was a formally endergonic process involving an input of some 7·9 kcal per mole. So, if 3 moles of ATP are synthesized from ADP and P_i, 3 × 7·9 or 23·7 kcal of energy are captured. The oxidation of NADH was shown in section 6.9 to be exergonic and to release some 52 kcal per mole. Hence, from the point of view of energy capture, the efficiency with which electrons are transported from NADH along the electron transport chain is

$$\frac{23\cdot7}{52\cdot6} \times 100 = 45\%$$

Similarly, when 2 moles of ATP are formed as electrons flow from $FADH_2$

to oxygen, the efficiency is

$$\frac{2 \times 7 \cdot 9}{36 \cdot 5} \times 100 = 43 \cdot 4\%$$

These calculations assume that the energy changes occur under standard conditions. In the living cell, the efficiency may well be somewhat different since non-standard conditions prevail (e.g., the solutions are not molar). It is interesting to note that the figures compare favourably with those for many types of man-made machines—steam engines and motor cars, for example, work at efficiencies well below 15 per cent.

It is found that mitochondria deprived of oxygen fail to show significant oxidative phosphorylation. Under anaerobic conditions, each component of the electron transport chain becomes and remains in the reduced state and hence both the flow of electrons along the chain and the synthesis of ATP ceases. Similarly, poisons such as carbon monoxide and cyanide are able to react with certain members of the chain (e.g., cytochrome oxidase) so as to stop electron transport, thus preventing oxidative phosphorylation.

As will be seen in the next section, the three molecules of ATP produced when the electrons from NADH pass down the electron transport chain are thought to be synthesized at three separate points in the chain. The exact mechanism by which this occurs is, however, not yet understood. There is some evidence that when the reduced form of a carrier C passes on its electrons, the energy of the oxidation is conserved by the simultaneous formation of a high-energy compound, which can be represented, for example, as $C^+ \sim X$. The decomposition of this compound is then probably coupled directly or indirectly with the $ADP \rightarrow ATP$ system, so that the reaction, in its simplest form becomes

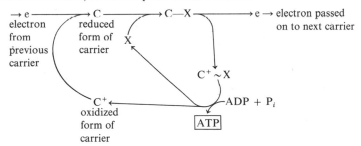

Certain poisons known as 'uncoupling agents' (e.g., 2,4-dinitrophenol and its homologues) cause an increase in the rate of electron flow down the chain and hence greater oxygen consumption, but ATP synthesis nevertheless ceases. The reason is thought to be that they cause the link

between C and X to break $(C^+ \sim X \rightarrow C^+ + X + \text{heat})$ so making C^+ available more quickly for further electron transport. In other words, the phosphorylation which accompanies oxidation is normally the pacemaker, acting as a friction brake on the whole process. An analogy is

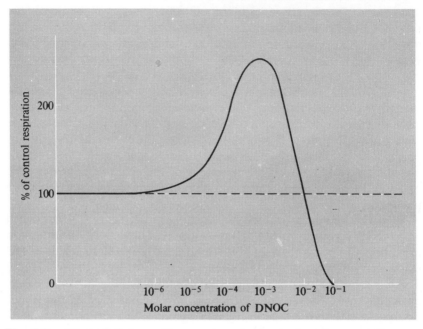

Fig. 6.8 Effect of dinitro-*ortho*-cresol on the respiration of starved *Chlorella vulgaris* (pH 6·4, temp. 30°C, in the dark)

offered by the case of a motor car being driven with a slipping clutch; the more the clutch is allowed to slip, the more the accelerator must be depressed to maintain constant speed, until eventually no amount of 'gas' will compensate for the fully depressed clutch, and motion ceases. Figure 6.8 illustrates how the herbicide 2,4-dinitro-6-methyl phenol (DNOC), stimulates oxygen uptake over a certain concentration range; above about 10^{-3}M, death of the plant ensues.

6.11 The site of oxidative phosphorylation in the cell

In chapter 1 it was seen that the mitochondrion consists of a folded inner membrane and a separate outer membrane. There is a body of evidence to suggest that the electron transport components from NAD to cytochrome oxidase are located in small particles (electron transport particles) forming an integral part of the inner membrane. By means of fractionation techniques it has been possible to break up these particles and thus to show that they consist of four sub-particles closely bound together. They are usually numbered I, II, III, and IV.

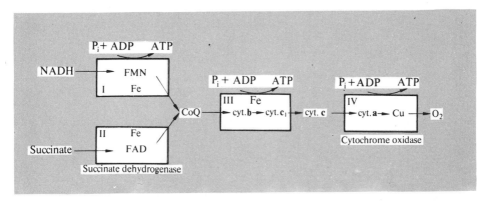

Fig. 6.9 Outline of the electron transport assembly showing the four particles. (Although drawn separately, they are closely linked together. Coenzyme Q and cytochrome **c** are mobile and shuttle the electrons between the particles at the points indicated. Note that ATP synthesis only occurs in particles I, III, and IV. Fe indicates non-haem iron protein)

Particle I contains FMN and is responsible for passing electrons from NADH to ubiquinone (Fig. 6.9). Particle II is succinate dehydrogenase, particle III contains several catalysts, including cytochromes **b** and c_1, and particle IV is cytochrome oxidase. Ubiquinone and cytochrome **c** are apparently not bound to any particle but act as mobile electron carriers between the particles. Oxidative phosphorylation occurs as a result of electron transport in particles I, III, and IV but not in II, in accordance with the observation that succinate oxidation produces one less molecule of ATP than does the oxidation of malate, isocitrate, or α-ketoglutarate, all of which pass their electrons, via NAD, to particle I. Each particle apart from IV also contains considerable quantities of a non-haem iron protein.

The following two chapters describe several important reaction sequences which produce $NADH$, $FMNH_2$ and $FADH_2$ in the mito-chondrion. These substances act in turn as electron donors to the transport system located at the inner mitochondrial membrane and so provide energy for the synthesis of ATP from ADP and inorganic phosphate.

7

The tricarboxylic acid cycle

Much of the NADH essential for the process of oxidative phosphorylation (chapter 6) is produced by a series of reactions known as the tricarboxylic acid (TCA) cycle. This cycle is often called the Kreb's cycle after Sir Hans Krebs whose work helped to elucidate its reactions. In addition to its energetic function, the cycle is important as a route allowing the interconversion of carbohydrates and certain amino acids (chapter 13).

7.1 Two important general reactions

Before considering the tricarboxylic acid cycle in detail it may help readers if we outline two types of reaction which occur not only in this cycle but also in other important biochemical processes.

The first of these (Fig. 7.1) has three steps and begins with the oxidation of a pair of carbon atoms ($-CH_2-CH_2-$) with the formation of a double bond $\left(\begin{array}{c} H \\ \diagdown \\ -C=C- \end{array} \diagup H \right)$. This process involves the removal of two

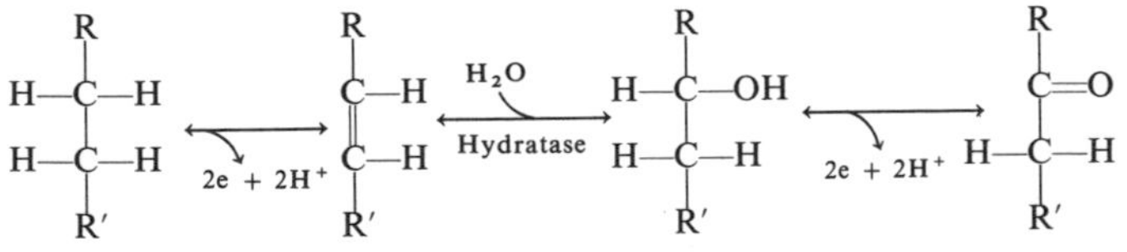

Fig. 7.1 Sequence of reactions occurring in several biochemical pathways

159

protons and two electrons (Fig. 6.3c) and is often catalysed by a flavo-protein enzyme, the FAD prosthetic group of which is reduced to $FADH_2$. In the second step a water molecule is added to the unsaturated compound to produce a hydroxylated derivative ($—CH_2—CHOH—$). Enzymes catalysing this type of hydration reaction are known collectively as *hydratases*. In the final step of this reaction sequence the hydroxyl compound is oxidized to a keto-compound. Again, two protons and two electrons are removed (Fig. 6.3b), the electrons reducing NAD^+ to NADH. This sequence of reactions occurs in the TCA cycle and also in both the catabolism and the synthesis of fatty acids (chapter 8).

The second type of reaction is called *decarboxylation*. This involves the removal of the —COOH group from a carboxylic acid, one of the products being carbon dioxide. Two versions of the reaction are frequently encountered in cell biochemistry and it is important to bear in mind that cleavage of a C—C bond is frequently an exergonic process even though the activation energy is often high. Sometimes, but not always, the cell is able to harness the energy released when such a link is broken. The converse is necessarily true—carboxylation reactions which link two carbon atoms to form a C—C bond are endergonic and therefore only occur when a supply of energy is available, either from external sources, or from simultaneous exergonic or entropy changes occurring elsewhere in the molecule.

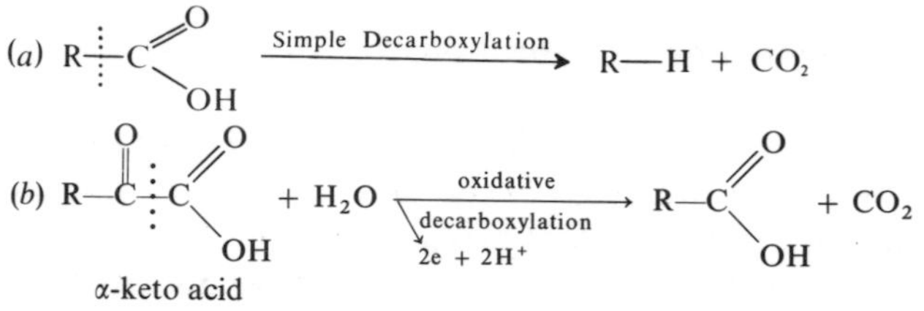

Fig. 7.2   Two types of decarboxylation reaction

The first type of decarboxylation reaction (Fig. 7.2a) requires no explanation but the second type is somewhat more complicated. α-Keto acids are readily decarboxylated because the keto carbonyl group on the α-carbon atom tends to weaken the terminal C—C bond. In the presence of water or some similar oxygen donor the products are a carboxylic acid and carbon dioxide; during the reaction two electrons and two protons are liberated and the former must be taken up by an electron acceptor (Fig. 7.2b). In other words, the reaction is an *oxidative decarboxylation*.

7.2 The TCA cycle—its fuel and function

The TCA cycle is an elegant catalytic device which brings about the total combustion of the acetyl group of acetyl coenzyme A which, as has been seen (Fig. 6.1), lies on the catabolic pathway of both fats and carbohydrates. The routes from the primary food materials—fats and carbohydrates—will be considered in the next two chapters, but the essential point to recognize now is that most of the energy produced as a result of the catabolism of these two types of materials actually *arises from the breakdown of acetyl coenzyme A*, and *not* from the various preliminary reactions which lead to the formation of acetyl coenzyme A.

The reactions which together form the machinery of the TCA cycle are represented by the circumference of the wheel of Fig. 6.1. The machinery is self-perpetuating; acetyl coenzyme A is fed in at the beginning of the cycle and, as the reaction sequence undergoes one cycle, two carbon atoms are eliminated as (respiratory) carbon dioxide. At the same time, reducing power, in the form of three molecules of NADH and one of $FADH_2$, is created; it is the *production of this reducing power*—from which ATP *is generated* by oxidative phosphorylation—which is the principal function of the cycle. Clearly, there is a close association, both functionally and locationally, between the reactions at the circumference of the wheel, which are now to be described, and the 'spokes' of the wheel which formed the subject of the last chapter.

Acetyl coenzyme A is a thioester, in which the acetyl group is attached through a sulphur bond to coenzyme A:

$$H—S—CoA \qquad CH_3—\underset{\underset{O}{\|}}{C}\sim S—CoA$$

Coenzyme A   Acetyl coenzyme A

Acetyl coenzyme A is a high energy compound and the C—S link is an example of a high energy bond (p. 131). This bond is of great importance, for the attachment of coenzyme A in this manner has the effect of making the acetyl group more reactive than it is in either the free acid or in ordinary esters such as $CH_3—\overset{O}{\overset{\|}{C}}—O—R$. The reactivity of the acetyl group of acetyl coenzyme A is, as we have already said, more comparable to that of the acetyl group of acetyl chloride—except that, whereas the latter is reactive in the presence of water and many other substances, the reactivity of acetyl coenzyme A is usually only evident in the presence of the

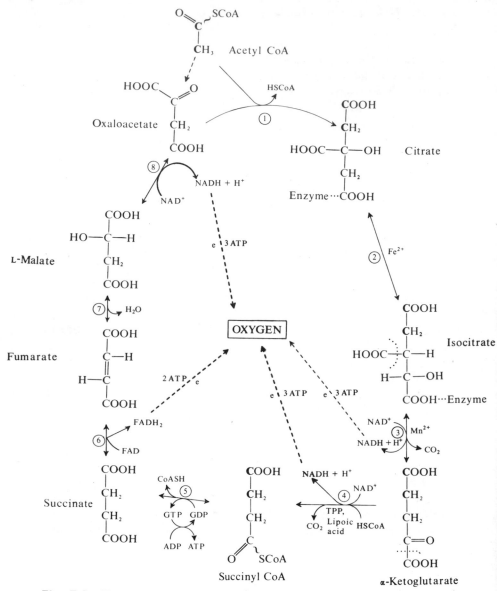

Fig. 7.3 The tricarboxylic acid cycle. (---$\xrightarrow{e}$--- represents the flow of electrons from NADH or FADH$_2$ along the electron transport chain. Enzyme attachment (top right) distinguishes two ends of the citrate molecule)

appropriate enzymes. As in other 'high-energy' compounds, the energy is not confined to a single bond but is spread over the whole acetyl group. A consequence of this, in the present instance, is that the hydrogen atoms of the —CH_3 group are often found to be more reactive than they would be if the coenzyme were not attached to the acetyl group.

7.3 Formation of citric acid

The complete cycle is shown in Fig. 7.3. The first reaction of the cycle results in the formation of a tricarboxylic acid known as citric acid, and for this reason the cycle is also sometimes called the citric acid cycle. The reaction is an addition or condensation reaction somewhat resembling the well known aldol 'condensation':

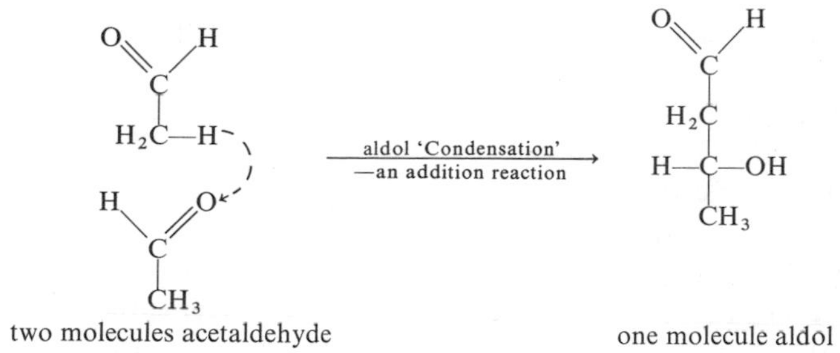

two molecules acetaldehyde   one molecule aldol

A mitochondrial enzyme known as citrate synthase catalyses the comparable 'condensation' of oxaloacetate and acetyl coenzyme A. The high energy associated with acetyl coenzyme A acts as a driving force ensuring that the reaction proceeds almost to completion. One of the C—H bonds of the methyl group of acetyl coenzyme A is particularly active (section 7.2) and 'condenses' with the keto-group of oxaloacetic acid with the formation of a C—C bond. The keto-group becomes a hydroxyl group, just as though a typical addition reaction were occurring, although the precise mechanism of the reaction under biological conditions is not fully understood. The reaction, an example of the one shown in Fig. 5.6c, is completed by the hydrolytic removal of coenzyme A.

Citric acid has six carbon atoms; a complete turn of the cycle results in the elimination of two carbon atoms as carbon dioxide, and oxaloacetate with four carbon atoms is again produced. This can then condense with more acetyl coenzyme A to re-form citrate. Hence with each

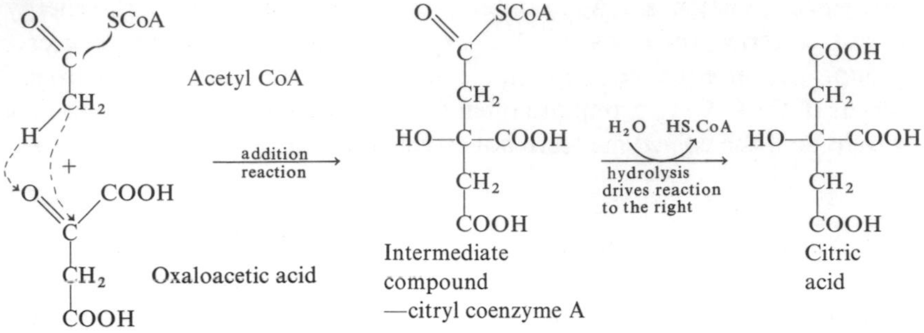

O SCoA
C
CH$_2$
Acetyl CoA
H

+

O COOH
C
CH$_2$ Oxaloacetic acid
COOH

addition reaction →

O SCoA
C
CH$_2$
HO—C—COOH
CH$_2$
COOH
Intermediate
compound
—citryl coenzyme A

H$_2$O   HS.CoA
hydrolysis
drives reaction
to the right →

COOH
CH$_2$
HO—C—COOH
CH$_2$
COOH
Citric
acid

turn of the cycle, one of the initial reactants, the oxaloacetate, is regenerated, making the cycle in this sense self-perpetuating.

7.4 Intra-molecular rearrangement of citrate to isocitrate

Citric acid, produced by the first reaction, is next converted to its isomer, isocitric acid. The reaction is catalysed by aconitate hydratase (aconitase), an enzyme dependent on ferrous ions; removal of these ions by dialysis inactivates the enzyme. In effect, (though probably not mechanistically) this hydratase catalyses the removal of the elements of water from the citrate, to produce an unsaturated intermediate, *cis*-aconitic acid, and isocitrate results from the subsequent hydration of the aconitate, the orientation of the entering water molecule being the opposite to that of the one initially removed. At equilibrium, there is a mixture of 90% citrate and roughly equal quantities of aconitate and isocitrate.

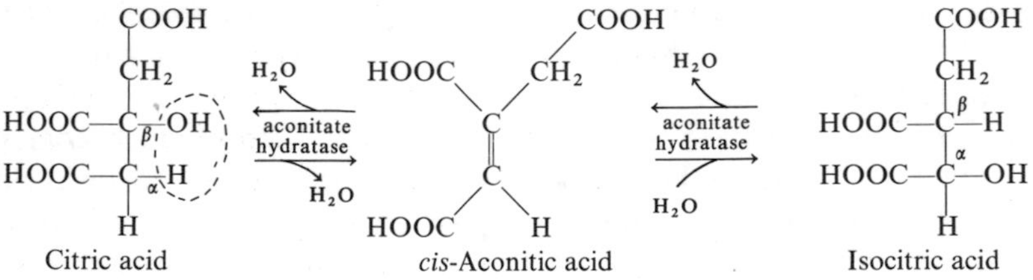

COOH
CH$_2$
HOOC—C—OH
β
HOOC—C—H
α
H
Citric acid

H$_2$O
aconitate hydratase
H$_2$O

HOOC   COOH
\   CH$_2$
 C
 ‖
 C
/   \
HOOC   H
cis-Aconitic acid

H$_2$O
aconitate hydratase
H$_2$O

COOH
CH$_2$
HOOC—C—H
β
HOOC—C—OH
α
H
Isocitric acid

The need for this isomeric change will be apparent if it is realized that here, as so often in biochemical oxidations, oxidation of a secondary alcohol group on an α-carbon atom is shortly to lead to the production of an α-keto-acid.

Cellular biochemistry and physiology

7.5 Isocitrate to α-ketoglutarate

The next reaction of the cycle is both a dehydrogenation and a decarboxylation. The whole conversion, from isocitrate to α-ketoglutarate, is catalysed by one enzyme, isocitrate dehydrogenase. It is one of the reactions of the cycle coupled with the reduction of NAD and hence leading to the generation of ATP by oxidative phosphorylation (chapter 6).

The reaction is best explained as occurring in two steps. The isocitric acid first undergoes an oxidation reaction whereby the secondary alcohol group is converted to an α-keto-group. In this reaction, two electrons and two protons are removed from the isocitrate (compare Fig. 6.3b) the electrons being transferred, directly or indirectly, to NAD*. The intermediate, oxalosuccinate, still firmly attached to the enzyme, is then decarboxylated on the *central* or β-carbon atom, a reaction which leads to the formation of α-ketoglutarate. As in many decarboxylation reactions, manganous ions are a necessary cofactor for the second stage of the reaction.

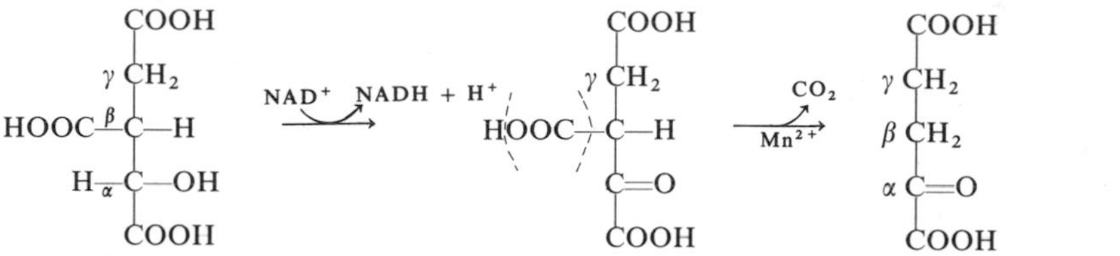

Isocitrate	Oxalosuccinate (Six carbon atoms)	α-Ketoglutarate (five carbon atoms)

7.6 The oxidative decarboxylation of α-ketoglutarate

The decarboxylation mentioned in the previous section accounts statistically (see section 7.10) for one of the two carbon atoms which enter the TCA cycle in the acetyl part of acetyl coenzyme A. We now come to a second decarboxylation which again leads to the production of carbon dioxide, accounting (statistically) for the second of the two carbon atoms

---

* Most organisms appear to have two mitochondrial isocitrate dehydrogenases—one specific for NAD and the other for NADP. In rat skeletal muscle and brain the NAD-linked enzyme predominates but in heart, kidney, and liver the NADP-coupled enzyme is present in greater amount. Nevertheless, transhydrogenation probably allows NADPH to pass its hydrogen on to NAD, thus justifying the statement in the text.

of acetyl coenzyme A.   The overall reaction can be represented as follows:

$$
\begin{array}{ccc}
\text{COOH} & & \text{COOH} \\
| & & | \\
\text{CH}_2 & \text{H}_2\text{O} \qquad \text{CO}_2 & \text{CH}_2 \\
| & & | \\
\text{CH}_2 & \longrightarrow & \text{CH}_2 \\
| & & | \\
\alpha\ \text{C}{=}\text{O} & 2e + 2\text{H}^+ & \text{COOH} \\
----|---- & & \text{Succinate} \\
\text{COOH} & &
\end{array}
$$

α-Ketoglutarate

The decarboxylation which occurs with α-ketoglutarate is quite different from that described for isocitrate, for it is an example of the second type of decarboxylation illustrated in Fig. 7.2b.   The reaction is oxidative as well as being a decarboxylation and results in carbon dioxide, two protons, and two electrons being eliminated in a reaction which is virtually *irreversible*.   The mechanism is far more complicated than indicated in the overall scheme above, because an ingenious device exists for capturing some of the energy involved in the cleavage of the C—C bond. The reactions are most simply explained by using the generalized formula for an α-keto acid, R—CO.COOH, where R— is —CH_2—CH_2COOH in the particular case of α-ketoglutaric acid; elsewhere, a similar reaction occurs with another keto acid, pyruvic acid as the substrate.

A series of reactions converts R—CO—COOH to the coenzyme A derivative, R—COS ∼ CoA.   The change is brought about by a mitochondrial complex, which, in the case of α-ketoglutarate, is called α-ketoglutarate dehydrogenase.   It is, however, not a single enzyme but several enzymes closely bound together.   Several cofactors are also intimately involved and include

(1) *Thiamine pyrophosphate*, TPP, the pyrophosphate of the vitamin, *thiamine*,

(2) *lipoic acid*, which contains two sulphur atoms:

$$
\begin{array}{l}
\qquad\qquad \diagup (\text{CH}_2)_4\text{COOH} \\
\text{S}-\text{CH} \\
\qquad\quad \diagdown \\
|\qquad\qquad \text{CH}_2 \qquad\qquad \text{Lipoic acid (oxidized form)} \\
\qquad\quad \diagup \\
\text{S}-\text{CH}_2
\end{array}
$$

(3) coenzyme A,
(4) FAD,
(5) NAD.

The sequence of reactions is summarized in Fig. 7.4 and begins with the linking of TPP with the α-keto acid.   At this step, in which carbon

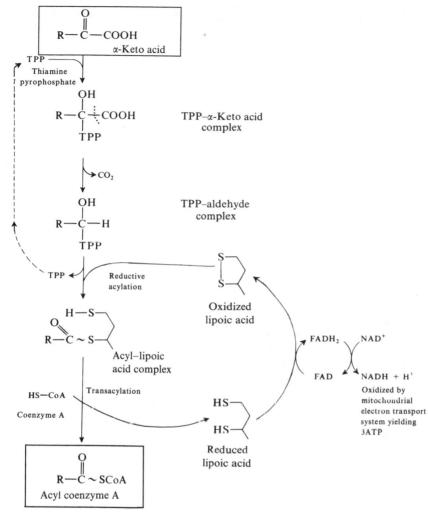

Fig. 7.4 The decarboxylation of an α-keto acid. (The α-ketoglutarate dehydrogenase complex of enzymes catalyses the conversion of α-*ketoglutarate* to *succinyl coenzyme A* (R = HOOC—CH₂—CH₂—). A similar enzyme, pyruvate dehydrogenase, converts *pyruvate* to *acetyl coenzyme A* (R— = H₃C—). At the reductive acylation stage, a hydrogen atom returns to the TPP as it splits off

dioxide is eliminated, the energy of the breaking C—C bond is retained in the TPP–aldehyde addition complex:

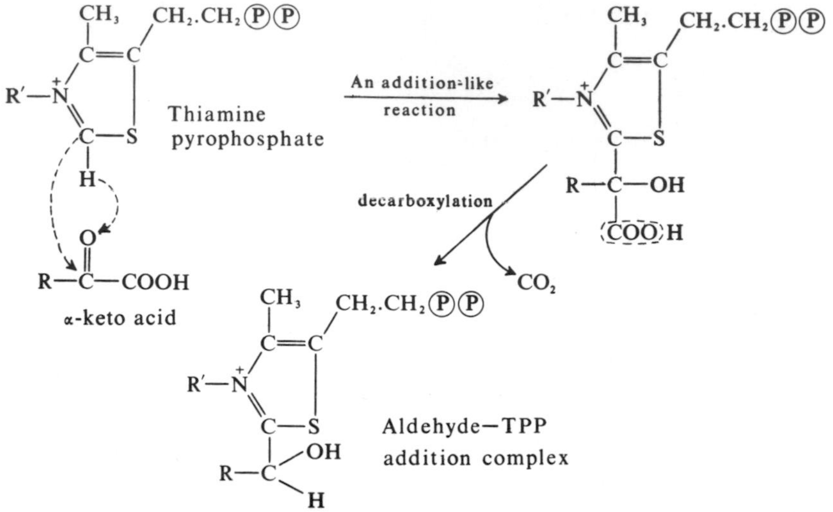

Next, the aldehyde–TPP complex reacts with the oxidized form of lipoic acid. TPP is released, ready to facilitate further decarboxylation, but at the same time the energy of the C—C bond broken initially is retained, and helps to forge the bond between the sulphur of lipoic acid and the acyl group in the R—CO ~ S–lipoic acid complex. Finally,

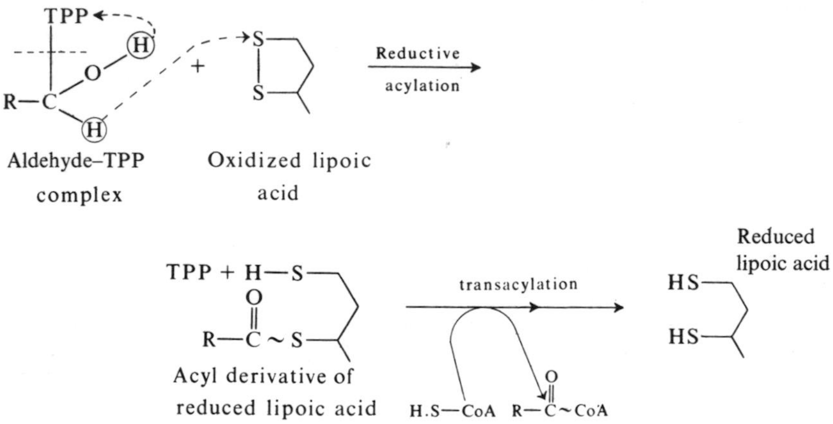

a transacylation reaction occurs, in which the acyl group is transferred from the acylated lipoic acid to coenzyme A. In this way, much of the energy released on cleavage of the original C—C bond is retained in the thiol linkage.

The lipoic acid has been changed significantly in this process since each sulphur atom is now part of a sulphydryl group. This means that the lipoic acid is now in a reduced form, possessing two protons and two electrons more than the oxidized form which reacted with the thiamine complex.

The dithiol is reoxidized to the disulphide (S—S) form of lipoic acid in the presence of a flavoprotein which acts as an electron acceptor, the electrons eventually reducing NAD^+ to NADH. The latter generates ATP by oxidative phosphorylation as the electrons pass along the electron transport chain.

7.7 Conversion of succinyl coenzyme A to succinic acid

When α-ketoglutarate is the substance undergoing oxidative decarboxylation, the product is *succinyl coenzyme A*. The next reaction of the TCA cycle converts this product to succinic acid. Before describing the mechanism involved, it must be stressed that some of the energy associated with the cleavage of the C—C bond of the α-ketoglutarate is carried over to succinyl coenzyme A, being preserved in the high energy C $\sim$ S bond. The structure and energetic characteristics of *succinyl* coenzyme A are very similar to those of *acetyl* coenzyme A:

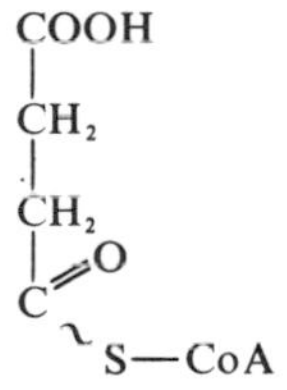

COOH
|
CH₂
|
CH₂
|
C⟋O
⟍S—CoA

Succinyl coenzyme A

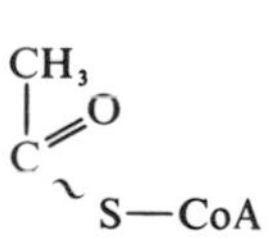

CH₃
|
C⟋O
⟍S—CoA

Acetyl coenzyme A

Many other acyl derivatives of coenzyme A are of great importance in biochemistry and a more complete picture of the high-energy 'quality' of this group of compounds can be obtained by digressing briefly to mention one aspect of fatty acid metabolism. A free fatty acid cannot combine directly with coenzyme A to form fatty acyl coenzyme A, as the reaction is endergonic, but enzymes known as 'thiokinases' can nevertheless bring about this synthesis by coupling the reaction to an exergonic reaction in which ATP is converted to AMP with the liberation of pyro-phosphate (PP):

(1) $ATP \rightarrow AMP + PP \quad (-\Delta G^0 = 7 \cdot 9 \text{ kcal})$

(2) $R—COOH + HS—CoA \rightarrow R—CO \sim SCoA + H_2O$

$$(+\Delta G^0 = 7 \cdot 5 \text{ kcal})$$

The coupled reaction catalysed by this thiokinase is thus

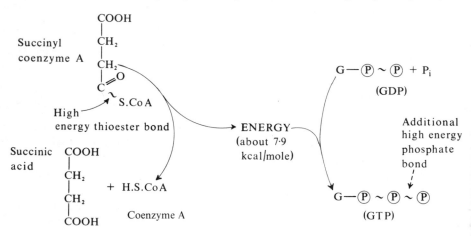

Since this coupled reaction is only slightly exergonic, it follows that acyl groups joined to coenzyme A possess, per mole, some 7 kcal of energy more than the uncombined acids. This is confirmed by the fact that the hydrolysis of succinyl coenzyme A and other acyl coenzyme A compounds is highly exergonic (equation (2) above working in reverse).

Within the mitochondrion the energy in succinyl coenzyme A is, however, not squandered by hydrolysis; instead, it is used to synthesize ATP. Two closely coupled reactions are responsible for the conservation of this energy. In one of them, succinyl coenzyme A is converted to free succinic acid and free coenzyme A. The energy made available by this reaction is used to drive the second reaction, in which a nucleotide diphosphate similar to ADP (it is actually *guanosine diphosphate*, GDP) unites with inorganic phosphate to give a nucleotide triphosphate (GTP).

Effectively, the consequence of this coupling is that the high energy of a (disappearing) thioester bond enables an extra 'high energy phosphate bond' to be built on to the nucleotide. ATP is now formed by a simple transfer of $\sim\!(\text{P})$ from GTP to ADP, making GDP available once more to accept energy from succinyl coenzyme A:

$$GTP + ADP \rightleftharpoons ATP + GDP$$

It follows from what has been said above, that some of the energy originating from the oxidative cleavage of the C—C bond of α-keto-glutarate leads to the generation of ATP by a process in which neither oxygen nor the electron transport system are involved. This type of phosphorylation is, in fact, another example of *substrate level phosphorylation* (compare pyruvate kinase, p. 134). In addition, of course, some of the energy of the cleavage is used to reduce lipoic acid, the reoxidation of which *does* involve oxidative phosphorylation (Fig. 7.4).

7.8 Succinate to oxaloacetate

The remaining steps of the cycle serve two purposes; they ensure that succinate is converted eventually to oxaloacetate so as to complete the cycle, and at the same time they create more ATP. Three reactions are, in fact, responsible for the remaining transformations; together they are a particular example of the generalized series of reactions to which reference has already been made (Fig. 7.1).

In the first reaction, succinate is oxidized to fumarate, the catalyst being succinate dehydrogenase and FAD the electron acceptor (section 6.7). This reaction occurs in a different location from the others of the cycle, for succinate dehydrogenase forms one of the electron transport particles (II) on the inner mitochondrial membrane. This implies that succinate formed elsewhere in the mitochondrion must diffuse to the inner membrane. The product (fumarate) presumably then diffuses back to the remaining TCA cycle enzymes. Note that succinate oxidation leads to the generation of 2 molecules of ATP by oxidative phosphorylation (section 6.10). The fumarate is then converted to a hydroxy acid, L-malic acid, by the addition of water in a reaction catalysed by the enzyme fumarate hydratase. The reaction is freely reversible *in vitro*, though it tends to move in the direction of L-malic acid synthesis *in vivo* because the TCA cycle is made uni-directional by the irreversible decarboxylation of α-ketoglutarate.

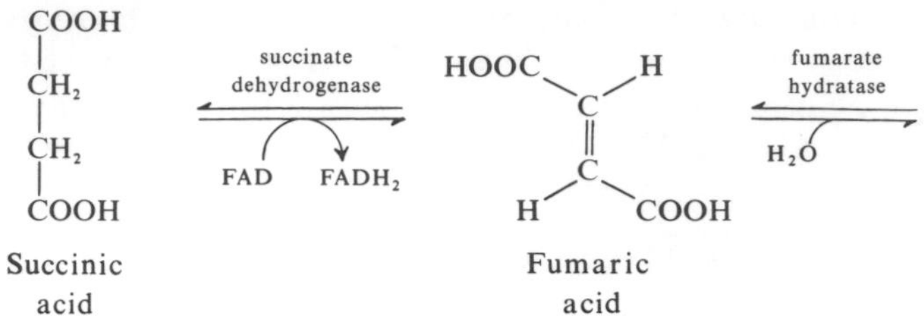

COOH
|
CH$_2$
|
CH$_2$
|
COOH

Succinic
acid

succinate
dehydrogenase

FAD   FADH$_2$

HOOC   H
   \   /
    C
    ‖
    C
   /   \
  H   COOH

Fumaric
acid

fumarate
hydratase

H$_2$O

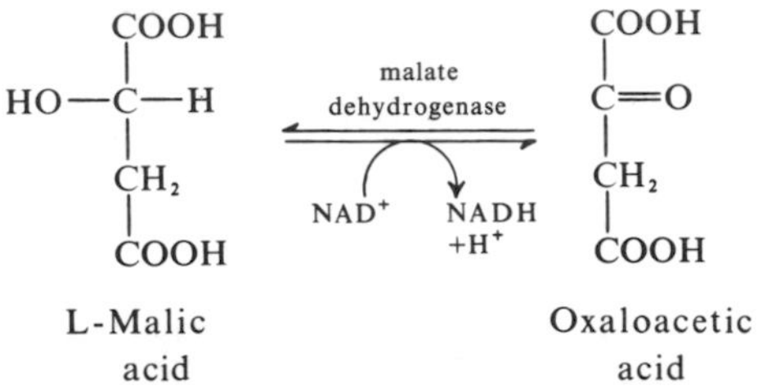

COOH
|
HO—C—H
|
CH$_2$
|
COOH

L-Malic
acid

malate
dehydrogenase

NAD$^+$ NADH
          +H$^+$

COOH
|
C=O
|
CH$_2$
|
COOH

Oxaloacetic
acid

Finally, oxaloacetic acid is formed in the presence of malate dehydrogenase, by the oxidation of the secondary alcohol group of malic acid. NAD is simultaneously reduced to NADH, so yielding three molecules of ATP by oxidative phosphorylation. As mentioned in the previous chapter (section 6.9), the standard redox potential E_0' for the reaction is such that malate dehydrogenase only reduces L-malate to oxaloacetate when there is a large excess of L-malate over oxaloacetate. However, the fact that oxaloacetate is very readily removed by formation of citrate (section 7.3) ensures that the ratio of malate to oxaloacetate is large enough to enable the oxidation of L-malate by NAD to proceed.

The standard redox potentials for the other two NAD-linked reactions of the cycle (i.e., the oxidation of isocitrate and α-ketoglutarate) are more negative than the standard redox potential of the system NAD → NADH (Table 6.1) so both reactions tend to move readily in the direction of substrate oxidation and NAD reduction.

7.9   Energy captured by the cycle

We are now in a position to look at the cycle as a whole and to calculate the net gain of ATP to the cell. In one complete turn of the cycle, one acetyl group is incorporated and two molecules of carbon dioxide are eliminated (Fig. 7.3); NAD is reduced to NADH at three stages, the enzymes catalysing the reductions being (1) isocitrate dehydrogenase,

(2) α-ketoglutarate dehydrogenase complex, and (3) malate dehydrogenase. In addition, FAD is reduced to $FADH_2$ by succinate dehydrogenase. Oxidative phosphorylation of ADP yields three molecules of ATP when one molecule of NADH is oxidized (Section 6.10). Moreover, two molecules of ATP are produced by the same process when the $FADH_2$ prosthetic group of succinate dehydrogenase is reoxidized. So the number of molecules of ATP produced by oxidative phosphorylation for each acetyl group incorporated into the cycle in the form of acetyl coenzyme A is $(3 \times 3) + 2$ or 11 ATP. In addition, one more ATP is produced by substrate level phosphorylation when succinyl coenzyme A is converted to succinate, making a *total of 12 ATP produced for each complete turn of the cycle.*

The TCA cycle and the associated electron transport chains together provide the machinery which produces most of the energy available to heterotrophic cells for endergonic processes. In plants, however, ATP is probably produced in photosynthesizing cells by photosynthetic phosphorylation.

7.10   The acids of the TCA cycle

Any of the acids of the cycle having six carbon (6-C) atoms can be readily converted to any of the others; however, the 4-C acids (succinate, L-malate, fumarate, and oxaloacetate) can be converted to the 6-C or 5-C acids only if acetyl coenzyme A is available to enter the cycle since the almost irreversible α-ketoglutarate decarboxylation makes the cycle a unidirectional process.

The cycle can be blocked by inactivating any enzyme in it. For example, fluoroacetate (FH_2C—COOH) prevents the cycle functioning because it unites with oxaloacetate in the same way as does acetyl coenzyme A. The fluorocitrate thus formed stops the cycle because aconitate hydratase cannot act upon fluorocitrate. Similarly, malonate stops the cycle at the succinate dehydrogenase step (p. 104). Both malonate and fluoroacetate have been used experimentally as inhibitors for studying the reactions of the cycle. Malonyl coenzyme A is an important cell metabolite—it is a precursor of long chain fatty acids (chapter 8)—but it is mainly extra-mitochondrial in location.

It is of some interest to note that the carbon dioxide eliminated in a single turn of the cycle is *not* derived from the acetyl coenzyme A incorporated in the same turn. This can be seen from Fig. 7.3, where the carbon atoms of the acetyl group are followed through the cycle in heavy type. At first sight this may seem surprising, because citric acid is a symmetrical molecule and therefore a differentiation between carbon atoms immediately derived from acetyl coenzyme A and those derived from oxaloacetate might not be expected. The explanation is that citrate and isocitrate are bound to the enzymes in such a way that the carbon atoms just incorporated are necessarily situated at the other end of the molecule from where decarboxylation subsequently occurs. By the time succinate is reached, however, no such distinction exists, because succinic acid is not only a symmetrical molecule but it also breaks away from the enzyme surface at which it is formed. Thus acetyl coenzyme A labelled with radioactive carbon (C^{14}) does not produce radioactive carbon dioxide until one turn of the cycle has been completed—the carbon dioxide is, in fact, derived from those carbon atoms of the citrate which originate from its oxaloacetate precursor.

8

Lipid metabolism

It was seen in chapter 1 that certain lipids, especially phospholipids and steroids, contribute to the architecture of the cells since they are present in lipoprotein membranes. The neutral glycerides may also be components of some membranes, but, more important, when broken down under oxidative conditions, they provide the cell with a large quantity of energy. A molecule of the long-chain fatty acid, palmitic acid, $CH_3.(CH_2)_{14}.COOH$, for example, yields eight molecules of acetyl coenzyme A. Some ATP is synthesized as a result of this process; far more, however, is formed by the subsequent oxidation of the acetyl coenzyme A by the TCA cycle (chapter 7).

In animals, adipose tissue is specially adapted for the storage of fats and oils. The insolubility of these lipids in aqueous cytoplasm keeps them isolated from the general metabolism, yet they can be mobilized readily when needed. In terms of their mass they represent a particularly efficient way to store energy; complete oxidation of 1 g of fat produces some 9 kcal of heat, while the same mass of carbohydrate or protein yields only 4–6 kcal. Nearly 10% of the weight of a normal man is fat, much of which can be drawn upon to provide energy during periods of starvation. Somewhat similarly, hedgehogs and other hibernating animals build up a considerable deposit of fat for the autumn, a deposit which provides thermal insulation up to the time it is catabolized and both energy and water thereafter. In plant cells, starch is frequently more common than fat as an energy reserve, but this is by no means always the case and sometimes, as in the castor oil seed, it is fat which is stored to provide the energy needed during germination.

Lipids should not be regarded as a static energy deposit, for, in animals, fats are constantly being mobilized and redeposited, as well as being drawn upon for catabolic purposes. Acetyl coenzyme A, formed when fat is catabolized, not only generates ATP (chapter 7), but may also take part in several important biosyntheses which lead to the production of other fats and lipids. In this chapter the mechanisms which result in the formation of acetyl coenzyme A from fat will be considered in detail and some consideration will also be given to the ways in which fats can provide the raw materials for the synthesis of new glycerides and certain other lipids.

8.1 Hydrolysis of glycerides

Glycerides are the fatty acid esters of glycerol and the initial step in their metabolism is almost invariably hydrolysis. In the germination of the castor oil seed the enzyme *lipase* catalyses this reaction, and a similar reaction occurs in many animal cells, especially in adipose tissue during starvation. As is well known, non-enzymic hydrolysis is possible *in vitro*; alkaline hydrolysis, for example, has long been employed as a method of making soap, the sodium and potassium salts of long-chain fatty acids.

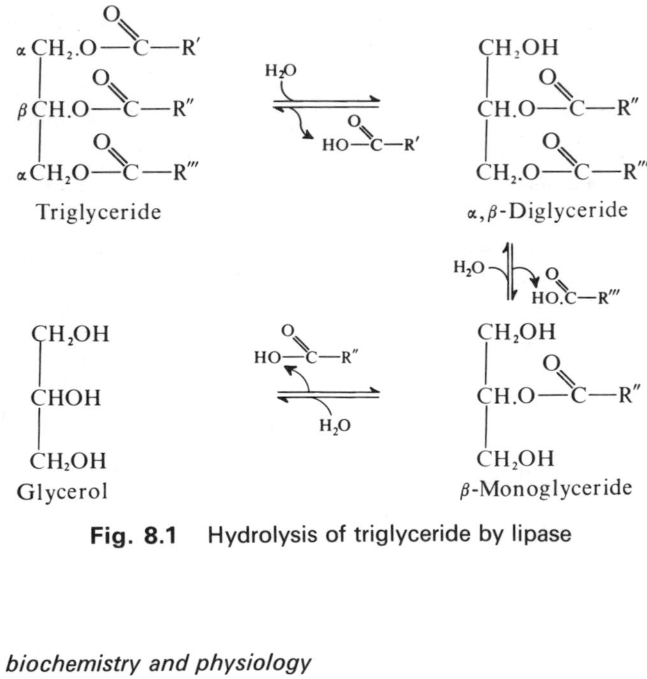

Fig. 8.1   Hydrolysis of triglyceride by lipase

Most lipases tend to attack the ester linkages situated at the two α-positions more readily than that in the central, or β-, position (Fig. 8.1). The two fatty acids attached at α-positions are removed one after the other to produce a β-monoglyceride. The latter is eventually converted to fatty acid and glycerol, in some cases by the action of another kind of lipase. The reactions catalysed by lipase are potentially reversible, but net synthesis of triglyceride from free fatty acid and glycerol (the reverse of hydrolysis) is not readily achieved unless one of these reactants is present in great excess; in practice, therefore, triglyceride biosynthesis follows a route which is different from that of its catabolism (section 8.7).

After fats have been hydrolysed by lipase, the glycerol and fatty acids so formed may undergo re-esterification to form other fats and phospholipids; under standard conditions of growth or nutrition, each species of animal and plant produces mixtures of triglycerides which are fairly constant and characteristic of that species and different from those which, if it is an animal, are present in the diet. Alternatively, the products of hydrolysis can be catabolized. In this case, the glycerol is phosphorylated and oxidized to form *triose phosphate*, which is then catabolized via the glycolytic pathway (chapter 9), while the *fatty acids* undergo the catabolic changes described in the next two sections.

8.2   The catabolism of long-chain fatty acids

The complete oxidation of a fatty acid is a highly exergonic reaction yielding carbon dioxide and water, e.g.,

$$H_3C(CH_2)_{16}COOH + 26O_2 \rightarrow 18CO_2 + 18H_2O \qquad (\Delta H = -2680 \text{ kcal})$$
Stearic acid

The ratio of the volume of carbon dioxide formed to the volume of oxygen taken up, whether by a whole organism or by a group of cells, may be measured experimentally and is known as the *respiratory quotient* (R.Q.). This is usually determined by measuring the volumes of the gases under the same conditions of temperature and pressure:

$$R.Q. = \frac{\text{volume of } CO_2 \text{ produced}}{\text{volume of } O_2 \text{ consumed}}$$

Since n moles of carbon dioxide occupy the same volume as n moles of oxygen under comparable conditions, the R.Q. for the complete oxidation of stearic acid is, from the equation above, $18n/26n$ or 0·69. This is a typical value for fatty acid oxidation. The R.Q. for carbohydrate oxidation (e.g., glucose) is approximately 1·0:

$$C_6H_{12}O_6 + 6O_2 \rightarrow 6CO_2 + 6H_2O$$

and that for protein oxidation is about 0·8. By measurement of the R.Q. it is therefore possible to obtain some idea of the organism's overall source of energy at the time the measurement is made. Thus the R.Q. of a rat fed with glucose is about 1·0, but in the fasting state the R.Q. is lower than this since the animal has to oxidize fats and proteins to maintain its metabolism. Similarly, the R.Q. of germinating barley seeds is approximately 1·0 while that of germinating castor oil seeds is 0·7. The *Warburg manometer* is a particularly useful instrument for measuring the release or uptake of gases by cells or tissues.

8.3 The fatty acid oxidation spiral

In animal and plant cells, the oxidation of fatty acids occurs exclusively in mitochondria. The oxidative process was known to involve β-oxidation long before the details of the individual reactions were known. Historic experiments by Knoop (1904) suggested that the overall process of stepwise degradation could be expressed in the following highly-simplified form:

$$R.(CH_2)_n.CH_2.CH_2.COOH \xrightarrow[\substack{\text{(now known to involve} \\ \text{four reactions)}}]{\beta\text{-oxidation}} R.(CH_2)_n.COOH + CH_3COOH$$
$$\,\beta\quad\alpha$$

$$R.(CH_2)_{n-2}.CH_2.CH_2.COOH \xrightarrow[\substack{\text{(same four reactions,} \\ \text{repeated)}}]{\beta\text{-oxidation}} R.(CH_2)_{n-2}.COOH + CH_3COOH$$
$$\phantom{R.(CH_2)_{n-2}.CH_2.CH_2.COOH}\,\beta\quad\alpha$$

The principal feature of β-oxidation is the removal of pairs of carbon atoms (as acetyl coenzyme A) from fatty acids to give shorter-chain fatty acids. It is now known that free fatty acids are not, in fact, metabolized. Instead, a fatty acid is first changed to a high energy form by conversion to an acyl coenzyme A derivative. The fatty acyl coenzyme A then acts as the substrate for the first of four consecutive reactions which collectively account for the process of β-oxidation. All the intermediates taking part in the remaining three reactions of the sequence are likewise coenzyme A derivatives.

All these coenzyme A derivatives are high-energy compounds (p. 161). When the free fatty acid is converted to fatty acyl coenzyme A, the energy needed to forge the thioester linkage is derived from ATP, although the synthesis actually occurs in two steps. In the first of these, a priming reaction, the free fatty acid reacts with ATP to give a fatty acyl-AMP compound, in which the carboxyl group of the fatty acid is linked to the phosphate of the mononucleotide. This is a high energy compound, for, in effect, much of the energy derived from the cleavage of pyrophosphate from the ATP is retained in the acyl-AMP linkage. This reaction

$$
\begin{array}{ccc}
\underset{\substack{\text{O} \\ \parallel}}{R-C}\!\!\!\diagdown\!\!\!\text{OH} + \text{(P)}\!\sim\!\text{(P)}\!\sim\!\text{(P)}\!-\!\text{ribose} & \xrightarrow{\qquad} & R-C\!\sim\!\text{(P)}\!-\!\text{ribose} \\
\text{Fatty acid} & \begin{array}{c}\text{ATP} \\ \text{PP} \\ \text{(pyrophosphate)}\end{array} & \text{Fatty acyl—AMP}
\end{array}
$$

is displaced to the right by the hydrolysis of a second product, pyrophosphate (p. 170), which is converted to inorganic orthophosphate by enzymes known as *pyrophosphatases*. Several similar reactions will be encountered later (p. 338).

The fatty acyl-AMP then reacts with coenzyme A with the formation of fatty acyl coenzyme A, AMP being released:

$$
\underset{\substack{\text{O} \\ \parallel}}{R-C}\!\sim\!\text{(P)}\!-\!\text{ribose} + \text{H}-\text{S}-\text{CoA} \longrightarrow \underset{\substack{\text{O} \\ \parallel}}{R-C}\!\sim\!\text{S}-\text{CoA} + \text{AMP}
$$

The formation of acyl coenzyme A has been described as a priming reaction, for although ATP is used initially, the overall process of β-oxidation has the function of generating ATP. The reaction is essential because coenzyme A has the effect of making the α and β carbon atoms of the fatty acid more reactive than might be expected from the saturated paraffin-like structure.

The overall effect of the four reactions constituting each turn of the spiral of fatty acid oxidation results in the formation of one molecule of acetyl coenzyme A and the release of four protons and four electrons. Two of these electrons reduce FAD and the other two, liberated at a later stage of the spiral, reduce NAD. When these coenzymes are reoxidized by the mitochondrial electron transport particles ATP is produced by oxidative phosphorylation (chapter 6).

After the initial activation leading to the formation of acyl coenzyme A, the whole catabolic process involves only four reactions which are essentially those shown in Fig. 7.1 and are very similar to those by which succinic acid is converted to oxaloacetate.

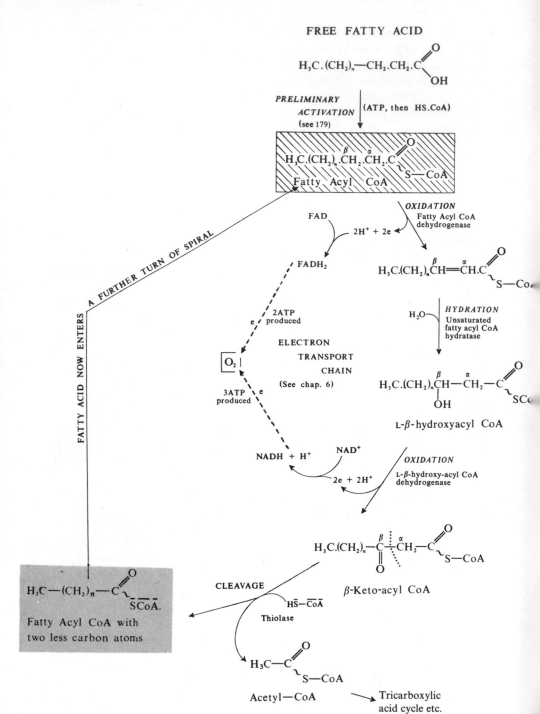

Fig. 8.2 Fatty acid oxidation spiral. (The reduced FAD and NAD pass the electrons to oxygen through the electron transport chain, producing ATP)

The first reaction is a dehydrogenation and is catalysed by a flavoprotein enzyme, *fatty acyl coenzyme A dehydrogenase*. This catalyses the removal of one proton and one electron from each of the α and β carbon atoms of the fatty acyl coenzyme A; these are transferred to FAD, so reducing it to $FADH_2$ (section 6.7). The reoxidation of the $FADH_2$ by the mitochondrial electron transport chain probably yields 2 ATP by oxidative phosphorylation. The similarity of this reaction to the succinate dehydrogenase reaction should be apparent. The product is the α, β-unsaturated fatty acyl coenzyme A.

The second reaction of the sequence is catalysed by a hydratase, *unsaturated fatty acyl coenzyme A hydratase*, which adds water to the unsaturated fatty acyl coenzyme A (Fig. 8.2). Although there are two possible stereoisomers only the L-form is produced. The formation of L-β-hydroxy-acyl coenzyme A recalls the stage in the TCA cycle where fumarate hydratase converts fumarate to L-malic acid. The hydratase has the high turnover number of about a million (p. 104).

The third enzyme, L-β-*hydroxy-acyl coenzyme A dehydrogenase*, is specific for both NAD and the L-form of the substrate. The products are the β-keto-acyl coenzyme A and reduced NAD (Fig. 8.2). The reaction is important energetically since 3 moles of ATP can be formed from each mole of NADH by oxidative phosphorylation.

The bond between the α and β carbon atoms is cleaved in the fourth and final reaction by the action of the enzyme β-*keto-acyl coenzyme A-thiolase* (Fig. 8.2), and it will be recalled that the cleavage of a C—C bond is generally an exergonic reaction (section 7.6). By coupling the cleavage with the acylation of coenzyme A, the energy released by the β-keto-acyl coenzyme A-thiolase reaction is not lost as heat but is conserved in a new thioester linkage. This is accomplished by the intervention of a molecule of free coenzyme A, the products being acetyl coenzyme A and a new (shorter) long-chain acyl coenzyme A. As the lines above the components of the formula of unbound coenzyme A show (Fig. 8.2), it is this molecule which helps to form the new *long-chain* fatty acyl coenzyme A, and is therefore the activator for the *next* turn of the spiral. The β-keto-acyl thiolase reaction thus shortens the carbon chain of the fatty acid by 2 carbon atoms. The new long-chain fatty acyl coenzyme A acts as the substrate for a further turn of the *catabolic spiral*, while the acetyl coenzyme A undergoes subsequent oxidation by the TCA cycle, to yield 12 molecules of ATP (p. 173).

The complete oxidation of a long chain fatty acid to acetyl coenzyme A yields considerable quantities of ATP. Palmitic acid

$$H_3C—(CH_2)_{14}—COOH$$

for instance, yields 8 molecules of acetyl coenzyme A through 7 turns of the catabolic spiral. This means that 14 molecules of ATP are produced as a result of the oxidation of 7 molecules of FAD by the mitochondrial electron transport chain and 21 molecules of ATP result from oxidation of 7 molecules of NADH. So (14 + 21) or 35 molecules of ATP are produced during breakdown to acetyl coenzyme A. Subsequent oxidation of the latter by the TCA cycle produces 12 molecules of ATP for every molecule of acetyl coenzyme A used up. Since 8 acetyl coenzyme A molecules are formed from palmitate, 96 more molecules of ATP originate in this manner. Hence, when one molecule of palmitic acid is completely oxidized to carbon dioxide and water, 131 molecules of ATP are produced, less the one molecule of ATP used up in the initial activation by thiokinase.

130 moles of ATP represent 7.9×130 kcal of energy (p. 123). Since the standard free energy change for the complete oxidation of palmitate to carbon dioxide and water is -2340 kcal, the overall efficiency of the energy capture is

$$\frac{7.9 \times 130}{2340} \times 100 = 44\%$$

8.4 The central metabolic position of acetyl coenzyme A

As has just been seen, the oxidation of fatty acids leads to the formation of acetyl coenzyme A within the mitochondrion, but this substance is also produced by other processes. In particular, and provided oxygen is available, it is produced by the decarboxylation of pyruvate (Fig. 6.1). Since pyruvate arises from the degradation of carbohydrates (chapter 9), acetyl coenzyme A possesses a very special biochemical significance, for it is thus a common product of carbohydrate and fat metabolism, a point of convergence of the two principal routes whereby energy is made available to the cell. It is the subsequent oxidation of acetyl coenzyme A by the machinery of the TCA cycle which leads to the production of a large proportion of the ATP needed by cells.

In addition to this catabolic function, acetyl coenzyme A also has important anabolic functions, for it plays a part in the synthesis of fatty acids and the synthesis of steroids. In animal and plant cells so far studied, it has been found that the biosynthesis of long-chain fatty acids occurs *outside* the mitochondrion in the cytosol. This biosynthesis is associated with large particles, with a molecular weight of a million or more, which in the intact cell form a part of the 'smooth' endoplasmic reticulum. They appear to convert acetyl coenzyme A to palmitate, $C_{15}H_{31}$.COOH, which may subsequently be *lengthened or shortened* as a result of a system of enzymes present *within* the mitochondrion.

The mitochondrial-based reactions leading to adjustment of carbon chain length (but *not* the synthetic processes located *outside* the mito-chondrion) can be regarded as being essentially the same as the fatty acid catabolism spiral shown in Fig. 8.2 working either in the forward or reverse direction. It is unlikely, however, that complete synthesis of long-chain fatty acids occurs by a reversal of the *intra-mitochondrial* catabolic route.

The acetyl coenzyme A needed for the *extra-mitochondrial* biosynthesis of long-chain fatty acids is unlikely to be the acetyl coenzyme A formed directly from the intra-mitochondrial catabolic processes, for the molecule of acetyl coenzyme A is too large readily to diffuse through the mito-chondrial membrane. Instead, intra-mitochondrial acetyl coenzyme A probably first condenses with oxaloacetate, exactly as it does in the first stage of the TCA cycle; the citrate so formed, having shed the coenzyme A, then diffuses into the extra-mitochondrial cytoplasm. Once in the cytosol, the citrate probably undergoes cleavage to form acetyl co-enzyme A and oxaloacetate once more. Extra-mitochondrial coenzyme A and also ATP are cofactors for this reaction. Figure 8.3 indicates the location within the cell of some major processes involving acetyl coenzyme A.

8.5 Nicotinamide adenine dinucleotide phosphate (NADP)

Before describing the biosynthesis of long-chain fatty acids it is necessary to refer to an important dinucleotide the structure of which is similar to NAD except that it possesses an additional phosphate group attached to ribose in position 2' of the formula on p. 84. NADP, like NAD, can assume an oxidized form ($NADP^+$) and reduced form (NADPH). Its importance rests in the fact that, like NAD, it is an electron acceptor. While NAD is electron acceptor for most dehydrogenases, there are some which *use only* NADP. It is of considerable interest that most dehydro-genases are *specific for the electron acceptor* as well as for the *substrate* (i.e., the electron donor).

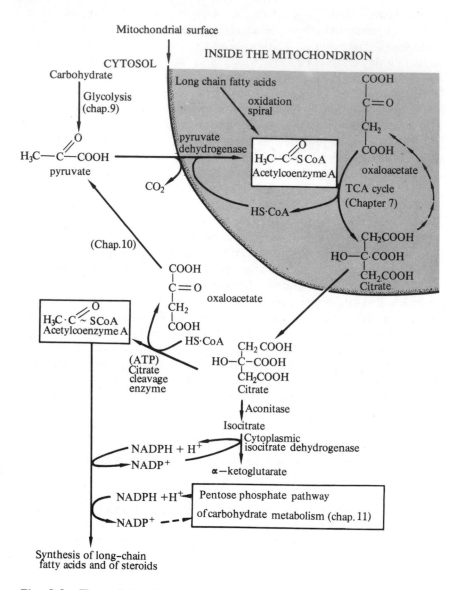

Fig. 8.3 The cellular location of some of the major processes involving acetyl coenzyme A

Most of the NADPH present in the cell is produced by a process known as the pentose phosphate pathway (chapter 11), but some is probably also produced by a NADP-dependent isocitrate dehydrogenase. The latter, like the pentose phosphate enzymes, is located outside the mitochondrion in the cytosol. It is quite distinct from the intra-mitochondrial isocitrate dehydrogenase yet, like its mitochondrial counterpart, it catalyses the oxidation of isocitrate to α-ketoglutarate.

The main functions of NADPH are discussed in chapter 11 but the most important appears to be to provide energy in the form of reducing power (electrons and protons) for certain extra-mitochondrial anabolic processes. Amongst such processes are the biosynthesis of steroids and (of particular importance in relation to the subject of this chapter) the biosynthesis of long-chain fatty acids. Figure 8.3 indicates some possible relationships between various mitochondrial and extra-mitochondrial enzyme systems, including those which generate NADPH.

8.6 The biosynthesis of long-chain fatty acids

It has been found that, *in vitro*, palmitate synthesis requires a number of cofactors, including NADPH, carbon dioxide, and biotin, none of which participates in the fatty acid oxidation spiral. Thus, even if the difference in cellular location were to be disregarded, the anabolic route cannot simply be the catabolic pathway operating in reverse. It is, in fact, not uncommon to find that biosynthesis occurs by a different route from catabolism. One reason for this is that the cell has to avoid a large endergonic step which would frequently have to be overcome if anabolism were the direct reverse of catabolism (cf. pp. 190, 221, 224).

Experiments using radioactive carbon (as $C^{14}O_2$) have shown that, while carbon dioxide is necessary for fatty acid synthesis, the carbon of this essential factor is not, in fact, incorporated into the final product. Moreover, *malonyl coenzyme A* proves to be a more potent precursor of palmitate than does acetyl coenzyme A. These facts were explained when it was realized that long-chain fatty acids are built up from shorter ones by successive addition of two-carbon units derived from malonyl coenzyme A with the release of carbon dioxide.

Malonyl coenzyme A is synthesized in two steps by the carboxylation of acetyl coenzyme A. In the first, carbon dioxide is activated by forming a derivative of *biotin*. This derivative is sometimes called 'active carbon dioxide' and the energy needed for its synthesis is supplied by ATP.

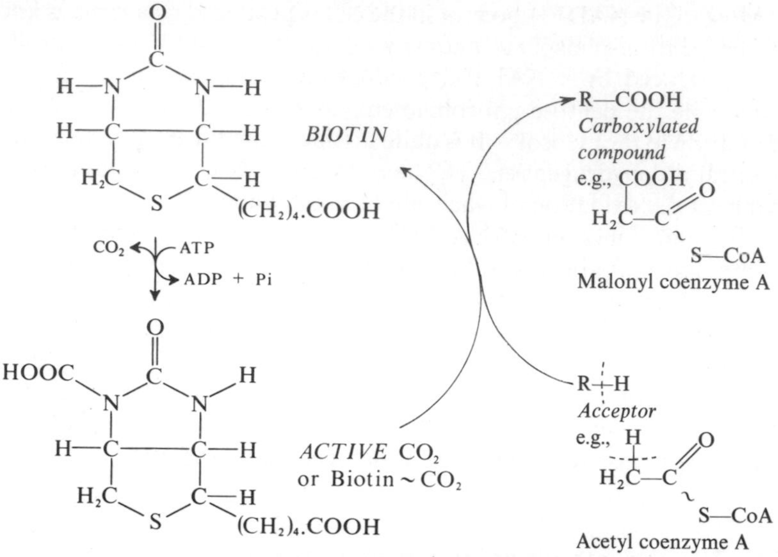

Fig. 8.4 Conversion of biotin to 'active CO_2' and the synthesis of malonyl coenzyme A. Other examples of similar carboxylation reactions involving 'active CO_2', considered in section 10.5

$$Pyruvate \rightarrow oxaloacetate$$
$$Propionyl\ CoA \rightarrow Methyl\ malonyl\ CoA$$

In both cases, the general reaction is R—H $\rightarrow$ R—COOH

The structures of biotin and of 'active carbon dioxide' are shown in Fig. 8.4. Secondly, malonyl coenzyme A is produced by a reaction in which the carbon dioxide is transferred from the biotin complex to a molecule of acetyl coenzyme A. Biotin is essential for several carboxylation reactions in addition to the one just described. Since animals cannot synthesize biotin, it is another example of a vitamin (p. 107). Raw egg-white contains avidin, a substance which combines with biotin and inactivates it.

The process of fatty acid synthesis consists essentially of the addition of two-carbon units *derived from malonyl coenzyme A* to a derivative of a fatty acid. However, the synthetic spiral differs from that which oxidizes fatty acids in a second important respect, for, in *E. coli* and other organisms so far investigated, the intermediates concerned in the anabolic sequence are not attached to coenzyme A, but instead are joined to a protein, *acyl carrier protein* or ACP. This protein has as its prosthetic group the sulphydryl compound, 4'-phosphopantetheine, a compound which, it will be recalled, also forms that part of the molecule of coenzyme A which possesses the sulphydryl group (p. 86). When a fatty acid forms a thioester with the sulphydryl group of ACP, the α and β carbon atoms are activated just as they are in the acyl coenzyme A compounds which take part in fat catabolism (section 8.3). Perhaps for this reason, although notable differences exist in cellular location and in coenzyme requirements, the principal reactions of the anabolic spiral strikingly resemble some of those of the catabolic pathway.

The initial reactants for the first turn of the anabolic spiral are derived from acetyl coenzyme A and malonyl coenzyme A. The malonyl coenzyme A, which can best be regarded as the *donor* of a two-carbon unit (Fig. 8.5) is converted to an ACP compound by the action of an enzyme called *transacylase*. The acetyl coenzyme A is similarly converted to

$$
\begin{array}{c}
\text{HOOC}-\text{CH}_2-\underset{\underset{\text{O}}{\parallel}}{\text{C}}\sim\text{S}-\text{CoA} \qquad\qquad \text{H}-\text{S}-\text{ACP} \quad \begin{array}{c}\text{acyl carrier}\\\text{protein}\end{array} \\[2mm]
\text{malonyl coenzyme A} \qquad\qquad \text{transacylase} \\[2mm]
\text{H}-\text{S}-\text{CoA} \qquad \text{HOOC}-\text{CH}_2-\underset{\underset{\text{O}}{\parallel}}{\text{C}}\sim\text{S}-\text{ACP} \\[2mm]
\text{malonyl–ACP complex}
\end{array}
$$

acetyl$\sim$S—ACP and acts, for the *first* turn of the spiral only, as the acceptor of a two-carbon unit—in other words, acetyl$\sim$S—ACP is the simplest of the fatty acyl$\sim$S—ACP compounds shown centrally in Fig. 8.5. This use of acetyl coenzyme A must not be confused with its general function of providing the malonyl coenzyme A donor for *all* turns of the spiral.

The 'condensation' of malonyl$\sim$S—ACP with acyl ACP (acetyl, butyryl, hexanoyl, etc.) is the first of four reactions of a sequence which constitutes any one turn of the cycle. It is at this stage that the carbon dioxide originally inserted into acetyl coenzyme A by 'active carbon dioxide' is split away from the malonyl$\sim$S—ACP. The process of 'condensation' would probably be endergonic, like many reactions which result in the formation of a new C—C bond, if it were not facilitated by the energy released by the simultaneous cleavage of a nearby C—C bond as decarboxylation occurs. It is this elegant 'carbon-gain by carbon-loss' mechanism which explains why *malonyl* rather than *acetyl* coenzyme A is the initial two-carbon donor; it also explains why radioactive carbon dioxide is not incorporated into the final product, even though carbon dioxide is an essential co-factor for fatty acid synthesis. The product

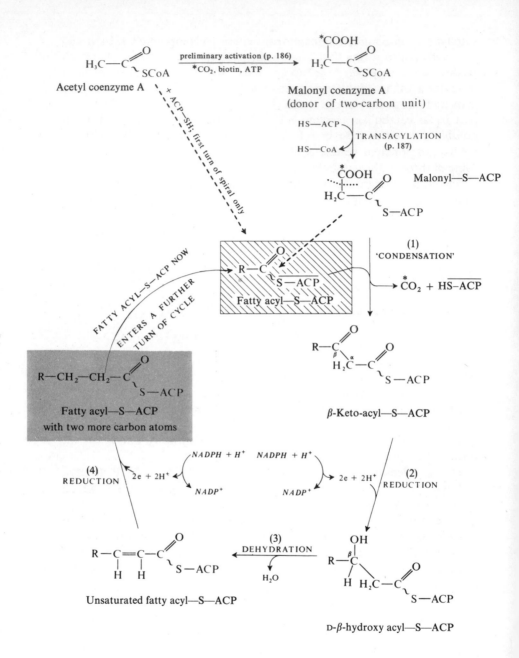

Fig. 8.5 Fatty acid synthetic spiral

of this 'condensation' is the β-keto-acyl derivative of ACP. In the remaining steps of the spiral this compound undergoes reactions of the type shown in Fig. 7.1, but moving from right to left.

In the second reaction of the spiral (Fig. 8.5), the β-keto compound is reduced to form a β-hydroxy-acyl-ACP. The reduction involves the transfer of two electrons and one proton from NADPH. It is important to note that this, and not NADH, is involved and that the product is D-β-hydroxy-acyl-ACP. In the catabolic route (Fig. 8.2), a L-β-hydroxy-acyl-coenzyme A participates. Thus, once again, a superficial similarity between the catabolic and anabolic routes should not be allowed to obscure the great differences between them. Indeed, the occurrence in one route of a L-reactant and in the other of a D-product of the same substance (present, of course, in different complexes) provides a remarkable example of the stereospecificity of enzymes (p. 89).

The third reaction of the anabolic spiral (Fig. 8.5) is a dehydration which · results in an α,β-unsaturated compound. It is analogous to the hydratase reaction of the catabolic route acting in reverse. Finally, in the fourth reaction the first turn of the spiral is completed with the formation of a fatty acyl-ACP compound with 2 carbon atoms more than the original fatty acid which took part in the first reaction. The reduction is achieved by the addition of two protons and two electrons donated by NADPH.

Thus, beginning with acetyl-ACP, each turn of the spiral adds a two-carbon unit so that after seven turns the product is palmityl-ACP,

$$H_3C.(CH_2)_{14}.C\overset{\displaystyle O}{\underset{\displaystyle \sim}{\diagup}}S—ACP.$$ Free palmitic acid is produced from this by hydrolysis of the thioester linkage.

The overall reaction for the synthesis of palmitic acid from acetyl coenzyme A may now be summarized:

$$8H_3C\ COSCoA + 7CO_2 + 14NADPH + 14H^+ + 7ATP \rightarrow$$
$$H_3C(CH_2)_{14}COOH + 8CoASH + 7CO_2 + 14NADP^+$$
$$+ 7ADP + 7P_i + 6H_2O$$

The stoichiometry shown in the equation indicates that one mole of palmitate is derived from 14 moles of reduced NADP and 7 moles of ATP. The process requires a great deal of energy since 14NADPH is equivalent to 14×3 high energy phosphate bonds. Moreover 8 acetyl coenzyme A, which would lead to 96ATP on oxidation, is also used up, and becomes unavailable for immediate catabolic purposes.

8.7 The biosynthesis of triglycerides

Free fatty acids do not usually accumulate in the cell to any great extent, but are largely incorporated into glycerides and phospholipids. It was noted earlier (p. 177) that the enzyme lipase does not catalyse net synthesis of triglycerides from glycerol and fatty acids, unless one of these is in great excess, because the equilibria shown in Fig. 8.1 favour hydrolysis rather than esterification. In the cell, triglyceride synthesis by-passes this unfavourable equilibrium by two devices; firstly, coenzyme A derivatives of the fatty acids, rather than free fatty acids, are involved, and secondly, instead of glycerol, it is glycerol phosphate which is esterified.

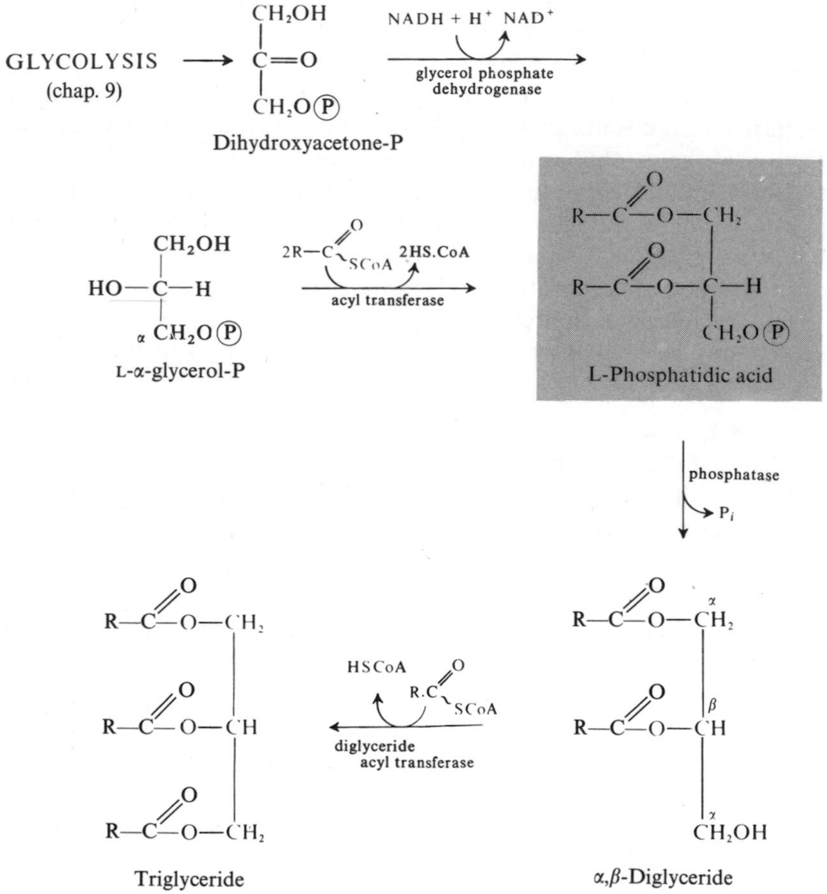

Fig. 8.6 The biosynthesis of phosphatidic acid and of triglyceride

Hence both reactants are at a higher energy level than in the lipase reaction (compare p. 214).

The first step, the synthesis of L-α-*glycerol phosphate*, usually occurs by a reduction of dihydroxyacetone phosphate, which is an important intermediate product of the glycolytic breakdown of glucose (chapter 9).

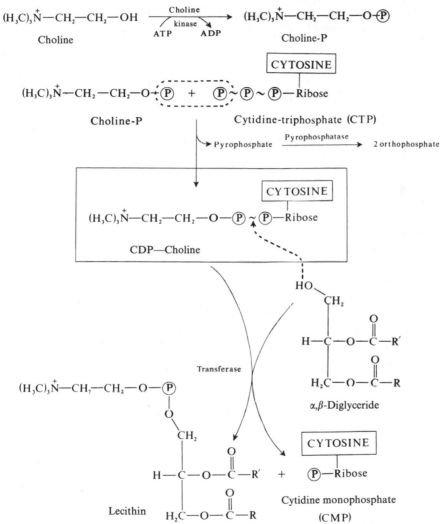

Fig. 8.7 The biosynthesis of lecithin. (Choline is converted to an 'active' form, CDP-choline, a form in which it is readily transferred to an α,β-diglyceride)

The reduction results from the passage of two electrons from NADH to the dihydroxyacetone phosphate (Fig. 8.6). The L-α-glycerol phosphate then reacts successively with two molecules of the acyl-coenzyme A derivatives of the appropriate fatty acids. The product is a *phosphatidic acid*. The synthesis of triglyceride is completed by two reactions; first, the phosphate group of the phosphatidic acid is hydrolysed, forming an α,β-diglyceride and orthophosphate (enzymes which catalyse the hydrolysis of organic phosphates in this way are termed *phosphatases*), and finally the diglyceride is esterified by the transfer of the acyl group from a fatty acyl coenzyme A (Fig. 8.6).

8.8 The biosynthesis of lecithin

The synthesis of phospholipids and of glycerides occurs predominantly in the microsomal fraction of the cell—probably in the 'smooth' regions of the endoplasmic reticulum (p. 30). Phospholipid formation may be illustrated by reference to lecithin synthesis. The lecithin molecule consists essentially of two parts, *choline phosphate* linked to an α,β-*diglyceride*. However, the synthesis does not result simply from the reaction of these two compounds but occurs by an energetically more favourable route; there is a preliminary activation which results in the conversion of choline phosphate to a more reactive choline phosphate-nucleotide derivative. In this case the nucleoside involved is cytidine—its triphosphate (CTP) reacts with choline phosphate to form cytidine-diphosphate-choline (CDP-choline). As was pointed out in chapter 7, compounds of this type are frequently encountered in anabolic reactions—fatty acyl-AMP, for example, is an intermediate in fatty acyl coenzyme A synthesis. The second product of the reaction, inorganic pyrophosphate, is itself rapidly hydrolysed to inorganic phosphate by the enzyme *pyrophosphatase*. Since this last reaction is highly exergonic, the hydrolysis of pyrophosphate is almost complete and ensures that the reaction of choline phosphate with CTP is also almost irreversible (Fig. 8.7). Lecithin finally results from the transfer of the choline phosphate moiety of CDP-choline to an α,β-diglyceride. Cytidine monophosphate (CMP), the second product of this reaction, is converted back to CTP by the transfer of phosphate groups from ATP:

$$2\text{ATP} + \text{CMP} = 2\text{ADP} + \text{CTP}.$$

It will be shown in chapter 10 that the formation of di- and polysaccharides from monosaccharides occurs by reactions very similar to those by which lecithin is synthesized from CTP and choline.

Glycolysis and the fate of pyruvate

Carbohydrates play a vital and central role in cellular biochemistry for in addition to their functions as structural units and as food reserves, their breakdown by one or other of several possible routes yields ATP, NADH, or NADPH. All of these, in different ways, offer a supply of energy which can be used by the cell to drive endergonic processes. Furthermore, carbon atoms derived from carbohydrate molecules can be incorporated into a variety of other biological compounds—the carbon frameworks of certain amino acids, for instance, can be synthesized from materials formed from carbohydrates.

9.1 The overall position of glycolysis

In most types of cells the principal catabolic pathway is one which converts various carbohydrates, via fructose diphosphate, to *pyruvate*. This route, which is termed *glycolysis* ('sugar-splitting'), involves a series of reactions which occurs almost universally in cells, even though the materials being fed into the process and those being obtained from it may sometimes appear dissimilar in different organisms or under different environmental conditions. This illustrates one of the most fascinating—and simplifying—aspects of biochemistry, for processes which, physiologically, look very different, may in fact proceed for a considerable part of their metabolic routes along pathways which are extremely similar. The reactions of glycolysis, for instance, are essentially the same, whether they occur in the setting of an anaerobic fermentation process or form a part of the mechanism by means of which aerobic respiration occurs. It is, very largely, *the fate of the pyruvate*, the end product of glycolysis, which distinguishes these different activities, and this complication will be considered towards the end of the chapter. For the moment, it is only necessary to recognize that pyruvate, once it is formed by glycolysis, can be readily converted under *aerobic* conditions to acetyl coenzyme A, the fundamental importance of which has already been emphasized (pp. 161, 182) in chapters 7 and 8.

When one mole of glucose undergoes complete oxidation to carbon dioxide and water, the overall free energy change is 691·2 kcal. As in the case of fat metabolism, this energy is not all released at once, but in successive small steps, some of which are of such a magnitude (8–12 kcal) that much of the energy liberated is conserved by the coupled formation of high energy bonds. In fact, glycolysis follows a very gradual downhill course as far as pyruvate and, in itself, provides the cell with little energy (Fig. 9.1). The great energetic potential of carbohydrates is only realized when, under aerobic conditions, pyruvate first forms acetyl coenzyme A and this then undergoes the oxidative changes that formed the subject of chapter 7.

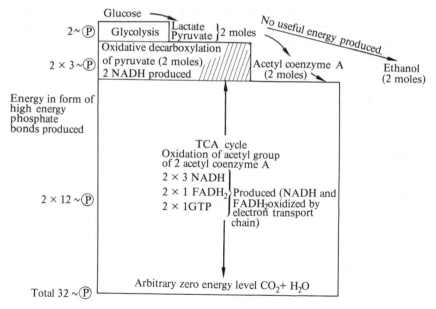

Fig. 9.1 Yield of energy during the catabolism of glucose. (Note the small energy change (vertical height) for the glycolytic pathway compared with the complete (aerobic) oxidation)

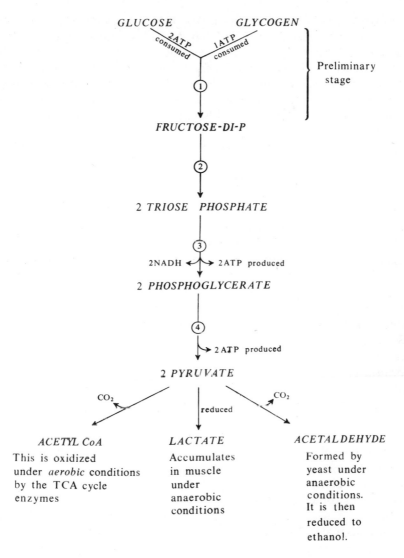

Fig. 9.2 The four stages of glycolysis and the fate of pyruvate

Glycolysis is often referred to as the E.M.P. pathway, in recognition of the classical researches of Embden, Meyerhof, and Parnas. Each of the enzymes participating in glycolysis occurs in the *cytosol*, or *soluble cell fraction*, so it is here, and not in the mitochondrion, that the initial steps of sugar catabolism take place. The glycolytic pathway can be subdivided into four stages (Fig. 9.2), and it will be discussed in general terms before each of these stages is considered in more detail (sections 9.3–9.6).

9.2 The four stages of glycolysis

Sugars almost always react biochemically in a phosphorylated form, and in *stage 1* of glycolysis the carbohydrate to be catabolized is converted to fructose diphosphate. In fact, this substance can, more properly, be regarded as the real starting point for the pathway. Its formation is a multi-step process and differs slightly according to whether it is derived from a carbohydrate store, such as glycogen or starch, or is formed directly from monosaccharide entering the cell (as occurs when yeast is grown in glucose solution or when glucose enters a muscle cell from the blood). When glycogen and starch are metabolized to fructose diphosphate, only one high energy bond is used up for each molecule of the diphosphate formed, but when glucose is the precursor, two molecules of ATP are used up per molecule of fructose diphosphate. The reason is that the polysaccharides are at a higher energy level than glucose, and ATP is used up in their synthesis (chapter 10).

In *stage 2*, each fructose diphosphate molecule splits into two molecules of triose phosphate, thus reducing to half the size of the molecules participating in subsequent steps. Incidentally, since these later steps commence with triose phosphate, it is inevitable that fructose diphosphate, and not the monophosphate, should be the precursor.

In *stage 3*, which is another multi-step process, the aldehydic form of triose phosphate (glyceraldehyde-3-P) is oxidized to phosphoglyceric acid, much of the energy so liberated being conserved by the coupled synthesis of ATP from ADP. Lastly, in *stage 4*, which again involves a sequence of reactions, the phosphoglyceric acid is converted to pyruvic acid with a further release of energy, much of which is again conserved by synthesis of ATP. As Fig. 9.2 shows, it is only after this point that the pathways of fermentation by yeast, anaerobic respiration in muscle, and the more common aerobic respiration diverge from one another.

9.3 Formation of fructose-di-P

The mechanism of glycolysis can now be discussed in more detail, stage by stage. In what may be regarded as preliminary reactions, carbohydrate is first converted to fructose-di-P (Fig. 9.3). When the carbohydrate is glucose, the initial reaction is the formation of glucose-6-P,

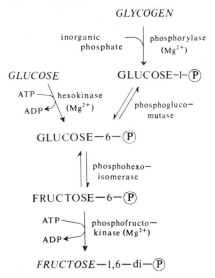

Fig. 9.3 Glycolysis—stage 1. Synthesis of fructose-di-P

the enzyme responsible, *hexokinase*, requiring magnesium ions as cofactor. There are many enzymes bearing the suffix *-kinase* and it is important to remember that this term indicates that the enzyme catalyses the transfer of a phosphate group from the high energy compound ATP (or some similar nucleoside triphosphate) to another compound—in the present case to the *hex*ose, glucose. The glucose-6-P is so named because it is the hydroxyl group on carbon atom 6 (i.e., in the straight-chain form, the atom farthest from the aldehyde group) which is phosphorylated.

Glucose   Glucose-6-P   Glucose-1-P

Glycolysis and the fate of pyruvate

When the carbohydrate to be catabolized is endogenous glycogen or starch, glucose-6-P is again formed but by a *two-step* process. The first step is the cleavage of the bond linking the terminal glucose unit, situated at the non-reducing end of the polysaccharide chain. The reaction is catalysed by the enzyme *phosphorylase*; phosphorolysis resembles hydrolysis except that it results in the attachment of an inorganic phosphate group to carbon atom 1 of the cleaved glucose molecule, and so leads to the formation of glucose-1-P. Note that the phosphate is not derived from ATP (phosphorylase is therefore not a kinase) and that inevitably, it is glucose-1-P, not glucose-6-P, which is produced by the reaction.

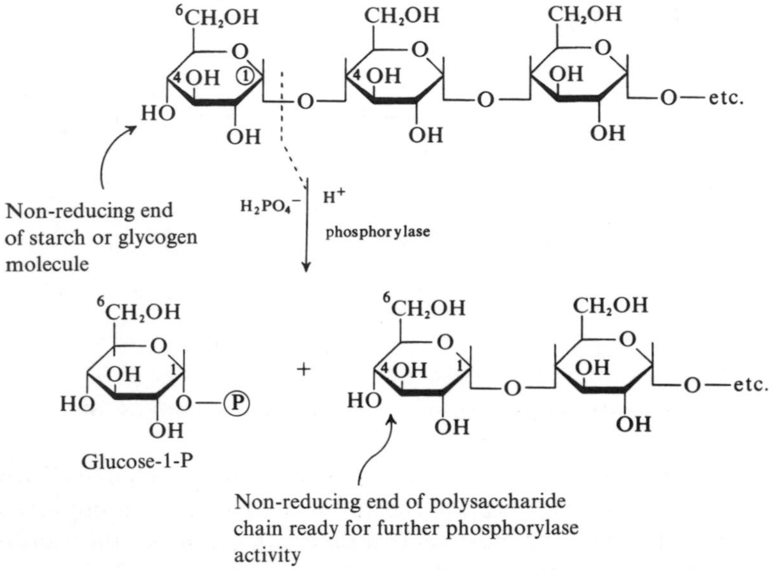

Next, the glucose-1-P is readily converted by the enzyme *phosphoglucomutase* to glucose-6-P (this is the same substance as is produced by the hexokinase reaction when glucose is the precursor). In this reaction the two isomeric glucose phosphates form an equilibrium mixture containing about 94·5% glucose-6-P. The name phosphogluco*mutase* implies that the enzyme transfers a phosphate group from one part of the glucose molecule to another. Glucose-1,6-di-P is a cofactor for the reaction.

Whether the carbohydrate to be metabolized by glycolysis is glucose, glycogen, or starch, these reactions ensure that it is converted to *glucose-6-P*. Isomerization then occurs with the conversion of the aldose, glucose-6-P, to the ketose, fructose-6-P. The enzyme responsible is termed *phosphohexoisomerase*, and at equilibrium about one third of the phosphohexose is in the form of fructose-6-P.

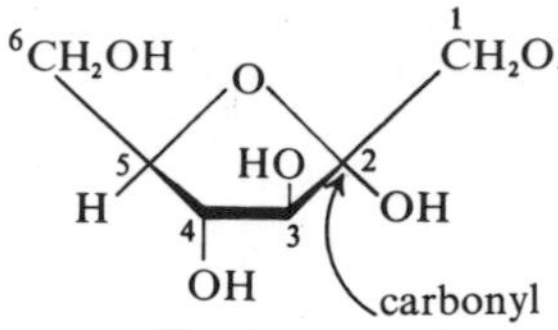

In the final reaction of stage 1 (Fig. 9.3), further phosphorylation takes place at the expense of ATP. The enzyme responsible is specific for fructose-6-P and is therefore termed *phosphofructo*kinase. The product is fructose-1,6-di-P, a fructose molecule phosphorylated at each end of the straight chain form.

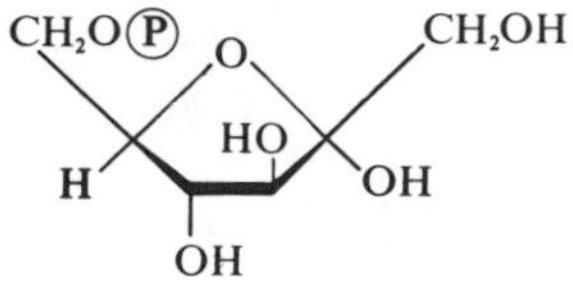

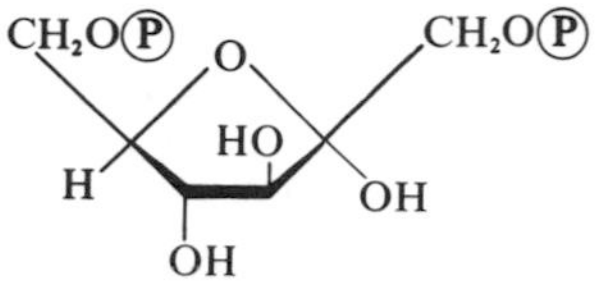

In stage 1 of glycolysis, there is a net loss of two molecules of ATP for every glucose molecule metabolized. If glycogen is the source of the fructose-di-P only one molecule of ATP is lost, since the second phosphate group is derived from inorganic phosphate through the phosphorylase reaction. This loss of ATP during a process whose primary function is to produce ATP need cause no concern since, as will be seen in stages 3 and 4, there is nevertheless a net overall synthesis of ATP. The ATP used up in stage 1 is analogous to the energy used in priming a gun— when the cap is detonated there is a much greater release of energy than was needed initially to pull the trigger. A similar situation was encountered earlier (p. 179) in relation to the 'priming' of fatty acids.

9.4 Formation of triose phosphates

Stage 2 of glycolysis is the cleavage of the fructose-di-P produced in the previous stage, to yield two molecules of triose phosphate (Fig. 9.4). The first enzyme involved is called *aldolase* because it catalyses a reaction known in organic chemistry as the aldol 'condensation'. The simplest

Glycolysis and the fate of pyruvate

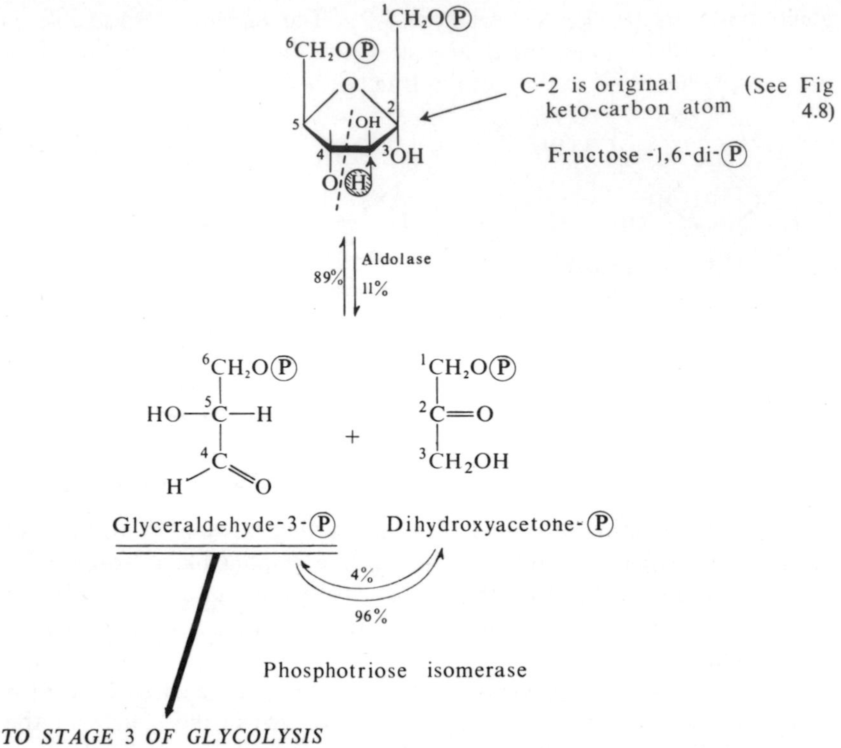

$^1CH_2O\text{ⓟ}$

$^6CH_2O\text{ⓟ}$

O

^{5}

^{4}

^{2}

^3OH

OH

O H

C-2 is original   (See Fig
keto-carbon atom   4.8)

Fructose -1,6-di-ⓟ

89%   Aldolase
11%

$^6CH_2O\text{ⓟ}$

HO—^5C—H

^4C

H   O

$^1CH_2O\text{ⓟ}$

^2C=O

$^3CH_2OH$

+

Glyceraldehyde-3-ⓟ   Dihydroxyacetone-ⓟ

4%

96%

Phosphotriose isomerase

TO STAGE 3 OF GLYCOLYSIS

Fig. 9.4   Glycolysis—stage 2.   Cleavage of fructose-1,6-di-ⓟ
(See also Fig. 4.8, p. 108)

example of this type of condensation—better thought of as an addition
reaction—occurs, when acetaldehyde is treated with alkali. Two molecules of acetaldehyde unite to form one molecule of aldol in a reaction
in which one of the two carbonyl groups is converted into a hydroxyl
group by reaction with a 'framework' hydrogen atom of the other
molecule:

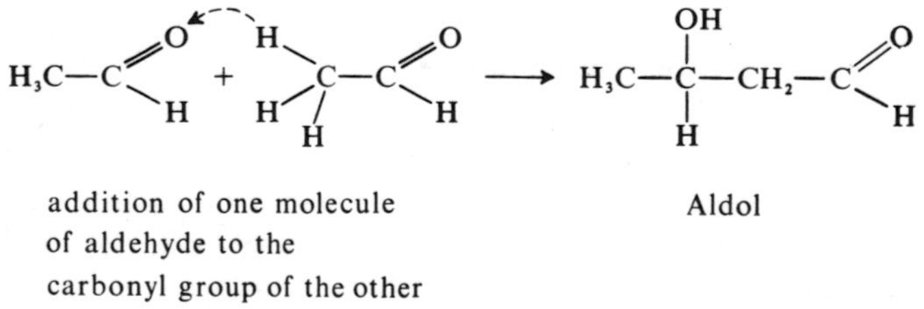

H_3C—C O   H
       H   C—C O
         H   H   H

+

→

OH
H_3C—C—CH_2—C O
     H   H

addition of one molecule
of aldehyde to the
carbonyl group of the other

Aldol

Cellular biochemistry and physiology   **200**

Aldolase catalyses a comparable reaction between a molecule of glyceraldehyde-3-P, acting as the hydrogen acceptor, and one of dihydroxyacetone phosphate. In other words, aldolase causes molecules of the two isomeric forms of triose phosphate to join together, the carbonyl group of the glyceraldehyde-3-P being changed to a hydroxyl group. The reaction is reversible but it tends to favour the formation of fructose-di-P rather than triose phosphate. At equilibrium, the mixture contains some 89% fructose-di-P but only 11% triose phosphate. This reversibility is of importance, since in some circumstances aldolase catalyses the synthesis of hexose from triose phosphate (chapter 10).

The aldolase reaction produces equimolecular quantities of glyceraldehyde-3-P and dihydroxyacetone-P; the final reaction of stage 2 allows inter-conversion of these two compounds. This is necessary because glyceraldehyde-3-P alone is the reactant for stage 3. The two trioses are readily inter-converted through the intervention of the enzyme *triose phosphate isomerase*, the equilibrium mixture containing about 96% dihydroxyacetone-P. The enzyme has a very high turnover number (p. 104) so that, in spite of the equilibrium favouring dihydroxyacetone-P, the concentration of glyceraldehyde-P is nevertheless sufficient to enable stage 3 of glycolysis to proceed unhindered.

9.5 Oxidation of glyceraldehyde-3-P

Stage 3 is the oxidation of the aldehyde, glyceraldehyde-3-P, and is the only step in the glycolytic pathway involving oxidation. It occurs in such a way that the energy released in the formation of the carboxylic acid is coupled to the synthesis of an energy-rich bond, instead of being dissipated as heat. The aldehyde group of glyceraldehyde-3-P undergoes an addition reaction with the sulphydryl group of the enzyme *phosphoglyceraldehyde dehydrogenase*, so attaching the aldehyde to the enzyme surface. Oxidation then occurs with the removal from the complex of two protons and two electrons. The electrons and a hydrogen ion combine with NAD^+, which is thereby reduced to NADH.

Since phosphoglyceraldehyde dehydrogenase couples oxidation of glyceraldehyde-3-P with reduction of NAD^+ it is better called D-glyceraldehyde-3-P:NAD^+ *oxidoreductase*, but the older name is still widely used. The high energy complex of enzyme and oxidized glyceraldehyde-3-P is next attacked, not by water but by inorganic orthophosphate, and is thereby split into free enzyme and a carboxyl phosphate of high energy, 1,3-diphosphoglyceric acid (Fig. 9.5).

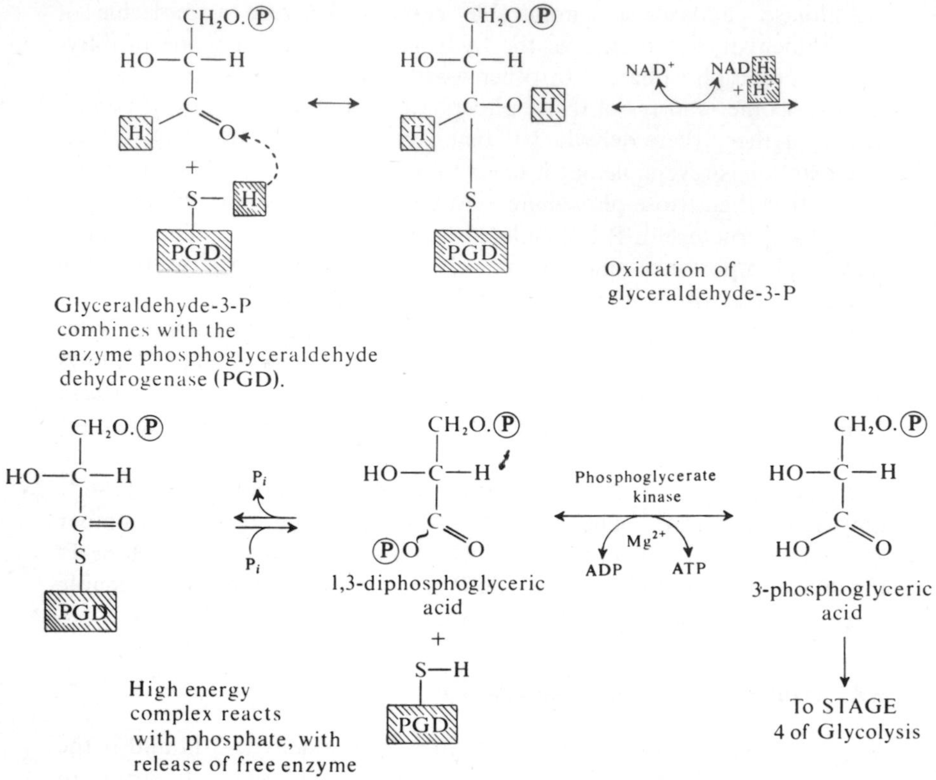

CH$_2$O.℗
HO—C—H
C
H   O
+
S—H
PGD

⟷

CH$_2$O.℗
HO—C—H
C—O H
H
S
PGD

NAD$^+$ NADH
+H

Oxidation of
glyceraldehyde-3-P

Glyceraldehyde-3-P
combines with the
enzyme phosphoglyceraldehyde
dehydrogenase (PGD).

CH$_2$O.℗
HO—C—H
C=O
S
PGD

P$_i$
P$_i$

CH$_2$O.℗
HO—C—H
C
℗O   O
1,3-diphosphoglyceric
acid
+
S—H
PGD

High energy
complex reacts
with phosphate, with
release of free enzyme

Phosphoglycerate
kinase
Mg^{2+}
ADP   ATP

CH$_2$O.℗
HO—C—H
C
HO   O
3-phosphoglyceric
acid

To STAGE
4 of Glycolysis

Fig. 9.5 Glycolysis—stage 3. Oxidation of glyceraldehyde-3-P and the several steps leading to 3-phosphoglyceric acid.   All the steps are reversible

1,3-Diphosphoglyceric acid is highly reactive, and, *in vitro*, it is found that its hydrolysis yields 3-phosphoglyceric acid, phosphoric acid, and much energy in the form of heat.   It is, therefore, a high-energy compound. In the living cell, a comparable reaction, catalysed by the enzyme *phosphoglycerate kinase*, couples the energy released with the transfer of phosphate to ADP.   This enzyme, like almost all known kinases, requires the presence of magnesium ions.   In the present reaction the enzyme is, in a sense, 'working in reverse', since ATP is being formed rather than destroyed.   It provides a further example of substrate level phosphorylation.

In terms of the hexose unit from which *two* molecules of triose phosphate were formed, there is, at this stage, a gain of two molecules of ATP.

9.6 Synthesis of pyruvate

Stage 4 of glycolysis is the conversion of 3-phosphoglyceric acid to pyruvic acid. In the first of three reactions the phosphate attached to the 3-phosphoglycerate is transferred to the central carbon atom, thus freeing the hydroxyl group on the terminal carbon in readiness for a subsequent reaction (Fig. 9.6). This transfer of a phosphate group from one part of the molecule to another is catalysed by the enzyme *phosphoglyceromutase*. It should be observed that the reaction is closely similar to that catalysed by *phosphoglucomutase* in stage 1 (p. 198). In just the same way as the latter enzyme only functions in the presence of glucose-1,6-di-P (1,6-diphosphoglucose), so phosphoglyceromutase likewise needs 2,3-diphosphoglycerate as cofactor.

The 2-phosphoglyceric acid is now dehydrated by the elimination of the elements of water from positions 2 and 3. The unsaturated product, phosphoenolpyruvate (PEP) is interesting since, like ATP, it possesses a high energy phosphate bond (p. 133). In this respect it should be contrasted with many phosphate esters of alcohols and of similar hydroxylated compounds; glucose-1-P and 2-phosphoglycerate, for example, liberate only a small amount of energy when the phosphate group is removed by hydrolysis. The difference may be related to the fact that the enol form of pyruvic acid is very unstable. The dehydration of 2-phosphoglycerate converts a low energy phosphate bond to a high energy bond—the energy of the exergonic dehydration, in a certain sense, is diverted to the phosphorus bond, the change in molecular configuration consequent upon dehydration providing the trapping device. The reaction is catalysed by an extensively investigated enzyme usually called *enolase*, though a better name is phosphopyruvate hydratase. It requires the presence of magnesium ions and is inhibited by fluoride ions. The fluoride forms a stable complex with magnesium and phosphate ions, thus effectively diminishing the availability of magnesium ions to the active centre of the enzyme.

The next reaction involves the removal of the phosphate group from the molecule of phosphoenol pyruvate (Fig. 9.6). The enzyme *pyruvate kinase* catalyses the transfer of the phosphate group to ADP—and with it is transferred much of the energy associated with the cleavage of this group from the phosphoenol pyruvate. This reaction is therefore a further example of substrate level phosphorylation. The two products are ATP and the *enol* form of pyruvic acid; the equilibrium lies very much in favour of the products of this reaction, so the process is virtually

Glycolysis and the fate of pyruvate

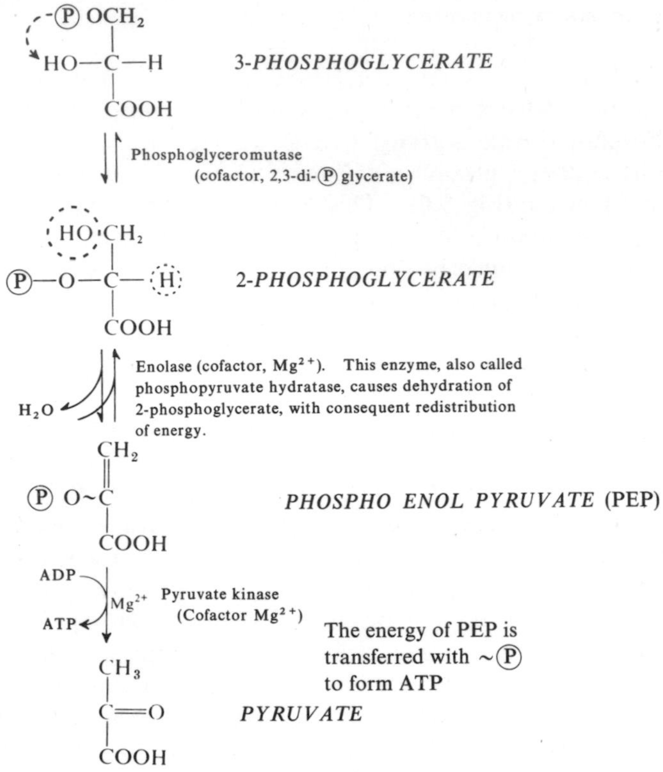

-Ⓟ OCH₂

HO—C—H   3-*PHOSPHOGLYCERATE*

COOH

     Phosphoglyceromutase
     (cofactor, 2,3-di-Ⓟ glycerate)

HO CH₂

Ⓟ—O—C—H   2-*PHOSPHOGLYCERATE*

COOH

     Enolase (cofactor, Mg^{2+}). This enzyme, also called
     phosphopyruvate hydratase, causes dehydration of
H_2O 2-phosphoglycerate, with consequent redistribution
     of energy.

CH₂

Ⓟ O~C   *PHOSPHO ENOL PYRUVATE* (PEP)

COOH

ADP

Mg^{2+} Pyruvate kinase
    (Cofactor Mg^{2+})

ATP

       The energy of PEP is
       transferred with ~Ⓟ
       to form ATP

CH₃

C=O   *PYRUVATE*

COOH

Fig. 9.6   Glycolysis—stage 4.   Synthesis of pyruvate.   (Note: in stage 4, 2 molecules of ATP are formed per original molecule of hexose)

irreversible. The enol form of pyruvic acid is unstable and tends to change spontaneously, by internal re-arrangement, to the more familiar keto-form of the molecule, pyruvic acid.

During the fourth stage of glycolysis, 2 molecules of ATP are thus produced for each original hexose unit. Therefore, when glucose is the source of energy, the overall synthesis of ATP from ADP per mole of hexose degraded is $(-2 + 0 + 2 + 2)$ or 2 moles ATP, while reducing power in the form of 2 moles NADH is also created. This NADH is in the *cytosol* but if its reducing power can, directly or indirectly, become available within the *mitochondrion*, each molecule is, of course, potentially able to produce 3 molecules of ATP by oxidative phosphorylation.

9.7 Oxidative decarboxylation of pyruvate to acetyl coenzyme A

Under normal *aerobic* conditions almost all cells convert pyruvate to acetyl coenzyme A. This is energetically a very rewarding process, for not only are 3 molecules of ATP formed for each molecule of pyruvate which is decarboxylated, but, even more important, the subsequent oxidation of each molecule of acetyl coenzyme A by the TCA cycle yields 12 molecules of ATP.

We come now to the important connecting link between glycolysis (and therefore a major pathway of sugar metabolism) and oxidation of acetyl coenzyme A. The whole series of reactions which constitute the glycolytic pathway takes place outside the mitochondrion in the cytosol, whereas the oxidation of acetyl coenzyme A is a mitochondrial event. The mitochondrion also contains a complex of enzymes, collectively known as *pyruvate dehydrogenase*. This contains very similar components to the α-ketoglutarate dehydrogenase system discussed earlier (section 7.6) and the reactions which convert pyruvate to acetyl coenzyme A are very similar to those which change α-ketoglutarate to succinyl coenzyme A. Thus the reactions indicated in Fig. 7.4 apply equally to pyruvate and to α-ketoglutarate; it should be noted that for pyruvate R— in the figure is —CH_3. The overall effect of pyruvate dehydrogenase is to catalyse the reaction

$$H_3C.\overset{\overset{\text{O}}{\|}}{C}.COOH + CoASH \longrightarrow CO_2 + H_3C - \overset{\overset{\text{O}}{\|}}{C}_{\diagdown SCoA} + 2e + 2H^+$$

The energy released when the C—C bond is broken is largely conserved as the high-energy C~S bond. Two electrons and one of the two protons removed from the pyruvate are passed on to NADH; they subsequently lead to the production of 3 molecules of ATP by the process of oxidative phosphorylation. The oxidative decarboxylation of pyruvate, like that of α-ketoglutarate, is virtually irreversible.

Under aerobic conditions we can thus envisage pyruvate being produced from carbohydrates by glycolysis in the cytosol, the pyruvate then diffusing into the mitochondrion where it is converted by pyruvate dehydrogenase to acetyl coenzyme A which then undergoes oxidation by the TCA cycle. The pyruvate dehydrogenase, like all but one of the TCA dehydrogenases, is coupled to the electron carrier NAD. The subsequent oxidation of NADH by the electron transport complexes of the *inner* mitochondrial membranes yields ATP by oxidative phosphorylation.

Since it is the carbon atom of the carboxylic acid group of pyruvate which is lost, and this group arises from the aldehyde group of glyceraldehyde-3-P, the carbon dioxide produced originates, in the first instance, from carbon atoms 3 and 4 of the original hexose unit undergoing catabolism. This has been shown by experiments using glucose, the carbon atoms of which were specifically labelled at positions 3 and 4. In this respect the catabolism of glucose by the glycolytic pathway differs from its metabolism by a second pathway, the pentose phosphate pathway, which is described in chapter 11.

9.8 The fate of pyruvate under anaerobic conditions

Even a temporary lack of oxygen can be a serious hazard for many types of cells; vertebrate brain cells, for example, are particularly susceptible to damage caused by *anoxia*. In the absence of oxygen, the mitochondrial electron transport system is unable to function, and so the TCA cycle, which generates ATP, comes to a halt. Furthermore, the fuel for this cycle is acetyl coenzyme A, itself often formed by the oxygen-dependent oxidative decarboxylation reaction.

Since fats also give rise to acetyl coenzyme A via the oxygen-dependent process of β-oxidation, fat catabolism also stops under anaerobic conditions and, if the cell is to survive, it must turn to glycolysis as the means for producing the bulk of its ATP. This, in effect, means that an anaerobic mechanism must somehow bring about the re-oxidation of the NADH produced by the activity of phosphoglyceraldehyde dehydrogenase in stage 3 of glycolysis; there is only a limited amount of NAD^+ in the cell, and glycolysis would also stop under anaerobic conditions if all this coenzyme were to be converted to the reduced form. Various types of cells, and especially different micro-organisms, can achieve the anaerobic oxidation of NADH in a variety of ways, and the mechanism will now be illustrated by considering the case of muscle undergoing strenuous exercise and that of yeast growing under anaerobic conditions.

In the *presence* of air, the electrons removed from an oxidizable substrate eventually reduce oxygen, the ultimate electron acceptor, to water. This reduced form of the electron acceptor is, paradoxically, inconspicuous owing to the abundance of water in all living cells. In the *absence* of air, however, some electron acceptor other than gaseous oxygen must be present and consequently some other reduction product than water must be formed. In muscle suffering oxygen lack, that product is lactate, while in yeast fermentation it is alcohol. They are both formed from pyruvate, the end-product of glycolysis, and their formation provides an elegant device which allows muscle and yeast to solve their redox problems when anaerobic conditions prevail.

When a muscle is heavily exercised, the blood is unable to supply oxygen rapidly enough for its requirements. Under these conditions of temporary anoxia, the muscle uses pyruvate as an electron acceptor, thus enabling NADH, formed in stage 3 of glycolysis, to revert back to NAD$^+$. In effect, the *oxidation* of aldehyde by the phosphoglyceraldehyde dehydrogenase reaction is coupled, via the NAD$^+$/NADH system, to the *reduction* of the ketone group of pyruvate (Fig. 9.7a). The enzyme responsible for the reduction of pyruvate is lactate dehydrogenase (p. 88). Thus the formation of lactic acid gives the muscle a temporary boost in the form of ATP derived from glycolysis, at a time when it is most needed.

Lactate, therefore, is the final product of anaerobic muscular activity, not because muscle has any need of this substance biochemically, but because it has an imperative need to *utilize pyruvate as an electron acceptor*. The system nevertheless has its limitations, as every athlete familiar with muscle fatigue will know; the muscle is said to be 'fatigued' when its activity is so prolonged that lactate accumulates at the site of its formation.

The process whereby alcohol is formed when yeast ferments sugar is essentially similar to the process occurring in muscle under anaerobic conditions. Yeast normally respires in the ordinary aerobic manner utilizing the electron transport system to oxidize NADH; fermentation begins when the carbon dioxide evolved during the aerobic process is unable to escape from the vessel and there is inadequate access of oxygen. Whereas animals in such circumstances would suffocate, yeast switches to the anaerobic process of fermentation.

In industrial fermentation processes, fructose-di-P, the primary reactant for glycolysis and hence for alcohol production, is formed from exogenous glucose (which is often made by man from starch before the yeast is added). Organisms faced with anaerobic conditions in nature may use exogenous glucose, or, of course, they can call upon internally stored carbohydrate. The pyruvate formed by the pathway of glycolysis is converted to ethanol for the same reason that it is converted to lactate by anoxic muscle. The difference rests in the fact that, whereas pyruvate in muscle is *directly* reduced by NADH, in yeast it is first *decarboxylated*, and the product of the reaction, acetaldehyde, acts as the electron acceptor.

Glycolysis and the fate of pyruvate

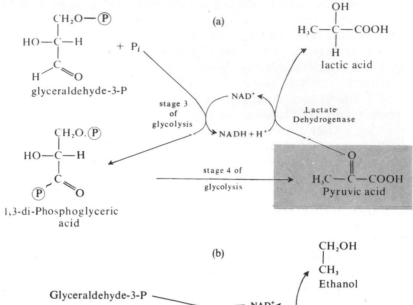

Fig. 9.7 The fate of pyruvate
(a) Muscle under anaerobic conditions
(b) Fermentation of yeast (anaerobic)

The (non-oxidative) decarboxylation of pyruvate by yeast growing under *anaerobic* conditions should be both compared to, and contrasted with, the oxidative decarboxylation which occurs under *aerobic* conditions and which leads to the formation of acetyl coenzyme A (p. 205). The anaerobic reaction is catalysed by *pyruvate decarboxylase*; this enzyme, like pyruvate dehydrogenase, requires thiamine pyrophosphate

Cellular biochemistry and physiology   **208**

(TPP) and manganous ions as cofactors. The reaction, in fact, resembles the dehydrogenase reaction in its initial stages, for an 'aldehyde–TPP complex' is formed, and this complex is then decarboxylated. However, unlike the pyruvate dehydrogenase reaction, the complex does not pass the aldehyde to lipoic acid, but releases it as acetaldehyde:

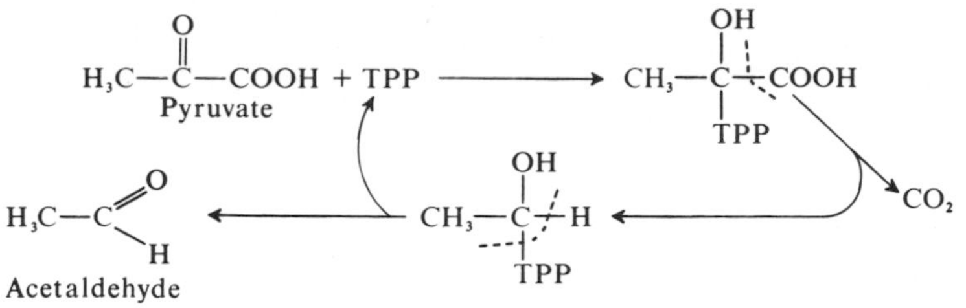

$$H_3C-\overset{\overset{O}{\|}}{C}-COOH + TPP \longrightarrow CH_3-\overset{\overset{OH}{|}}{\underset{\underset{TPP}{|}}{C}}-COOH$$

Pyruvate

$$H_3C-\overset{\overset{O}{\diagup}}{\underset{H}{C}} \longleftarrow CH_3-\overset{\overset{OH}{|}}{\underset{\underset{TPP}{|}}{C}}-H \longleftarrow \quad CO_2$$

Acetaldehyde

Since the reaction does not involve the reduction of lipoic acid, no ATP is formed at this stage (in contrast to the pyruvate dehydrogenase reaction). However, in the presence of alcohol dehydrogenase, the acetaldehyde which is produced acts as electron acceptor for the NADH produced by the phosphoglyceraldehyde dehydrogenase reaction. By maintaining the balance between NAD^+ and NADH, the reduction of acetaldehyde thus fulfils the same function as the reduction of pyruvate in anaerobic muscle—it keeps the glycolysis pathway functioning, and so allows continued anaerobic production of ATP.

Needless to say, the yeast has no use for the alcohol formed by fermentation. As in the case of lactate in muscle, it is a *waste product* and can actually be disadvantageous; as its concentration increases it acts as a poison, eventually preventing further growth.

9.9 Energy capture resulting from glucose catabolism

It is now possible to assess the energy captured by the system $ADP + P_i \rightarrow$ ATP as a result of glucose catabolism. The *anaerobic* catabolism of glucose to ethanol or to lactate will be considered first. In both cases, 4 moles of ATP result from the conversion of 1 mole of fructose-di-P to pyruvate. Two triose phosphate molecules arise from 1 molecule of fructose-di-P but 2 molecules of ATP are required for the synthesis of fructose-di-P from glucose. Thus when 1 mole of glucose is metabolized either to lactate in muscle or to ethanol in yeast, 4 minus 2 moles of ATP are synthesized.

When 1 molecule of glucose is metabolized under *aerobic* conditions, the number of ATP molecules produced is much larger than under anaerobic conditions. As each pyruvate molecule is converted to acetyl coenzyme A, 3 molecules of ATP are produced at the pyruvate dehydrogenase step. Moreover, when each acetyl coenzyme A molecule is subsequently metabolized, 12 molecules of ATP are generated. But two pyruvate molecules arise from one of glucose and so $(2 \times 3) + (2 \times 12)$, or 30 molecules, of ATP are produced as a result of the pyruvate oxidation. Since, in addition, 2 molecules of ATP are formed during the glycolytic production of pyruvate from 1 molecule of glucose, a total of 32 molecules of ATP result from each glucose molecule catabolized. In other words, 16 times more energy is captured in the aerobic than in the anaerobic degradation of a molecule of glucose. In addition, 2 molecules of NADH are produced in the cytosol (pp. 201, 204).

The cell is able partially to compensate for the lower yield of ATP which occurs when the oxygen supply is inadequate. This compensating device is known as the *Pasteur effect*. Under anaerobic conditions yeast catabolizes more glucose and produces more carbon dioxide than when oxygen is freely available. This implies that the rate of glycolysis is greater under anaerobic conditions. A somewhat similar effect (increased metabolism of carbohydrate) is observed in anaerobic muscle.

The mechanism which induces the Pasteur effect is still poorly understood. However, it is possible that the availability of phosphate in the cytosol may be one of the factors responsible. Under aerobic conditions, phosphate tends to accumulate in the mitochondria in the form of ATP (produced by oxidative phosphorylation). This tendency disappears when oxygen deprivation occurs. Inorganic phosphate and ADP may then move from the mitochondria to the cytosol where the glycolytic enzymes are situated. Since glycolysis requires the presence of inorganic phosphate and ADP, this movement may increase its rate, i.e., under normal aerobic conditions, a relative shortage of inorganic phosphate in the cytosol may act as a brake on the process of glycolysis. This may therefore be another example of the way biochemical reactions can be regulated or controlled by the availability in different parts of the cell of the various metabolites and cofactors upon which they are dependent.

10

Glycogenesis and gluconeogenesis

The present chapter describes a number of processes which are related to those discussed in chapter 9; glycogenesis is the synthesis of glycogen from glucose, and gluconeogenesis is the formation of glucose from *non-carbohydrate* precursors such as oxaloacetate and lactate. These conversions are in several important respects different from the reactions of the glycolytic pathway operating in reverse.

The earlier parts of this chapter describe the synthesis of starch, glycogen, cellulose, and two typical disaccharides. All of these are formed from *glucose-1-P*; di- and polysaccharides are not derived directly from free hexoses. Next, the formation of monosaccharides from simpler non-carbohydrate precursors is considered commencing with the synthesis and further reactions of *oxaloacetic acid*, for in many instances (Fig. 10.1) gluconeogenesis takes place by the following route:

precursor → oxaloacetate → glucose-1-P → polysaccharide

In animals, glycogen is typically synthesized from the three principal monosaccharides formed when they digest their food. These are D-*glucose*, D-*fructose*, and D-*galactose*. In addition, certain amino acids are important precursors of glycogen. In all higher animals, for example, the carbon skeletons of most amino acids can be catabolized in the liver and kidneys to yield *oxaloacetate, α-ketoglutarate, fumarate, pyruvate*, or *propionate*. Each of these is readily converted to glucose or glycogen. The process whereby carbohydrates are formed from precursors of non-carbohydrate *origin* may involve not only recently absorbed amino acids, but also, under conditions of starvation, the amino acid residues of the animal's own tissue proteins.

Lactate is a third major precursor of glycogen in animals. Strenuous muscular exercise results in the temporary accumulation of lactate in the

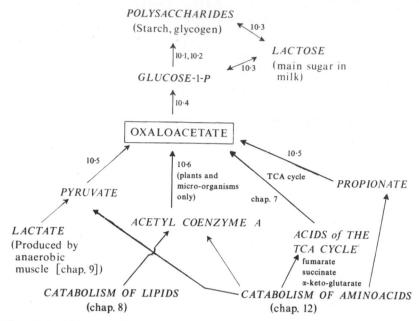

POLYSACCHARIDES
(Starch, glycogen)

10·3

LACTOSE
(main sugar in
milk)

10·3

GLUCOSE-1-P

10·1, 10·2

10·4

OXALOACETATE

10·5

10·6
(plants and
micro-organisms
only)

TCA cycle

chap. 7

10·5

PROPIONATE

PYRUVATE

ACETYL COENZYME A

ACIDS of THE
TCA CYCLE
fumarate
succinate
α-keto-glutarate

LACTATE
(Produced by
anaerobic
muscle [chap. 9])

CATABOLISM OF LIPIDS
(chap. 8)

CATABOLISM OF AMINOACIDS
(chap. 12)

Fig. 10.1   The overall pattern of polysaccharide synthesis from various precursors.   (Numbers refer to the sections where the steps are indicated)

muscles and the bloodstream (chapter 9).   During the period of recovery from exercise, the concentration of lactate in the animal's tissues is found to decrease.   This is largely occasioned by the uptake of the lactate by the liver, where, in the presence of lactate dehydrogenase it is reoxidized to pyruvate.   The pyruvate is then converted to glycogen by a series of reactions described later in the chapter, and thus the animal's supply of fuel is conserved.   The relationship between these reactions leading to conservation of carbohydrates, and those responsible for carbohydrate breakdown, are summarized in Fig. 10.2.

Plants and micro-organisms are even more versatile in that they can build up storage carbohydrates from a wide range of simple precursors. The major carbohydrate produced in photosynthesis in the green plant is D-fructose-6-P (chapter 11) and this is readily converted to starch and other compounds.   Many micro-organisms are able to grow on any one of the major food materials—provided sufficient trace elements and, in some cases, appropriate vitamins—are available.   This implies that such organisms possess the ability to interconvert lipids, carbohydrates, and

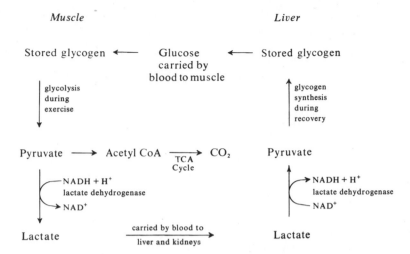

Fig. 10.2   The relationship between glycogen breakdown and its resynthesis during muscular exercise and the recovery phase

proteins. Animals differ from plants and micro-organisms in being *unable to convert long chain fatty acids* or the acetyl part of acetyl coenzyme A to carbohydrate and this considerably limits the range of their biochemical activities (section 10.6).

10.1   Polysaccharide biosynthesis from glucose-1-P

Before considering gluconeogenesis from compounds such as pyruvate and other simple organic acids it is necessary to establish the mechanism of the final stages of polysaccharide formation from hexose derivatives such as glucose-1-P.

When starch or glycogen undergoes glycolysis, the first reaction, catalysed by phosphorylase, involves the removal of glucose residues situated at the ends of the branches of polysaccharide molecules.   In this reaction, inorganic phosphate is incorporated in such a way that the glucose residue is released as glucose-1-P (p. 198).

$$[\text{Glucose}]_n + \text{P}_i \xrightarrow{\text{phosphorylase}} \text{glucose-1-P} + [\text{Glucose}]_{n-1}$$

| starch or glycogen | starch or glycogen with one less unit of glucose |

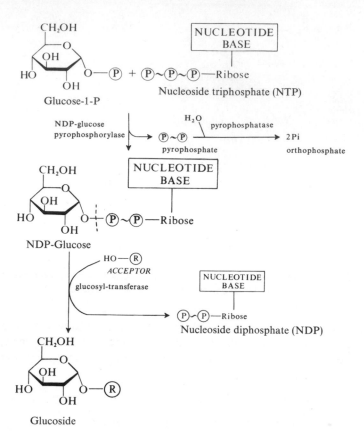

Fig. 10.3   The role of nucleoside diphosphate glucose in the synthesis of a glucoside, e.g., synthesis of starch or glycogen

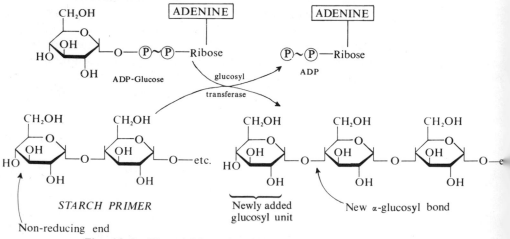

Fig. 10.4   The addition of a glucose unit to a starch molecule

It is unlikely that significant quantities of polysaccharide are synthesized from glucose-1-P by the reverse of the phosphorylase reaction. The position of the equilibrium, for concentrations of the reacting substances present under *biological* conditions, lies very much in favour of glucose-1-P formation. Net overall *synthesis* occurs, in fact, by a quite different mechanism and so does not involve the participation of phosphorylase.

An important new type of compound must now be introduced. In the synthesis of polysaccharides constructed from glucose units, a compound takes part which comprises glucose linked to a nucleoside diphosphate. These compounds—for there are several—can be represented as NDP-glucose, where N stands for nucleoside and will in due course be replaced by a letter representing the particular nucleotide base (p. 81) which is actually present. Such a compound is formed by the reaction of the corresponding nucleoside triphosphate (NTP) with glucose-1-P (Fig. 10.3). The reaction is freely reversible but, in practice, it favours the formation of NDP-glucose because *the second product*, inorganic pyrophosphate, is removed by hydrolysis to orthophosphate, the enzyme responsible being pyrophosphatase. Finally, the glucose is transferred from NDP-glucose to a suitable acceptor. The result is that a glucose residue is added to the acceptor and the nucleoside diphosphate (NDP) is released. This may subsequently undergo rephosphorylation to form the triphosphate (NTP) which is then available to repeat the process. Some of these reactions should be compared with those by means of which lecithin is synthesized (p. 192).

Figure 10.4 shows the final reaction whereby a glucose residue is transferred to a *starch* 'primer' molecule in starch synthesis. In this case the enzyme involved, ADP-glucose : starch *glucosyl transferase*, is specific for *adenosine* diphosphate glucose (ADP-glucose), which is formed from ATP and glucose-1-P via the reactions shown in Fig. 10.3. *Glycogen* synthesis in animals and fungi is very similar except that the nucleoside involved is *uridine*; uridine triphosphate (UTP) reacts with glucose-1-P, the UDP-glucose so formed eventually transferring the glucose to a glycogen 'primer' molecule. Attachment occurs, as in the case of starch, to carbon atom number 4 of the terminal glucose unit of a side-chain; the other product is UDP.

$$\text{UTP} + \text{glucose-1-P} \rightleftharpoons \text{UDP—glucose} + \text{pyrophosphate}$$

$$\text{pyrophosphate} + H_2O \longrightarrow 2 \text{ orthophosphate}$$

$$\text{UDP—glucose} + \underset{\substack{\text{glycogen 'primer'}\\ \text{(a partly-built}\\ \text{glycogen molecule)}}}{[\text{Glucose}]_n} \longrightarrow \text{UDP} + \underset{\substack{\text{glycogen with}\\ \text{additional}\\ \text{glucose unit}}}{[\text{Glucose}]_{n+1}}$$

Note that, for both starch and glycogen, attachment occurs to a smaller pre-existing molecule or 'primer'; as the molecule becomes larger it starts to branch (p. 217), and then parts of it break away to form new 'primer' molecules.

Cellulose is present in plant cell walls and its synthesis resembles those of starch and glycogen except that the nucleoside of the NDP-glucose is in this case *guanosine*. Moreover, the final glucosyl transferase reaction results in the formation of β-glucoside linkages, for unlike starch and glycogen, cellulose comprises β-glucoside residues.

Most poly- and disaccharides are synthesized from a nucleoside diphosphate derivative of the appropriate monosaccharide, the latter being transferred to a suitable acceptor. Although the nucleotide base, the monosaccharide unit being transferred, and the unit which accepts that monosaccharide, can all vary from one system to another, the overall pattern of the reaction is similar. Sucrose, for example, is formed in leaves from UDP-glucose, fructose-6-P acting as acceptor (Fig. 10.5).

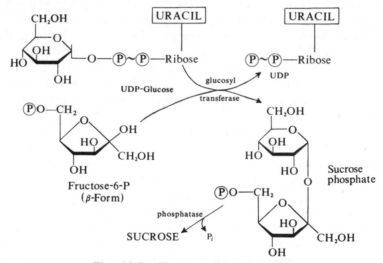

Fig. 10.5 The synthesis of sucrose

The product of the reaction, sucrose phosphate, is finally dephosphorylated in a reaction catalysed by a phosphatase.

In leaves, the synthesis of starch, sucrose, and cellulose may at certain times be going on simultaneously, but the NDP-glucose compound is different in each case (ADP-glucose for starch, UDP-glucose for sucrose, and GDP-glucose for cellulose). Since the glucosyl transferases are probably specific for the different nucleosides, it is not unreasonable to suppose that the various bases of the nucleosides act as a 'code' determining which carbohydrate is to be formed. If this is indeed the case an interesting analogy exists between polysaccharide and protein biosynthesis (chapter 14).

10.2 Branching enzyme

If the starch grains of plants are supplied with glucose-1-P and ATP they are able to elongate existing starch molecules by the repeated addition of glucose units, each step of the elongation involving the reactions shown in Fig. 10.4. This is probably the way in which amylose, the unbranched form of starch, is formed. Most plant starches, however, contain a branched chain form, amylopectin (p. 78), as well as amylose.

The branching is produced by an enzyme which acts upon a branch of amylopectin which has previously elongated, without further branching, until it contains some 42 glucose units. The enzyme hydrolyses an α-1,4-glucoside bond and transfers the shorter section of the glucose chain to another part of the amylopectin molecule, attaching it to a glucose residue so as to form a 1,6-glucoside bond (Fig. 10.6). The enzyme, sometimes known as Q-enzyme, is a glucosyl transferase, since it transfers a section comprising glucoside units from one part of the amylopectin molecule to another. The extent of branching and the number of glucose residues between branches are both determined by the enzyme.

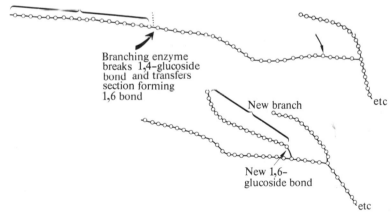

Branching enzyme
breaks 1,4-glucoside
bond and transfers
section forming
1,6 bond

New branch

etc

New 1,6-
glucoside bond

etc

Fig. 10.6 The branching of amylopectin by 'branching enzyme'. (The circles represent glucose units)

Glycogenesis and gluconeogenesis

Glycogen is a more highly branched molecule than amylopectin but here again the branching is achieved by a glucosyl transferase which removes a straight chain section of glucose units by breaking a 1,4α-glucoside linkage, and then attaches the sections so removed to another part of the glycogen molecule by a 1,6-glucoside linkage. Thus, both when glycogen is synthesized in animal tissues and when starch is made in plants, the glucosyl transferases which catalyse the formation of the NDP-glucose compounds and the branching enzymes co-operate in the task of forming these important polysaccharides.

10.3 Galactose and lactose metabolism

When certain micro-organisms such as yeasts are grown in a medium containing galactose, they are able to absorb this sugar and convert it to glucose and glycogen. In mammals, too, galactose is readily converted to glucose or to lactose.

In both yeast and liver cells galactose metabolism usually commences with phosphorylation, the reaction being catalysed by *galactokinase*. This enzyme transfers a phosphate group from ATP to the 'carbonyl' hydroxyl group of galactose, with the result that galactose-1-P is formed:

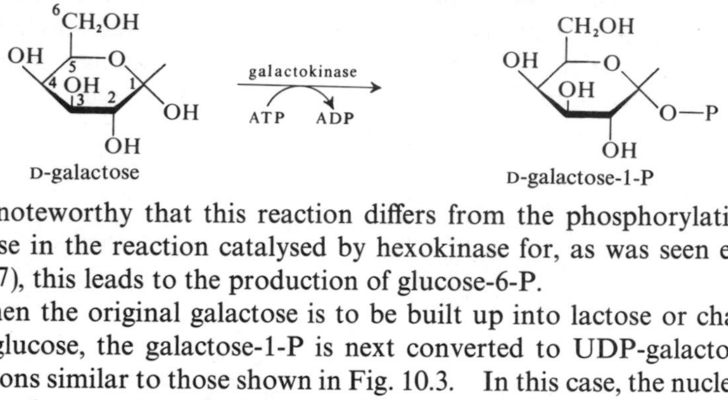

D-galactose   galactokinase   D-galactose-1-P
ATP → ADP

It is noteworthy that this reaction differs from the phosphorylation of glucose in the reaction catalysed by hexokinase for, as was seen earlier (p. 197), this leads to the production of glucose-6-P.

When the original galactose is to be built up into lactose or changed into glucose, the galactose-1-P is next converted to UDP-galactose by reactions similar to those shown in Fig. 10.3. In this case, the nucleoside triphosphate contains the base *uracil:*

$$\text{UTP} + \text{galactose-1-P} \rightleftharpoons \text{UDP-galactose} + \text{PP}$$

This is a necessary preliminary for the stereochemistry of galactose to be altered so that it is eventually converted to *glucose.* Once in this 'active' form, an enzyme called *UDP-galactose-4-epimerase* catalyses a rearrangement of the groups around carbon atom number 4 of the galactose ring, producing UDP-glucose (Fig. 10.7a). Such a rearrangement is called *epimerization*; it will be seen from the figure that a hydroxyl group written 'up' before epimerization occurs is written 'down' afterwards—i.e., the mirror image arrangement of the groups around this *one particular* carbon atom is somehow adopted.

The epimerase has NAD^+ bound to it, present in equimolecular amount to the protein, and one possible mechanism of epimerization is that it involves oxidation at carbon atom number 4 (with the formation of a

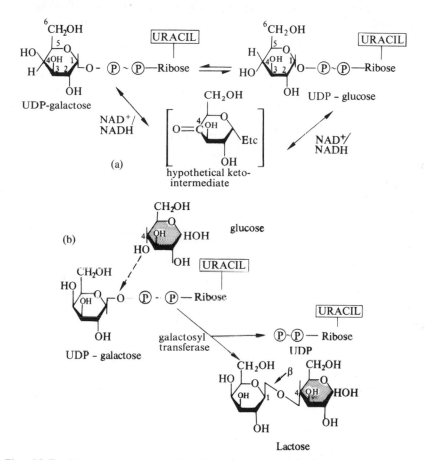

Fig. 10.7   Two important reactions in galactose metabolism
      (a) The interconversion of UDP-galactose and UDP-glucose, catalysed by UDP-galactose-4-epimerase
      (b) Synthesis of lactose from glucose and UDP-galactose

carbonyl group) followed by reduction of this *unsaturated, and therefore optically inactive*, group.

Operating in the opposite direction to that just described, UDP-galactose-4-epimerase is important in the mammary gland since it participates in the synthesis of the lactose present in milk. The gland in lactating animals absorbs *glucose* from the blood and converts it to glucose-1-P. This involves two reactions, both of which have been encountered already, namely, the phosphorylation of glucose to glucose-6-P and the conversion of glucose-6-P to glucose-1-P. The first is catalysed by hexokinase and the second by phosphoglucomutase. Glucose-1-P is then converted to UDP-glucose (section 10.1) which undergoes epimerization to UDP-galactose.

Lactose is finally formed by a reaction similar to that described for the synthesis of sucrose phosphate in plants (Fig. 10.5), for the UDP-galactose reacts with glucose to form the β-galactoside, lactose, the enzyme responsible being galactosyl transferase (Fig. 10.7b). UDP is the other product. Lactose plays an important role in infant nutrition for it is the major carbohydrate of milk. Dietary lactose is readily hydrolysed to galactose and glucose by the enzyme *lactose* present in the intestine.

10.4 Origin of oxaloacetate and its conversion to glucose-1-P

It was mentioned at the beginning of this chapter that plants and certain micro-organisms can convert most, and perhaps all, amino acids and fatty acids to carbohydrates. The ability to do this is more limited in animals. *Glucogenic* compounds are substances which can be converted, in animals, to *glucose* and hence to starches or glycogen.

Oxaloacetate plays a central role in metabolism. Besides being a member of the TCA cycle (chapter 7), it is an intermediate compound in several pathways which lead eventually to carbohydrate synthesis from amino acids and fatty acids. From this an important generalization emerges—any compound which can be converted by an organism to oxaloacetate is glucogenic in that organism.

Oxaloacetate is converted to *phosphoenol pyruvate* by phosphorylation and decarboxylation. Phosphoenol pyruvate is, as was seen in chapter 9, an intermediate in the glycolytic pathway and it is converted to glucose-1-P by reactions which are, with one notable exception, the *reverse* of those which occur in glycolysis. The glucose-1-P so produced can then serve as a precursor of either di- or polysaccharides by the reactions described in section 10.1.

Phosphoenol pyruvate is produced from oxaloacetate in a reaction catalysed by *phosphoenol pyruvate carboxykinase.* As the word 'kinase' suggests, the enzyme transfers a high energy phosphate group—in the present case from guanosine triphosphate (GTP)—to the oxaloacetate; at the same time it catalyses decarboxylation with the release of carbon dioxide:

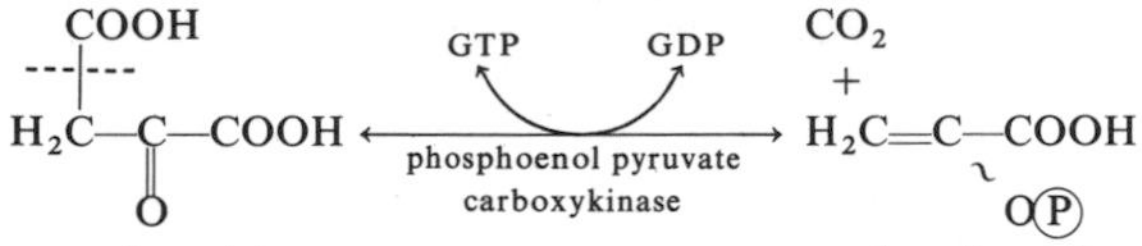

$$
\underset{\text{oxaloacetate}}{
\begin{array}{c}
\text{COOH} \\
\text{---}|\text{---} \\
\text{H}_2\text{C}\text{---}\underset{\underset{\text{O}}{\|}}{\text{C}}\text{---COOH}
\end{array}
}
\quad
\underset{\underset{\text{carboxykinase}}{\text{phosphoenol pyruvate}}}{
\overset{\text{GTP} \qquad \text{GDP}}{\rightleftharpoons}
}
\quad
\underset{\underset{\text{pyruvate}}{\text{phosphoenol}}}{
\begin{array}{c}
\text{CO}_2 \\
+ \\
\text{H}_2\text{C}\!=\!\text{C}\text{---COOH} \\
\overset{\backsim}{\text{O}\text{\textcircled{P}}}
\end{array}
}
$$

The enzyme is, in fact, named after the reverse reaction in which phosphoenol pyruvate is carboxylated and dephosphorylated to form oxaloacetate, GTP being simultaneously formed from GDP.

The phosphoenol pyruvate produced from oxaloacetate is next converted to glyceraldehyde-3-P by a series of reactions identical to those occurring in stages 2 and 3 of glycolysis (but operating in the reverse direction). Indeed, the enzymes involved are the same as those catalysing the glycolytic reactions. Triose phosphate isomerase ensures that sufficient dihydroxyacetone phosphate is produced from glyceraldehyde-3-P to enable aldolase to convert the two triose phosphate molecules to fructose-1,6-di-P. As was pointed out previously (p. 201), the position of equilibrium of the aldolase reaction favours hexose synthesis.

We now come to one major difference between the pathways of gluconeogenesis and glycolysis. If glycolysis were simply reversed in the conversion of fructose-1,6-di-P to fructose-1-P, the following reaction would be somewhat endergonic from left to right and hence the equilibrium mixture would contain little fructose-6-P:

$$\text{Fructose-1,6-di-P} + \text{ADP} \underset{\text{phosphofructokinase}}{\rightleftharpoons} \text{fructose-6-P} + \text{ATP}$$

$$(+\Delta G^\circ = 3\cdot0 \text{ kcal per mole})$$

This reaction is, however, by-passed by an exergonic and therefore, an energetically more favourable process. A phosphatase catalyses the hydrolytic removal of a phosphate group from the fructose-di-P, producing inorganic orthophosphate and fructose-6-P.

$$\text{Fructose-1,6-di-P} \overset{\text{fructose di-P phosphatase}}{\rightleftharpoons} \text{fructose-6-P} + \text{P}_i$$

$$(-\Delta G^\circ = 4\cdot0 \text{ kcal per mole})$$

Glycogenesis and gluconeogenesis

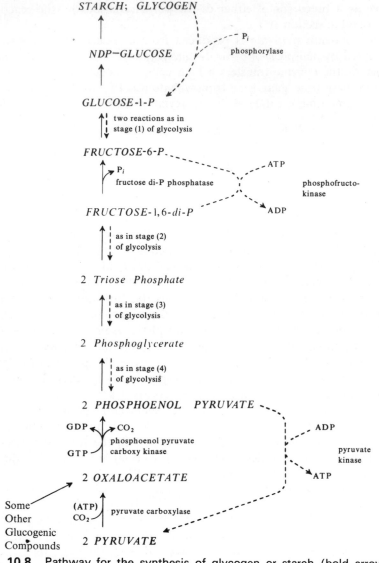

STARCH; GLYCOGEN

NDP–GLUCOSE

P_i
phosphorylase

GLUCOSE-1-*P*

two reactions as in
stage (1) of glycolysis

FRUCTOSE-6-*P*

P_i
fructose di-P phosphatase

ATP

ADP

phosphofructo-
kinase

FRUCTOSE-1,6-*di*-*P*

as in stage (2)
of glycolysis

2 *Triose Phosphate*

as in stage (3)
of glycolysis

2 *Phosphoglycerate*

as in stage (4)
of glycolysis

2 *PHOSPHOENOL PYRUVATE*

GDP CO_2

phosphoenol pyruvate
carboxy kinase

GTP

ADP

ATP

pyruvate
kinase

2 *OXALOACETATE*

Some
Other
Glucogenic
Compounds

(ATP)
CO_2

pyruvate carboxylase

2 *PYRUVATE*

Fig. 10.8 Pathway for the synthesis of glycogen or starch (bold arrows)
contrasted with the glycolytic pathway (broken arrows)

Lastly, the fructose-6-P is converted to glucose-6-P and eventually to glucose-1-P by reactions which are exactly the reverse of those described for stage 1 of glycolysis. Figure 10.8 summarizes the reactions leading to synthesis of glycogen or starch. They should be compared and contrasted with those of the glycolytic pathway (Fig. 9.2). Note that when the glucogenic compound is pyruvate three major differences exist:

1. Oxaloacetate is an intermediate in phosphoenol pyruvate formation.
2. A phosphatase converts fructose-1,6-di-P to fructose-6-P.
3. An NDP-glucose is finally involved in the synthesis of glycogen or starch.

10.5 Synthesis of oxaloacetate from amino acids, pyruvate, and propionate

Many amino acids are *glucogenic* because when the amino-group is removed, a carbon skeleton remains which readily gives rise to oxalo-acetate or one of the other intermediates of the TCA cycle. L-Aspartic acid is the most obvious example, for, on removal of the amino group (p. 269), the nitrogen-free compound so formed is *oxaloacetate* itself. However, many other amino acids give rise to oxaloacetate indirectly— some, for example, give α-*ketoglutarate* and others give *fumarate*, and both of these are oxidized by the enzymes of the TCA cycle to yield oxaloacetate.

In animals, *pyruvate* can arise from lactic acid during the period of recovery from exercise (Fig. 10.2) and it is also a product of the catabolism of certain amino acids (alanine, serine, cysteine—chapter 13). It can be converted to oxaloacetate by a reaction which involves carbon dioxide 'fixation'. The process closely resembles that by which malonyl coenzyme A is formed from acetyl coenzyme A in the course of fatty acid biosynthesis (p. 186). As in that reaction, carbon dioxide is first activated by forming a derivative of *biotin*, ATP being also essential. The carbon dioxide is then transferred from the 'active carbon dioxide' to pyruvate, so producing oxaloacetate:

$$
\overset{*}{C}O_2 + ATP \quad \searrow \quad \text{Biotin} \quad \nwarrow \quad
\begin{array}{l}
\overset{*}{C}OOH \\
CH_2 - C - COOH \\
\quad\quad\; \| \\
\quad\quad\; O \quad \text{oxaloacetate}
\end{array}
$$

$$
P_i + ADP \quad \nearrow \quad \text{Biotin} \sim \overset{*}{C}O_2 \quad \nearrow \quad
\begin{array}{l}
H \\
CH_2 - C - COOH \\
\quad\quad\; \| \\
\quad\quad\; O \quad \text{pyruvate}
\end{array}
$$

'active carbon dioxide'

The whole reaction is catalysed by the enzyme *pyruvate carboxylase*. This enzyme should not be confused with pyruvate decarboxylase which, in yeast, converts pyruvate to acetaldehyde (p. 208), nor with the complex of enzymes called pyruvate dehydrogenase, which is responsible for the oxidative decarboxylation of pyruvate (p. 205).

At this point, it may well be asked why pyruvate, arising, perhaps, from the lactate formed during vigorous exercise, should pass by this circuitous route (pyruvate → carboxylation to oxaloacetate → decarboxylation of oxaloacetate) to phosphoenol pyruvate, rather than undergoing direct phosphorylation. As is so often the case, energetic aspects of the reaction must be taken into consideration, for it will be recalled that when *pyruvate kinase* catalyses the conversion of phosphoenol pyruvate to pyruvic acid, the reaction is highly exergonic (p. 133):

$$\text{Phosphoenol pyruvate} + \text{ADP} \rightarrow \text{pyruvate} + \text{ATP};$$
$$(-\Delta G^\circ = 5 \cdot 7 \text{ kcal per mole})$$

The reverse reaction—the phosphorylation of pyruvate by ATP—is therefore correspondingly endergonic. The cell by-passes this unfavourable energy barrier by the two exergonic reactions described above:

$$\text{Pyruvate} + \text{CO}_2 + \text{ATP} \underset{\text{(biotin)}}{\overset{\overset{\text{pyruvate}}{\text{carboxylase}}}{\rightleftarrows}} \text{oxaloacetate} + \text{ADP} + \text{P}_i$$

$$(-\Delta G^\circ = 1 \cdot 08 \text{ kcal per mole})$$

$$\text{Oxaloacetate} + \text{GTP} \underset{\text{carboxykinase}}{\overset{\overset{\text{phosphoenol}}{\text{pyruvate}}}{\rightleftarrows}} \text{phosphoenol pyruvate} + \text{GDP} + \text{CO}_2$$

$$(-\Delta G^\circ = 1 \cdot 34 \text{ kcal per mole})$$

$$[\text{ATP} + \text{GDP} \rightleftharpoons \text{ADP} + \text{GTP}]$$

Note that in the single-reaction 'direct' route from phosphoenol pyruvate to pyruvate only one ATP is produced, whereas in the 'indirect' reverse direction the energy of two high energy phosphate groups is involved. The energetic principles we have just considered are applicable in more or less modified form to other instances where anabolic and catabolic routes differ, and for this reason they are very important.

Propionate can arise from the catabolism of several aminoacids, including methionine, threonine, and isoleucine. In ruminants such as cattle and sheep, propionate is produced by the gastric micro-organisms. These digest cellulose and other components of fodder; propionate, a major product, is absorbed by the alimentary tract and is an important precursor of carbohydrates in these animals.

For propionic acid to be converted to oxaloacetate, free propionic acid must first be converted to the active form, propionyl coenzyme A. The reactions concerned are similar to those occurring when a free fatty acid is 'primed' prior to its catabolism (p. 179). The next step is similar to the reaction in which pyruvate is carboxylated. It involves the catalytic action of *propionyl coenzyme A carboxylase*, working in conjunction with biotin (which acts as a carrier of 'active' carbon dioxide). Carbon dioxide is thus transferred to the propionyl coenzyme A, so producing methyl malonyl coenzyme A (Fig. 10.9). The last-named substance then undergoes a remarkable isomerization to succinyl coenzyme A in a reaction catalysed by an enzyme whose co-factor is vitamin B_{12}; the enzyme is called *methyl malonyl coenzyme A mutase*. The succinyl coenzyme A so formed is then converted to oxaloacetate by the machinery of the TCA cycle in the manner described in chapter 7.

10.6   The synthesis of oxaloacetate from acetyl coenzyme A

In *plants and many micro-organisms* fatty acids derived from glycerides or phospholipids can act as precursors of oxaloacetate and are therefore potentially glucogenic. The fatty acids are first broken down to acetyl coenzyme A by the pathway described in chapter 8. The acetyl coenzyme A is then fed into a cycle known as *the glyoxylate cycle*, the enzymes of which occur in the mitochondria of plants and in the equivalent structures of bacteria. The cycle resembles the TCA cycle (chapter 7) in many ways except that, instead of the two carbon atoms of acetyl coenzyme A being oxidized to carbon dioxide, reactions leading to a *net synthesis of oxaloacetate* take place. Net synthesis of oxaloacetate from acetyl coenzyme A is impossible in the TCA cycle, since for each molecule of acetyl coenzyme A fed into the cycle, two carbon atoms are always lost as carbon dioxide.

The first reaction of the glyoxylate cycle, as in the TCA cycle, is the 'condensation' of acetyl coenzyme A and oxaloacetate to form citric acid. This is catalysed by citrate synthase. The citric acid is now converted to isocitric acid by aconitate hydratase. The next step is quite different from the TCA cycle for, instead of being oxidized, the isocitrate undergoes cleavage. The enzyme responsible is called *isocitrate lyase* and the products are succinate and *glyoxylic acid* (Fig. 10.10). The last-named substance gives the cycle its name, for it does not occur in the TCA cycle.

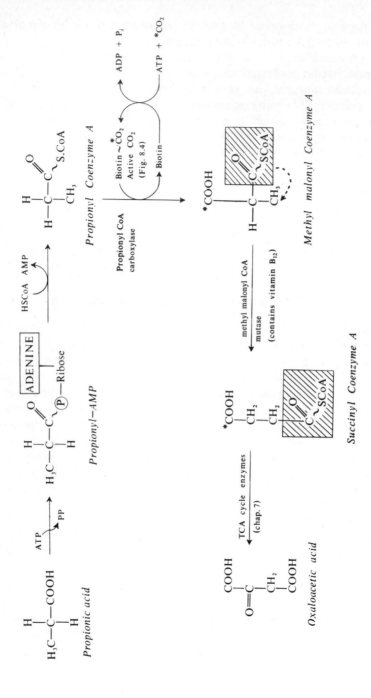

Fig. 10.9 Conversion of propionate to oxaloacetate

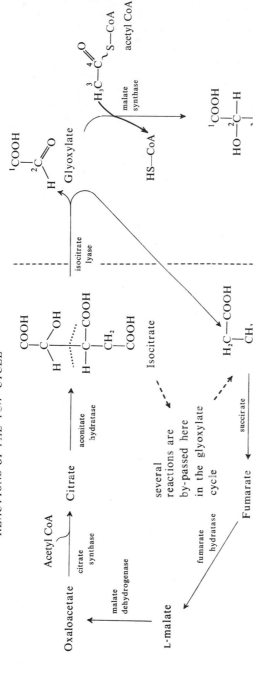

REACTIONS UNIQUE TO THE GLYOXLATE CYCLE

^1COOH
$^2C=O$
H
Glyoxylate

malate
synthase

H_3C-^4C
^{3} $\begin{matrix} O \\ \parallel \\ \end{matrix}$ S—CoA
acetyl CoA

HS—CoA

^1COOH
HO—^2C—H
$^3CH_2$
^4COOH
L-Malate

REACTIONS OF THE TCA CYCLE

Acetyl CoA

citrate
synthase

Oxaloacetate

malate
dehydrogenase

Citrate

COOH
H—C—OH
H

aconitate
hydratase

H—C—COOH
CH₂
COOH
Isocitrate

isocitrate
lyase

several
reactions are
by-passed here
in the glyoxylate
cycle

$H_2C—COOH$
CH₂
COOH
Succinate

succinate
dehydrogenase

malate dehydrogenase

L-malate

fumarate
hydratase

Fumarate

OXALOACETATE

Fig. 10.10 The glyoxylate cycle. (In this cycle, the overall reaction is 2 acetyl CoA → 1 oxaloacetate. Most of the reactions occur also in the TCA cycle (chapter 7). Two enzymes, *isocitrate lyase* and *malate synthase*, occur in the glyoxylate cycle but not in the TCA cycle)

Glycogenesis and gluconeogenesis

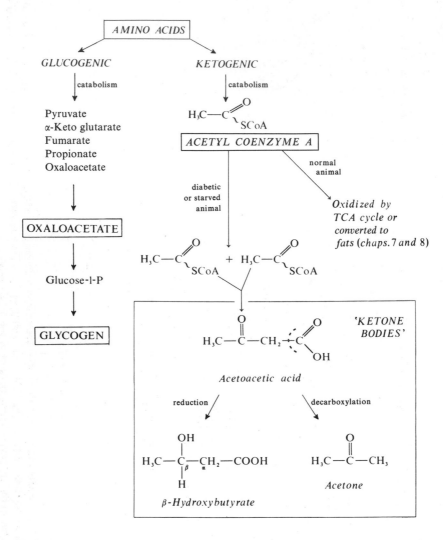

Fig. 10.11 The contrast between the catabolism of glycogenic and ketogenic amino acids

An enzyme called *malate synthase* now catalyses an aldol-type 'condensation' (p. 163) between a second molecule of acetyl coenzyme A and the glyoxylate, the product of the reaction being L-malate. Thus two molecules of acetyl coenzyme A and the original molecule of oxaloacetate are converted to one molecule each of succinate and L-malate. These are now both converted to oxaloacetate, as in the TCA cycle, by reactions catalysed by succinate dehydrogenase, fumarate hydratase, and malate dehydrogenase (only the latter being necessary, of course, for the L-malate). One of the oxaloacetate molecules effectively replaces the one used initially to form the citrate but the *overall effect* of the glyoxylate cycle is to convert two acetyl coenzyme A molecules to one molecule of oxaloacetate.

Animal tissues lack the two enzymes which are peculiar to the glyoxylate cycle (isocitrate lyase and malate synthase) and so they are *unable to catalyse a net conversion of acetyl coenzyme A to oxaloacetate*. This has an important consequence, for it means that substances such as fatty acids and certain amino acids (especially *leucine*), whose catabolism yields acetyl coenzyme A, cannot be converted to oxaloacetate. Consequently, such substances are *not glucogenic*. Normally acetyl coenzyme A resulting from their catabolism is oxidized to carbon dioxide by the TCA cycle. However, under certain abnormal circumstances—namely during severe starvation, or when there is an insufficiency of the hormone, insulin—the acetyl coenzyme A is converted to acetoacetate, β-hydroxybutyrate and acetone by reactions summarized in Fig. 10.11. These substances are usually known collectively as 'ketone bodies'. When these are formed in appreciable quantities in animal tissues, they accumulate in the blood and are excreted in the urine where they are easily detected by tests for ketones. Amino acids which, when fed to diabetic animals, produce ketone bodies are known as *ketogenic amino acids*, so distinguishing them from the glucogenic ones (chapter 13). Some amino acids are both *glucogenic and ketogenic* because their catabolism produces some substances capable of being converted to oxaloacetic acid (and hence to glucose and glycogen) but, in addition, acetyl coenzyme A is also formed. Tyrosine and phenylalanine are two important examples.

10.7 A general outline of carbohydrate metabolism in green plants

In the present and previous chapters several aspects of carbohydrate biochemistry in plants have been described and at this point it is

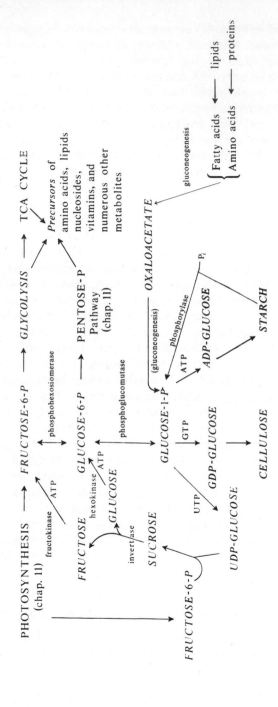

Fig. 10.12 Outline of carbohydrate metabolism in the green plant

convenient to indicate some of the connections between them as well as some other metabolic processes to be considered later.

In many plants the fructose-6-P produced by photosynthesis (chapter 11) is first converted to the highly soluble sugar, sucrose, which acts as a *temporary* carbohydrate store and is also frequently the major form in which carbohydrate is translocated from one part of the plant to another. Fructose-6-P is converted to sucrose by the reactions indicated in Fig. 10.5 but, for this to be possible, a proportion of it must first be made into glucose-1-P, the immediate precursor of UDP-glucose. Two enzymes, previously mentioned in chapter 9, phosphohexoisomerase and phosphoglucomutase, are responsible for the interconversion of fructose-6-P and glucose-1-P (Fig. 10.12).

Once sucrose has been formed in the leaf it is often transported (in the phloem) to other parts of the plant. Upon arrival, the sucrose is hydrolysed by invertase to glucose and fructose, each of which must be phosphorylated by appropriate kinases before undergoing further metabolism. The resulting fructose-6-P and glucose-6-P are easily interconverted, and, once formed, numerous other substances can be synthesized from them which can subsequently produce not only ATP but a number of other important compounds as well. These include acetyl coenzyme A which, in turn, is a precursor of fatty acids, and acids of the TCA cycle from which many amino acids for protein synthesis arise (chapters 12 and 13). Glucose-6-P is the initial reactant for a cycle known as the *pentose phosphate cycle*, a metabolic pathway which has numerous anabolic functions (chapter 11). Alternatively, the glucose-6-P may be converted to glucose-1-P from which cellulose or starch can be formed (section 10.1).

Clearly, the formation of fructose-6-P by photosynthesis initiates a large number of biochemical reactions throughout the plant. This is true also of the breakdown of storage compounds such as the food materials in germinating seeds, potato tubers, tap roots, corms, and bulbs. Stored carbohydrates are usually hydrolysed to their component hexose units which, after phosphorylation, are converted to sucrose or some other soluble sugar for transportation to the growing parts of the plant. The fatty acids of stored lipids are converted by the fatty acid oxidation spiral (chapter 8) to acetyl coenzyme A and this, like the breakdown products of amino acids, can be converted to oxaloacetate and hence to glucose-1-P. This is then able to initiate any of the numerous processes indicated in Fig. 10.12.

Glycogenesis and gluconeogenesis

10.8 Regulation of blood glucose levels in the mammal

Normally, the blood of fasting human beings contains about 80 mg glucose per 100 ml. If this level were not maintained the consequences could be serious, since the brain and other vital organs would be denied their principal source of energy. The blood glucose level is controlled by a complex system of hormones, a detailed description of which would take us beyond the scope of this account. Nevertheless, two hormones are particularly effective in bringing about a rise in the blood glucose concentration. They are *glucagon*, a hormone secreted by certain cells of the pancreas, and *adrenalin*, which is secreted by the medulla of the adrenal gland. They do this by activating phosphorylase in liver and muscle cells. Phosphorylase breaks down stored glycogen to form glucose-1-P, which is then converted to glucose-6-P by phosphoglucomutase. Glucose-6-phosphatase hydrolyses the glucose-6-P to form inorganic orthophosphate and glucose, the latter being released into the blood, thus increasing the glucose concentration there.

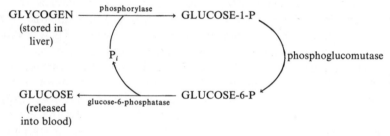

While glucagon and adrenalin have the effect of tending to raise the blood glucose levels by breaking down stored glycogen enzymically, another pancreatic hormone, *insulin* (p. 50), has the opposite effect. Absorption of glucose, after a meal containing it, causes the blood glucose level to increase to values which may be as high as 180 mg glucose per 100 ml. Insulin stimulates the uptake of glucose by many cells and enhances the activity of the enzymes responsible for glycogen synthesis. This results in an increased glycogen store and a consequent reduction of the blood glucose level.

Patients with the disease *diabetes mellitus* lack insulin so the level of glucose in their blood increases far more than normal after consuming glucose. The extremely high blood glucose concentration (it may be as high as 400 mg per 100 ml) results in loss of glucose in the urine. Moreover, the synthesis of glycogen is lower than normal. Untreated diabetic subjects thus have a reduced glycogen store and, between meals, they catabolize lipids and proteins to provide energy.   This is why the catabolism of lipids and protein is enhanced and why more acetyl coenzyme A is produced than the mitochondria can cope with.   Under these conditions ketone bodies are formed as a result of the breakdown of lipids and ketogenic amino acids (Fig. 10.11).

11
Photosynthesis and the pentose-phosphate cycle

The major pathway for the catabolism of carbohydrates in almost all organisms is glycolysis (chapter 9). There is, however, a second pathway present in most, if not all, organisms; it is known as the pentose-phosphate pathway, or sometimes as the hexose-monophosphate 'shunt'. A considerable part of this pathway is similar to many of the reactions of the carbon cycle of photosynthesis and it is tempting to imagine this alternative system of metabolism as being evolutionarily primitive. Like the photosynthetic machinery, its functions appear to be closely associated with anabolic processes. Perhaps for this reason it tends to be most highly developed in certain specialized cells which have particular anabolic functions.

11.1 The overall process of photosynthesis

Many of the earlier theories about the remarkable process which ultimately supports all life, the reduction of carbon dioxide to sugar by green plants and by certain photosynthetic bacteria, postulated various types of reactions which had little or no connection with any other facet of biochemistry. Recent advances, made possible by the introduction of radio-isotopes and chromatography, have however shown that photosynthesis fits into the same framework as other known biochemical processes. In particular, the reduction of carbon dioxide is now known to be brought about by the same two driving forces encountered in other anabolic processes—ATP and reducing power. It is the *origin* of these energy-yielding substances, and not their chemical nature, which is different in photosynthesis in that they are formed at the expense of the energy of sunlight.

The reactions of photosynthesis can conveniently be divided into two groups:

1. The conversion of light energy into chemical energy
2. The use of that chemical energy to reduce carbon dioxide.

The sum total of all the reactions of photosynthesis is often represented by the following equation

$$6CO_2 + 6H_2O + \text{light} \xrightarrow[\text{enzymes}]{\text{chlorophyll,}} C_6H_{12}O_6 + 6O_2$$

This overall equation gives no clue as to how photosynthesis actually works. But the advent of radioactive carbon (C^{14}) made it possible to label the carbon of carbon dioxide and so to investigate which cellular constituents became radioactive after plants had been exposed to labelled carbon dioxide for very short periods of time. Much of the work was carried out by Calvin and his team at Berkeley, California—the carbon cycle of photosynthesis is consequently sometimes known as the Calvin cycle. The principal test organisms were the algae *Chlorella* and *Scenedesmus*, although leaves of higher plants were also employed. When either genus of unicellular algae was employed, the suspension of $C^{14}O_2$-treated cells was subjected to exposure to light for a few seconds, and then plunged into boiling absolute alcohol to terminate the reaction. After concentrating by low-temperature evaporation and after clean-up procedures which included centrifugation and fat extraction, the remaining aqueous layer was investigated by paper chromatography in order to identify the labelled compounds produced.

Experiments of this type revealed much of what we know about the main carbon cycle of photosynthesis (at least in some plants, alternative pathways exist). It consists of some twelve separate enzyme-catalysed reactions, each of which can occur in darkness. Clearly, therefore, the contribution of light to these syntheses must be vicarious. In fact, as will be shown in section 11.3, the reactions of the cycle can be carried out individually in the laboratory, provided the appropriate enzyme and, where applicable, suitable cofactors and an energy supply, are present. In its simplest form, the cycle can be summarized in the way shown in Fig. 11.1 where C_3, C_4, C_5, etc., represent sugar phosphates containing three, four, five, etc., carbon atoms. The contribution of the light reaction is to provide energy to drive the cycle; ATP raises the energy of reactants by phosphorylation at two places in the cycle, and reducing power, in

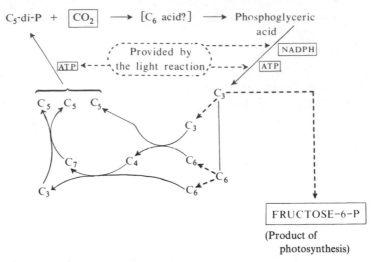

Fig. 11.1 The carbon cycle of photosynthesis in outline. (C_3, C_4, C_5, C_6, C_7, monophosphorylated sugars; C_5-di-P, ribulose diphosphate: square brackets indicate compounds entering the carbon cycle and (lower right) leaving it. The cycle morks out statistically, so stoichiometric relationships are not shown: the dotted arrows indicate that some part of the material C_5 or C_6 takes the pathway shown)

the form of NADPH, is involved in one step, namely, the reduction of phosphoglyceric acid to triose phosphate (C_3). The light reaction is the only part of the total photosynthetic process which may have any claim to mechanistic uniqueness.

11.2 The light reaction of photosynthesis

The 'light reaction' of photosynthesis undoubtedly involves the sequential operation of a very considerable number of enzymic reactions, and the vast majority of these are almost certainly 'ordinary' biochemical reactions in the sense that, once they have been provided with their substrates, they are in no way dependent on light. The precise nature of each step of the light reaction has not yet been fully elucidated either in green plants or in photosynthetic bacteria, and the subject is therefore only considered here in the most general way.

Photosynthesis takes place in the chloroplasts (p. 28). In higher plants, but not universally throughout the plant kingdom, the lipoprotein membranes which provide the ultrastructure of the chloroplast, and with which the chlorophyll is associated, come together in certain areas and separate again in others. This has the effect of producing regions of greater and lesser concentration of chlorophyll; the greener areas are called grana and the lighter, less membranous regions are the stroma (Fig. 11.2). Typically, the grana are cylindrical in shape, consisting of stacks of platelets, or lamellae. Isolated lamellae, when provided with suitable cofactors, are not only capable of trapping light energy but can carry out the whole carbon cycle of photosynthesis. The exact orientation of pigment molecules within or around the lipoprotein is unknown, and the nature of some of the molecules which donate and receive electrons in the presence of light remains uncertain. Pigment molecules are undoubtedly implicated as catalysts, but there may be other participants as well.

Electron micrographs which show a face-on view of a lamella often depict a mesh-like sub-structure. Individual sub-units are some 10 mμ thick and 16 mμ diameter, with a molecular weight of about two millions, half of which can be accounted for as protein. These units, which are not necessarily identical chemically, are termed *quantasomes* and are postulated to be the ultimate structural units responsible for the photon-trapping process which is the unique attribute of the chloroplast.

Chlorophyll **a** is almost universlly present in chloroplasts of plants in all phyla of the plant kingdom, but the accessory pigments which accompany it vary considerably. Slightly different chlorophylls, designated **b**, **c**, and **d** have been recognized in various green plants, as have several carotenoids and hydroxylated carotenoids. All the chlorophylls have a porphyrin-like structure with a central magnesium ion. When chloroplasts are ashed, 1 % of the residue is iron; magnesium, copper, and manganese are also present. In intact chloroplasts, the iron occurs, in part, as non-haem iron protein, but some is also complexed in haem groups of cytochromes **b**$_6$ and **f**. These cytochromes, which are different from those concerned with electron transport in mitochondria, take part in the transport of photo-activated electrons.

The light reaction of photosynthesis involves the participation of an electron transport system analogous to that of the mitochondrion, but the details of the electron carriers are less well understood. Nevertheless, a general picture has emerged in recent years and this is presented now in its simplest form. In the green plant, although perhaps not in photosynthetic bacteria, there are, in fact, *two* elegantly linked photochemical

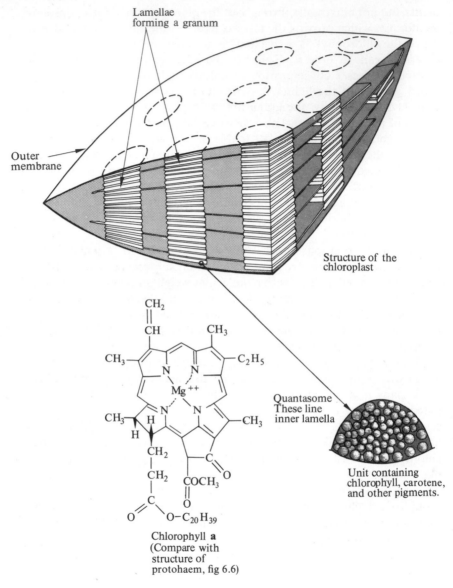

Lamellae
forming a granum

Outer
membrane

Structure of the
chloroplast

Quantasome
These line
inner lamella

Unit containing
chlorophyll, carotene,
and other pigments.

Chlorophyll **a**
(Compare with
structure of
protohaem, fig 6.6)

Fig. 11.2   Structure of the chloroplast and of chlorophyll

reactions. There is an energetic reason for this; 24 electrons are needed to reduce 6 molecules of carbon dioxide as the following equation shows:

$$6CO_2 + 24H^+ + 24e \rightarrow C_6H_{12}O_6 + 6H_2O$$

Consequently, the minimal quantity of energy needed per equivalent of reducing power to reduce 6 moles of carbon dioxide to 1 mole of sugar is

$$\frac{\Delta G^\circ \text{ for glucose formation}}{24} = \frac{691,000}{24} = 28 \cdot 8 \text{ kcal}$$

Light with a wavelength of 700 mμ is associated with 40·5 kcal per 6×10^{23} photons, and so, if only one photon energized one electron, the conversion of light energy into reducing energy would need to operate at an efficiency of 28,800/40,500, or 71%. This high level of efficiency is not achieved in cellular processes (compare, for example, the efficiency of energy capture during oxidative phosphorylation, p. 154), and, in somewhat different ways, two photons contribute energy to drive the carbon dioxide reduction process.

In the first light reaction (system I), which is most effective when illumination is with red light, chlorophyll **a** is activated by photons in such a way that an electron is raised to a high reducing potential. The energized electron is then able to reduce a non-haem iron protein, *ferredoxin*. Either directly, or via other electron carriers, the ferredoxin passes the electron to NADP which is thereby reduced to NADPH (Fig. 11.3). Ferredoxin is considered further on p. 265.

The *first* photochemical reaction therefore results in the loss of an electron from chlorophyll **a**. This must be replaced, and it is one of the functions of the *second* light reaction (system **II**) to do this. Light of a shorter wavelength, upon striking a second chloroplastic component, raises electrons to a high reducing potential. The electrons pass along an electron transport system, comprising carriers known as plastoquinone and plastocyanin, as well as the cytochromes b_6 and **f** mentioned earlier, and they eventually reach the oxidized form of chlorophyll **a** produced by the first light reaction. As the electrons move down the redox gradient, they raise ADP and inorganic phosphate to the higher energy level of ATP. The process is clearly analogous to the oxidative phosphorylation which occurs at the inner membrane of the mitochondrion. Here, however, the original energy is derived from light, and the process is known as *photosynthetic phosphorylation*.

The oxidized fraction from the second light reaction, which is the only redox fraction not yet accounted for, probably reacts with water and then

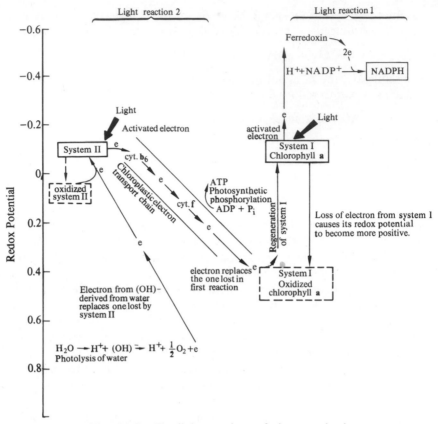

Fig. 11.3   The light reactions of photosynthesis

splits up to give the oxygen evolved in photosynthesis, the second chloro-plastic component mentioned being restored to its non-oxidized form (Fig. 11.3).   The outcome of the total light reaction is that water is used up and that oxygen and a reducing agent are formed.   The process is in some ways very similar to electrolysis, where light energy, instead of the energy of the electric current, acts as the driving force.   For this reason, the process has been termed *photolysis*.   Water ionizes automatically to produce hydrogen ions and hydroxyl ions:

$$H_2O \rightleftharpoons H^+ + OH^-$$

In electrolysis, the ions move under the influence of the current to produce hydrogen (a more reduced product than water) and oxygen (more oxidized):

$$4H^+ + 4OH^- \rightarrow 2H_2 + (O_2 + 2H_2O)$$

In photolysis, the *hydrogen* ions and electrons are harnessed to reduce $NADP^+$ to NADPH, while the actual movement of other electrons along an electron transport chain generates ATP. The *hydroxyl* ions, via unstable intermediate compounds, behave like the hydroxyl groups in electrolysis, the oxygen produced being merely a by-product of the photosynthetic reaction.

$$4H^+ + 4e + 2NADP^+ \rightleftharpoons 2NADPH$$
$$4OH^- \text{ minus } 4e \rightarrow 2H_2O + O_2$$

11.3   The photosynthetic carbon cycle

Unlike the light reaction, the mechanism whereby carbon dioxide is converted to sugars, and eventually to starch, fats, proteins, and nucleic acids, is for the most part well established. Only one reaction in step 1 of the sequence below is rather dubious. The carbon cycle can be divided into three steps in the following way:

  Step 1. The fixation of carbon dioxide, a process which converts a five carbon sugar, *ribulose-5-P*, to two molecules of the three carbon compound, 3-phosphoglyceric acid.
  Step 2. A series of reactions, essentially the same as some of those of glycolysis operating in the reverse direction, which converts the 3-phosphoglyceric acid to fructose-6-P.
  Step 3. A series of sugar phosphate transformations, similar to those of the pentose-phosphate pathway, regenerating ribulose-5-P from the fructose-6-P.

Step 1. Carbon dioxide fixation

The most convenient point at which to start considering the carbon cycle is with the reaction in which the keto-pentose phosphate, ribulose-5-P, reacts with ATP. The reaction is comparable to other kinase reactions, including that which leads from fructose-6-P to fructose-1,6-di-P (p. 197). The enzyme *phosphoribokinase* transfers a phosphate group from the high-energy donor, ATP, and places it upon carbon atom number one (C-1) of ribulose-5-P (Fig. 11.4). The reaction is entirely 'conventional', only the *origin* of the ATP possibly being different from usual, for, under good lighting conditions, the ATP in plant chloroplasts largely arises from photosynthetic phosphorylation and not from oxidative phosphorylation.

We come now to the reaction in the carbon cycle which results in the 'fixation' of carbon dioxide. 'Fixation', or the union of carbon dioxide with a pre-existing organic molecule, is, in itself, not unusual; carboxylation reactions dependent upon biotin as cofactor have already been referred to (p. 226), and many different carbon dioxide acceptors are known in both plants and animals. But in the particular case of photosynthesis it is the ribulose-1,5-di-P formed by the kinase reaction which acts as *the acceptor of the carbon dioxide*. The first identifiable product of this 'fixation' reaction is *3-phosphoglyceric acid*, but it is possible that the acid indicated in brackets in Fig. 11.4 is a transient intermediate.

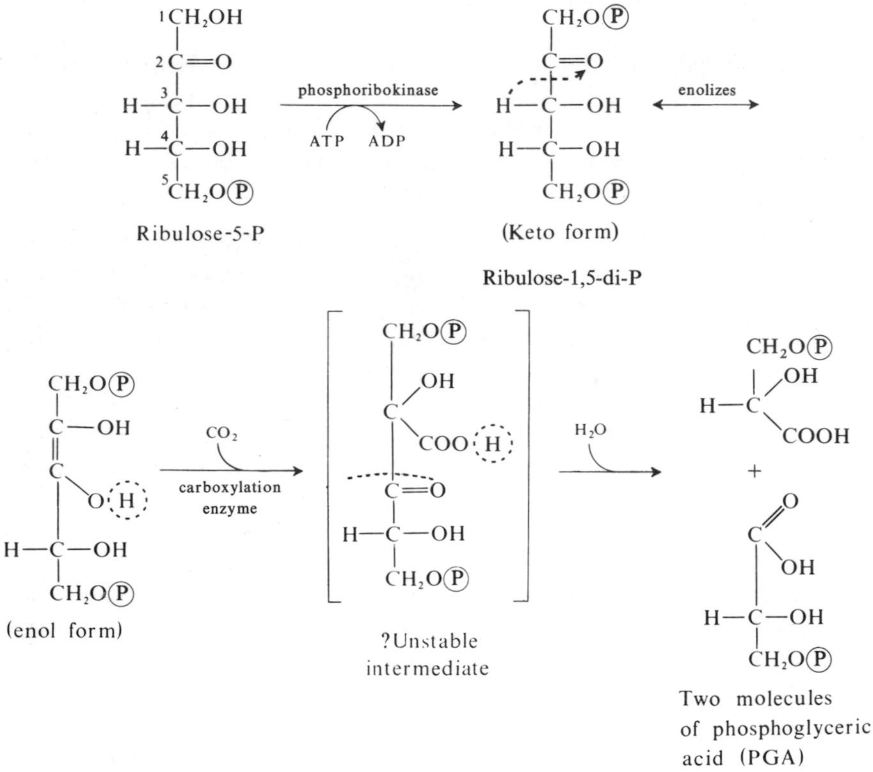

Fig. 11.4 Step 1 of the photosynthetic carbon cycle; the fixation of carbon dioxide

The reaction is not fully understood, and is catalysed by an enzyme, or enzyme complex, known simply as 'carboxylation enzyme'.

Step 2. Conversion of 3-phosphoglyceric acid to fructose-6-P

Each of the reactions in this step has been considered before, a fact which again illustrates a general simplifying feature of biochemistry. The cell shows great economy in that it often uses the same set of reactions for more than one purpose. Thus the 3-phosphoglyceric acid is converted to *fructose-1,6-di-P* by the reactions of stages 2 and 3 of glycolysis (Figs. 9.4 and 9.5). There are only two differences; the reaction proceeds in the reverse *direction* from that of glycolysis in this chloroplast-located reaction, and the chloroplast enzyme, phosphoglyceraldehyde dehydrogenase, uses NADP as cofactor instead of NAD. Since, in this instance, 1,3-diphosphoglyceric acid is reduced to glyceraldehyde-3-P, the reaction requires a supply of *reduced* NADP. NADPH is, however, a product of the light reaction and so the reduction is able to proceed very readily. The involvement of the $NADP^+/NADPH$ system rather than NAD is quite a common feature of *anabolic* as opposed to catabolic reaction sequences.

The fructose-1,6-di-P produced by the reactions mentioned above is dephosphorylated to form fructose-6-P, the initial reactant of step 3 of the carbon cycle. The dephosphorylation is catalysed by *fructose-di-P phosphatase*. This enzyme, and not phosphofructokinase, takes part in the reaction for the energetic reasons discussed in section 10.4.

The fructose-6-P has many possible fates including conversion to sucrose and other carbohydrates (Fig. 10.12), but much of it must be fed back into the cycle to regenerate ribulose-5-P so that step 1 of the cycle can take place again. This regeneration involves a series of inter-conversions of sugar phosphates and these together form step 3 of the carbon cycle.

Step 3. The inter-conversion of fructose-6-P and ribulose-5-P

All the reactions about to be considered are freely reversible, allowing fructose-6-P to be converted into ribulose-5-P, as occurs in photosynthesis, but equally allowing ribulose-5-P to be converted to fructose-6-P. The enzymes catalysing the reactions are water-soluble; they are present in the soluble fraction of the cell and in the aqueous phase of the chloroplasts.

It is best not to consider these reactions from a stoichiometric viewpoint; the direction of change and the quantity of sugar phosphate produced by

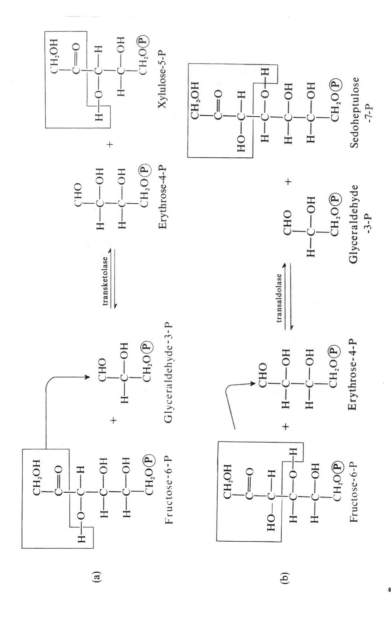

Fig. 11.5 Reactions catalysed by transketolase and transaldolase

any particular reaction is determined by environmental factors such as the nature and quantity of the sugar phosphate fed into the cycle, and by the cellular demand for individual sugar phosphates. Such factors result in more or less of the products of any one reaction being 'siphoned off' at any time. In other words, the best mental picture of this process is of a group of reactions feeding one another with reactants, or removing products, and therefore proceeding at rates determined by mass action considerations. The result is that the system settles down to a steady state (p. 129). The *statistical* outcome, in *photosynthesis*, is that the removal of ribulose-5-P by step 1 of photosynthesis draws the sugar transformation reactions in the direction of ribulose-5-P synthesis, whereas, in the pentose-phosphate oxidative pathway (p. 249), the reverse conditions drive the cycle in the direction of fructose-6-P synthesis.

In following the reactions about to be discussed, the reader is advised to refer to the general schemes shown in Fig. 11.1 and (in more detail) in Fig. 11.6. It should be observed that all the reactions involve sugar phosphates in which the phosphate group is attached to the terminal, highest-numbered, carbon atom. The main enzymes concerned, *transketolase* and *transaldolase*, dislodge respectively a two-carbon unit and a three carbon unit from a *ketone* sugar and place it on the carbonyl group of an *aldehyde* sugar.

Transketolase first removes a two-carbon fragment from fructose-6-P and attaches it to glyceraldehyde-3-P (Fig. 11.5). The reaction, like that already encountered for aldolase (p. 108), can be regarded as one in which addition occurs on to the carbonyl double bond of the glyceraldehyde-3-P. Figure 11.5a demonstrates how the reaction may be envisaged, although it does not imply any particular mechanism for the reaction *in vivo*. In the present case, the products of the reaction are erythrose-4-P and xylulose-5-P:

$$\overset{\overset{\textit{minus 2C}}{\frown}}{C_6} + C_3 \xrightarrow[]{\text{transketolase}} C_4 + C_5$$

Ketose	Aldose		Residue	Addition
			(an aldose)	product
				(a ketose)

The active centre of transketolase contains thiamine pyrophosphate (TPP) and it is to this that the two-carbon (2-C) fragment is temporarily attached.

In the presence of an *epimerase*, the xylulose-5-P is converted to another keto-pentose, ribulose-5-P; this involves interchanging the positions of the hydrogen atom and the hydroxyl group around one of the carbon atoms.

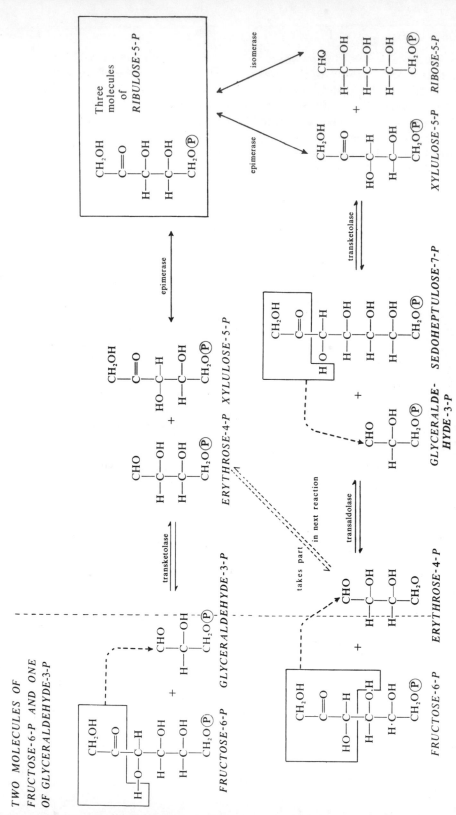

Fig. 11.6 The interconversion of fructose-6-P and ribulose-5-P. (The sequence of reactions from left to right represents step 3 of the carbon cycle of photosynthesis. In the reverse direction, it forms step 3 of the pentose-phosphate cycle)

Next, the erythrose residue from the transketolase reaction accepts a three-carbon fraction from fructose-6-P. This results in the formation of a triose phosphate and a sugar phosphate with a molecule containing seven carbon atoms. It is called sedoheptulose-7-P, and the enzyme is a *transaldolase*. It is easy to remember that the enzyme catalysing the transfer of a 3-C unit is transaldolase by comparison with the aldolase reaction (p. 108). The transaldolase reaction is represented in Fig. 11.5b and can be summarized by the equation:

$$\text{C}_6 \;+\; \text{C}_4 \;\xrightleftharpoons{\text{transaldolase}}\; \text{C}_3 \;+\; \text{C}_7$$

Ketose	Aldose		Residue	Addition
			(an aldose)	product
				(a ketose)

It should be observed that in this and all other reactions involving sugar phosphates for which transaldolases or transketolases act as catalysts, the aldehyde acceptor of the transferred group becomes a ketone, while the ketone donor of that group becomes an aldehyde—and thus ready for the next step in the reaction sequence. On this occasion and in the presence of transketolase, sedoheptulose-7-P reacts with glyceraldehyde-3-P. By the transfer of a 2-C unit, the sedoheptulose-7-P is converted to ribose-5-P, while glyceraldehyde-3-P is changed to xylulose-5-P. Specific enzymes then change both pentose phosphates to ribulose-5-P (Fig. 11.6). All these reactions, working together, have therefore produced three molecules of ribulose-5-P; these are converted to ribulose-1,5-di-P in the first step of the next turn of the cycle (step 1, p. 241).

The statistical result of these equilibrated transformations, shown in Fig. 11.7, is that, for three molecules of carbon dioxide fed into the cycle by reaction with three molecules of ribulose-1,5-di-P, *three molecules of ribulose-di-P and one molecule of triose phosphate* are produced:

$$3\text{C}_5 \;+\; 3\text{C}_1 \rightarrow 3\text{C}_5 \;+\; 1\text{C}_3$$

Since two molecules of triose phosphate can unite to give fructose-1,6-di-P (the aldolase reaction), this equation can also be written in the following way:

$$6\text{C}_5 \;+\; 6\text{C}_1 \rightarrow 6\text{C}_5 \;+\; 1\text{C}_6$$

It must, however, be remembered that *any* of the intermediate sugar phosphates can equally well be 'siphoned off' if a cellular demand for any one of them exists. For the production of nucleic acids, for example, pentoses are needed, and in this case the equation would read:

$$5C_5 + 5C_1 \rightarrow 5C_5 + 1C_5$$

For each molecule of carbon dioxide fixed in step 1 of the carbon cycle *one ATP molecule* is required for the phosphoribokinase reaction. In

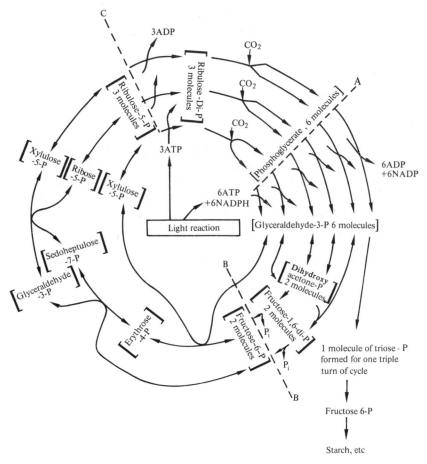

Fig. 11.7 Photosynthetic carbon cycle. (The scheme from C to A is stage 1, from A to B resembles the glycolytic pathway working in reverse, and B to C completes the cycle. The part of the cycle C to B (anticlockwise) represents the third step of the pentose-phosphate pathway)

addition, *two molecules of ATP* and *two of NADPH* are required in step 2 to convert the two molecules of phosphoglyceric acid, formed in step 1, to glyceraldehyde-3-P. The ATP and NADPH are produced by the light reaction in the living plant, although non-photosynthetic sources of these compounds can be used in *in vitro* studies, showing that, given these sources of energy, the whole carbon cycle can be carried out in darkness. Some of the ATP could arise from oxidative phosphorylation *in vivo*, but since photosynthesis can take place under good lighting conditions at twenty times the rate of respiration, the respiratory contribution is normally small.

11.4   The pentose-phosphate, or hexose-monophosphate, pathway

The principal catabolic route for carbohydrates is the glycolysis pathway, followed, under aerobic conditions, by the TCA pathway. The glycolytic enzymes are present in the soluble fraction of the cell but the route from pyruvate is located in the mitochondria.

There is, however, an alternative route by which the oxidation of carbohydrates can take place. Quantitatively, it usually accounts for only a very small percentage of the oxygen taken up by the cell but, as will be seen later, it serves a qualitatively different purpose from the main pathway, and at least one of the products, NADPH, is of vital importance in cell anabolism. The enzymes concerned are present in the cytosol; it should be noted that the mitochondrion is not directly involved in any part of the cycle. The pentose-phosphate pathway consists of two oxidative steps, followed by the same series of interconversions of sugar phosphates encountered as step 3 of photosynthesis.

The pathway is also known as the hexose-monophosphate pathway (or 'shunt') because the initial reactant is *glucose-6-P*. As in glycolysis (p. 197), this substance can be formed from glucose by the hexokinase reaction, or from starch or glycogen by the action of phosphorylase and phosphoglucomutase (chapter 9).

Step 1.   Oxidation of glucose-6-P to 6-P-gluconic acid

In this process, the secondary alcohol grouping associated with carbon atom 1 of glucopyranose-6-P is oxidized to a carbonyl group (Fig. 11.8). Because of the proximity of the pyrane ring's oxygen atom, this oxidation has an interesting consequence—the sugar phosphate is oxidized to the oxidation level of a carboxylic acid, yet the 6-membered ring structure

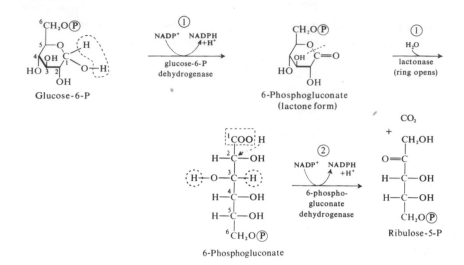

Fig. 11.8   Steps 1 and 2 of the pentose-phosphate pathway.   (Note that in both of the steps NADP is reduced to NADPH)

is nevertheless retained.   In other words, the oxidation produces an internal ester, formed between a carboxyl group and a hydroxyl group in the *same* molecule; such an internal ester is called a *lactone*.   The enzyme catalysing the production of the lactone form of 6-P-gluconic acid is *glucose-6-P dehydrogenase*, and a very important feature of this enzyme is that its cofactor is NADP.   In this respect it differs from the enzymes of glycolysis and the TCA cycle, for almost all the dehydrogenases of these pathways are dependent on NAD.   Consequently, the glucose-6-P dehydrogenase reaction, together with the reaction of step 2 below, provides the cell with *most of its NADPH*.

The dehydrogenase reaction just described is reversible *in vitro*, but *in vivo* the oxidation is followed at once by a hydrolytic reaction catalysed by *lactonase*.   By opening the ring system to give the straight chain form of 6-phosphogluconic acid (Fig. 11.8), the position of the equilibrium is pulled towards the right.

Step 2.   Oxidative decarboxylation of 6-phosphogluconate

In this reaction the enzyme *6-phosphogluconate dehydrogenase* brings about oxidation of the secondary alcohol group on carbon atom number 3 of the 6-phosphogluconic acid molecule at the same time as the elements of carbon dioxide are removed from the unphosphorylated end of the molecule. Consequently, as the original carbon atom number 1 is released as carbon dioxide, the third carbon atom of the original molecule becomes the ketone carbon atom of a ketopentose (Fig. 11.8). This ketopentose is *ribulose-5-P*, and, since 6-phosphogluconate dehydrogenase is NADP-linked, its formation is coupled with the reduction of $NADP^+$ to NADPH.

Thus, in these two steps of the pentose-phosphate pathway, glucose-6-P is converted to ribulose-5-P with the simultaneous reduction, outside the mitochondrion, of two molecules of $NADP^+$ to NADPH. It is of interest to note that it is carbon atom number 1 of the original glucose molecule which appears as (respiratory) carbon dioxide when its oxidation follows this pathway, whereas, in the glycolytic pathway, the carbon dioxide released when pyruvate is decarboxylated arises from carbon atoms 3 and 4 of the original hexose (p. 206). Glucose labelled either at C-1 or at C-3 has been used to distinguish between the two routes and so to determine the contribution of each route to the respiration in any particular cell or tissue. Although this technique has provided useful information, its overall reliability has been questioned.

Step 3. Interconversion of ribulose-5-P and glucose-6-P

The ribulose-5-P produced by the first two steps of the pathway may be the precursor of ribose and deoxyribose. However, for the pentose pathway to be a complete cycle, much of the ribulose-5-P must be converted back to glucose-6-P.

Apart from its localization in the cytosol, the mechanism whereby ribulose-5-P is converted back to glucose-6-P differs in only very minor respects from step 3 of the carbon cycle of photosynthesis, operating in the opposite direction. This is best followed by reference to Fig. 11.6 and to Fig. 11.7 commencing at C and proceeding anticlockwise to B. An epimerase first converts two molecules of ribulose-5-P to xylulose-5-P, while an isomerase converts a third molecule to ribose-5-P. *Transketolase* then enables the ribose-5-P to accept a two-carbon fragment from xylulose-5-P, the products being sedoheptulose-7-P and glyceraldehyde-3-P:

$$\overbrace{C_5 \ + \ C_5}^{minus\ 2C} \ \underset{\text{transketolase}}{\rightleftarrows} \ C_3 \ + \ C_7$$

Ketose Aldose	Residue Addition
	(an aldose) product
	(a ketose)

The products of the last reaction next react, in the presence of *transaldolase*, to give erythrose-4-P and fructose-6-P:

$$\underset{\text{Ketose}}{C_7} + \underset{\text{Aldose}}{C_3} \xrightleftharpoons{\text{transaldolase}} \underset{\substack{\text{Residue} \\ \text{(an aldose)}}}{C_4} + \underset{\substack{\text{Addition} \\ \text{product} \\ \text{(a ketose)}}}{C_6}$$

minus 3C

The sugar with four carbon atoms in its molecule, erythrose-4-P, next accepts two carbon atoms from the remaining molecule of xylulose-5-P, in a reaction catalysed by *transketolase*, The products are another molecule of fructose-6-P and a molecule of glyceraldehyde-3-P. The fructose-6-P is readily converted to glucose-6-P by the enzyme *phosphohexoisomerase* (p. 197). Similarly, the glyceraldehyde phosphate can readily give glucose-6-P via fructose-1,6-di-P, the enzymes responsible including aldolase and fructose-di-P-phosphatase (Fig. 10.8).

Not only can ribulose-5-P be formed from hexose by steps 1 and 2 of the pentose-phosphate cycle, but any excess ribulose-5-P can be converted back to hexose phosphate without loss of the reducing power of NADPH formed during the oxidative steps. The outcome of six turns of the pentose-phosphate cycle is that, statistically, 1 molecule of hexose is completely metabolized to 6 molecules of carbon dioxide in such a way that 12 molecules of NADPH are formed from 12 molecules of $NADP^+$ at the same time. Naturally, $NADP^+$ must be regenerated from NADPH or the process would soon stop (compare NADH, p. 207), and it is therefore necessary to consider now a matter of considerable importance—the ways in which NADPH can function within the cell as a reducing agent.

11.5 Differences between the pentose-phosphate and main pathway of sugar degradation

The pentose-phosphate pathway differs in several important respects from the glycolytic pathway (followed by the TCA cycle) for not only are the products different, but so are the cofactor requirements and the functional characteristics of the two systems. Glycolysis, like the pentose-phosphate pathway, is located in the cytosol, but whereas the major functions of glycolysis are to produce ATP and pyruvate (the oxidation of which, via acetyl coenzyme A, yields additional ATP by mitochondrial oxidative phosphorylation), the pentose-phosphate pathway produces no ATP directly but instead results in the reduction of NADP. It is unlikely that NADPH, produced in the cytosol, acts as a direct source of electrons for mitochondrial ATP production, for, apart from the difficulty relating to location, a prior trans-hydrogenation reaction of the type

$$NADPH + NAD^+ \rightleftharpoons NADP^+ + NADH$$

would need to occur. While such systems are known, it appears likely that the principal functions of the pentose-phosphate pathway are (a) to produce various non-hexose sugars and (b) to produce NADPH, which provides reducing power needed for a number of anabolic processes.

The pentose-phosphate pathway has few cofactor requirements. NADP is necessary, of course, and thiamine pyrophosphate is a prosthetic group for transketolase. In contrast, as was seen in earlier chapters, the main catabolic route requires numerous cofactors including inorganic phosphate, NAD, lipoic acid, thiamine pyrophosphate, and coenzyme A.

Differences in enzyme and coenzyme requirements of the main and pentose-phosphate oxidative systems are reflected in their different vulnerability to several poisons. The glycolytic pathway is highly sensitive to fluoride ions, because of the effect of this poison on enolase (p. 203). Similarly, the TCA cycle is inhibited by fluoroacetate and malonate (p. 173). Again, the principal energetic function of the main pathway—the production of ATP—is prevented by dinitrophenols (which uncouple oxidation from phosphorylation) and by such substances as cyanide ions and hydrogen sulphide (which inactivate cytochrome oxidase). The pentose-phosphate pathway, on the other hand, operates in the presence of $10^{-1}M$ fluoride and is also rather insensitive to cyanide ions. On the other hand, the pentose-phosphate system appears to be sensitive to copper ions, and both of the dehydrogenases of the cycle are inhibited by thioperazine.

The pentose-phosphate pathway is often well developed in lower organisms and plants, but most vertebrate tissues also possess some activity. Up to 40% of the carbohydrate oxidation which takes place in the lactating mammary gland and 20% in the liver occurs via the pentose-phosphate pathway. On the other hand its activity in brain and muscle is often very small, although these tissues contain some of the enzymes of the cycle.

11.6 Utilization of sugars produced by the pentose-phosphate pathway

One of the main functions of the pathway is to produce pentoses which are components of nucleotides and are therefore essential for the synthesis

Photosynthesis and the pentose-phosphate cycle

of nucleic acids, NAD, ATP, GTP, and of other similar electron and energy carriers.

The pentose phosphates occurring in the cycle can be converted readily to *ribose-5-P*. This does not directly combine with the nucleotide bases but undergoes a prior activation. In this process a pyrophosphate group is added from ATP to carbon atom 1 of the furanose ring, so producing AMP and 5'-phosphoribose 1'-pyrophosphate (PRPP). The latter reacts with nucleotide bases by a reaction catalysed by specific pyrophosphorylases, pyrophosphate being released. This synthesis of the nucleoside monophosphate is drawn towards completion by the hydrolysis of the pyrophosphate to orthophosphate.

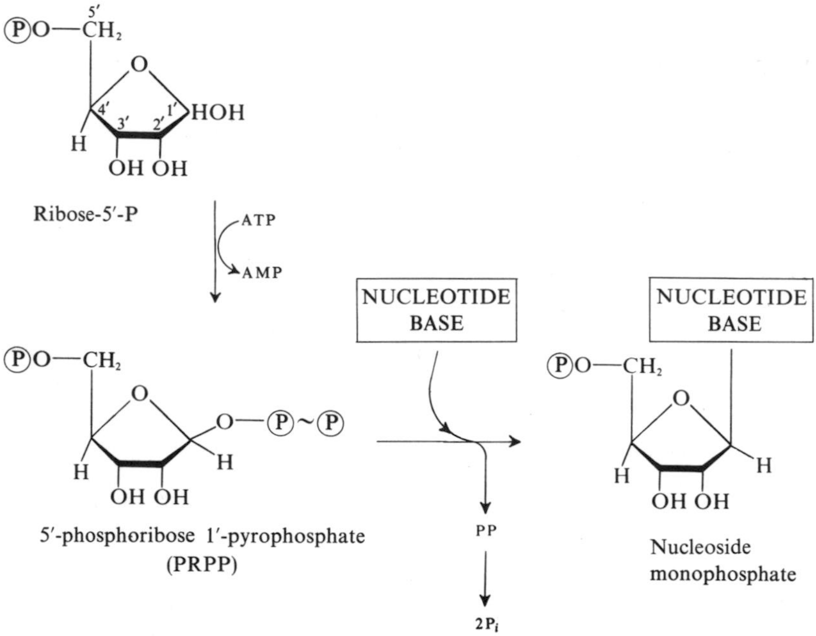

Ribose-5'-P

5'-phosphoribose 1'-pyrophosphate (PRPP)

NUCLEOTIDE BASE

NUCLEOTIDE BASE

Nucleoside monophosphate

Deoxyribose mononucleotides for DNA synthesis are largely formed from pre-formed *ribo*nucleotides. In other words, deoxyribose is not synthesized from free ribose but the reduction occurs at the nucleotide level. The process of reduction takes place as indicated below, the parent ribonucleoside being in the form of the diphosphate. The electrons and protons involved in the reduction are derived from NADPH—a further instance where this cofactor provides reducing power for an anabolic process.

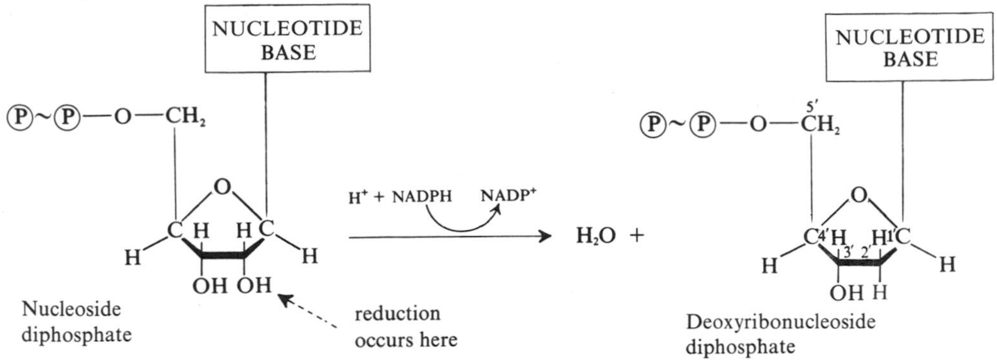

NUCLEOTIDE BASE

$P \sim P - O - CH_2$

$H^+ + NADPH \qquad NADP^+$

$H_2O +$

Nucleoside diphosphate

$C H \quad H C$

H \qquad H

OH OH

reduction occurs here

NUCLEOTIDE BASE

$P \sim P - O - \overset{5'}{C}H_2$

$C4'H \quad H1'C$

H \qquad H

3' \quad 2'

OH H

Deoxyribonucleoside diphosphate

Erythrose-4-P, another sugar produced by the pentose-phosphate pathway, is of considerable importance as a precursor of part of the carbon skeleton of three amino acids, phenylalanine, tyrosine, and tryptophan. However, these can only be synthesized by plants and certain micro-organisms, for animals lack the necessary enzymes. In plants, 6-phosphogluconic acid is a precursor of L-ascorbic acid (vitamin C).

11.7 Some processes requiring a supply of reduced NADP

Since both the pentose-phosphate pathway and the light reaction of photosynthesis are principal routes for producing NADPH, it is appropriate that an indication should now be given of some of the major reactions, other than the reduction of carbon dioxide by plants, which require this form of reducing power.

NADPH is of particular importance in the anabolism of lipids for, in addition to its use in the formation of *long-chain fatty acids* (chapter 8), considerable quantities are needed for the synthesis of *steroids* and *carotenoids*. The steroids (and, in plants, the carotenoids and rubber latex) are synthesized from *isopentenyl pyrophosphate*. This 5-carbon compound is formed by a series of reactions which are outlined in Fig. 11.9. Three molecules of acetyl coenzyme A give rise to one molecule of β-hydroxyl-β-methyl-glutaryl coenzyme A (HMG—CoA) with the elimination of two molecules of coenzyme A. The acyl linkage of HMG—CoA undergoes *reduction* reactions which require four equivalents of reducing power derived from two molecules of NADPH. The acyl

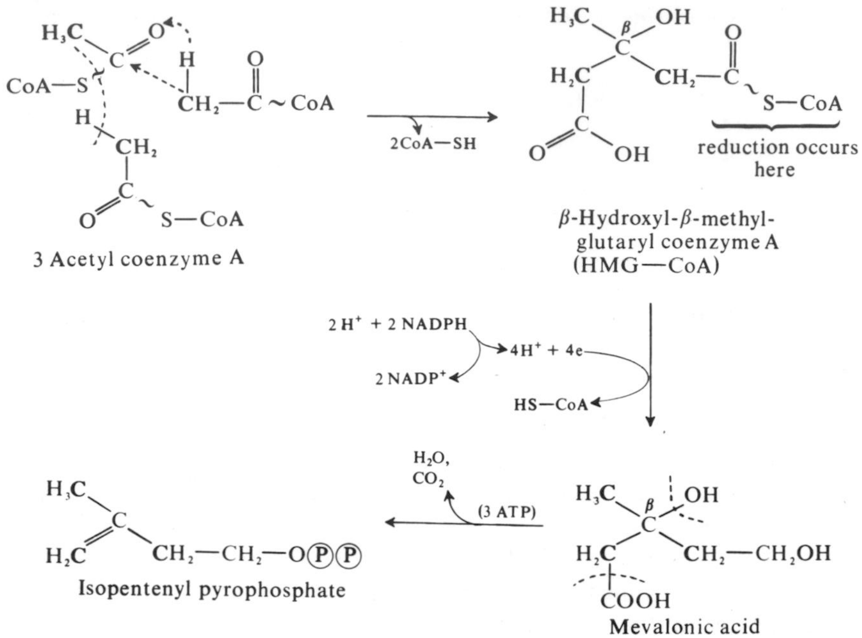

Fig. 11.9 Synthesis of isopentenyl pyrophosphate from acetyl coenzyme A

linkage of acyl coenzyme A is thereby converted to an alcohol group with the release of free coenzyme A. The di-hydroxy acid so formed, *mevalonic acid*, then undergoes reactions which involve the loss of the elements of water and carbon dioxide. Three molecules of ATP provide the necessary energy and also provide the pyrophosphate contained in the product, isopentenyl pyrophosphate. Three of the carbon atoms of this compound have been shown by the use of radioactive carbon to arise from the methyl group of acetate (shown in heavy type in Fig. 11.9) while the other two are derived from the carboxyl group.

Isopentenyl pyrophosphate is itself the precursor of numerous lipids. Rubber latex is a polymer consisting of thousands of isopentenyl groups linked together (Fig. 11.10). The side chain of ubiquinone (Fig. 6.5) is also derived from it.

Cholesterol contains 27 carbon atoms and each of these has been shown to arise from acetyl coenzyme A; labelling techniques have shown that mevalonic acid and isopentenyl pyrophosphate are again intermediates in the reaction sequence. Fifteen of the carbon atoms are derived from the methyl group of acetate (shown in heavy type in Fig. 11.10), the other

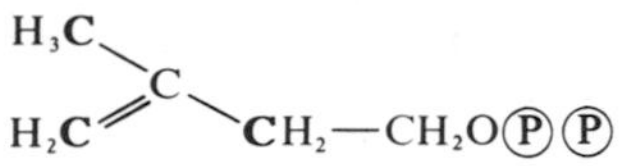

Isopentenyl pyrophosphate
(carbon atoms in heavy type derived from CH_3 group of acetyl CoA)

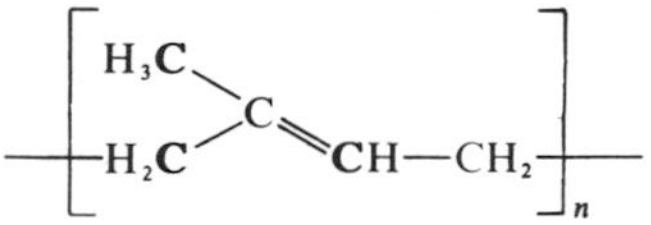

RUBBER

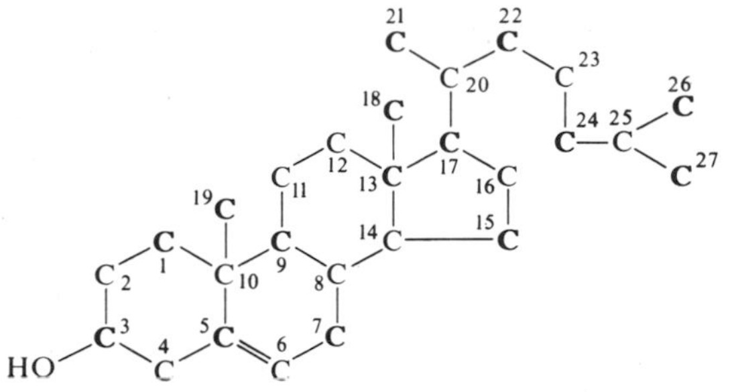

CHOLESTEROL
(hydrogen atoms omitted)

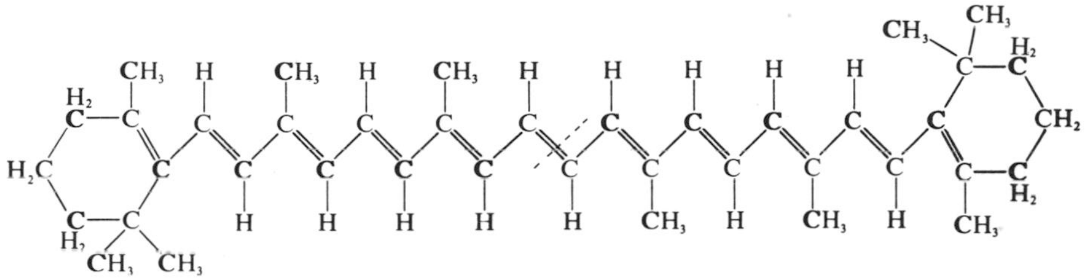

β-CAROTENE

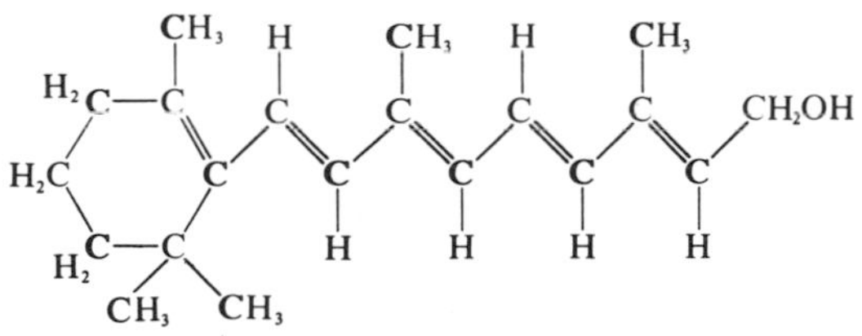

VITAMIN A_1

Fig. 11.10   Four derivatives of isopentenyl pyrophosphate

Photosynthesis and the pentose-phosphate cycle

12 originating from the carboxyl group. The cholesterol molecule (and no doubt those of other steroids) is formed by the combination of 6 molecules of isopentenyl pyrophosphate, and the formation of 6 molecules of this precursor requires 12 molecules of NADPH. In addition, other reactions in the sequence by which cholesterol is synthesized appear to use the reducing power of NADPH.

The carotenoids are common plant pigments; β-carotene (Fig. 11.10) is a typical example and contains 40 carbon atoms which arise by the combination of 8 isopentenyl units. These, of course, require 16 molecules of NADPH for their synthesis. Higher animals cannot synthesize carotenoids and for this reason β-carotene is an important factor they must acquire in their diet. Animals are able to convert β-carotene to vitamin A by splitting the molecule at the point of symmetry. Vitamin A is a precursor of the photo-sensitive pigment of the eye (chapter 16, p. 406).

11.8   The microsomal electron transport system

It has become apparent in recent years that, in addition to the mito-chondrial and chloroplastic electron transport systems, a *third electron transport* system exists, often associated with the membranes of the smooth endoplasmic reticulum. This electron transport system may prove to be as important in its own way as the other two systems; it merits a mention at the end of this chapter since it cannot function unless the cytosol provides a continuous supply of NADPH.

The principal role of the microsomal electron transport system is to bring about a variety of oxidative changes in molecules of a lipophilic nature. These oxidative changes are sometimes simple conversions of the type: R—H → R—OH, but they can also be much more complex. Because the range of possible substrates is so wide and the type of oxidation so varied, the enzyme complexes present in the microsomes which catalyse these changes have been called *mixed function oxidases*. It is particularly noteworthy that the *reduced* nucleotide, NADPH, is an essential cofactor for these *oxidative* changes.

Lipophilic substances, both endogenous and exogenous in origin, are known to be oxidized by the microsomes under aerobic conditions. Two examples will illustrate some of the activities of the mixed function oxidases, and the intriguing diversity of the substrates upon which they act.

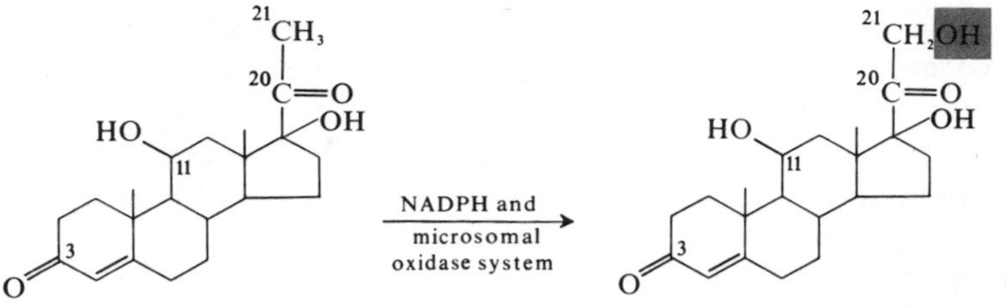

21-deoxycortisol

NADPH and microsomal oxidase system →

Cortisol cf. Figs. 1.6 and 11.12)

Similarly, 11-deoxycorticosterone is oxidized to corticosterone

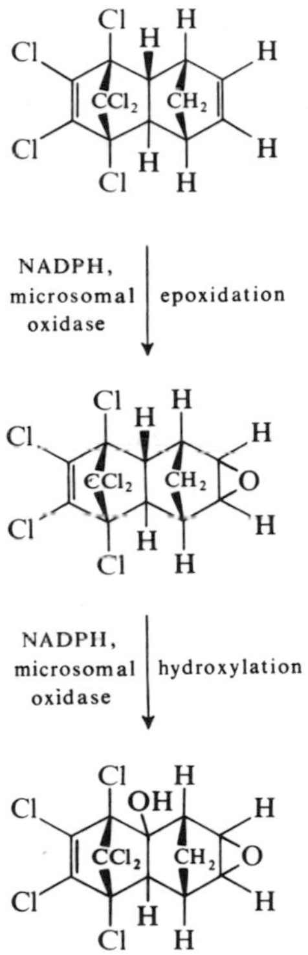

Aldrin

NADPH, microsomal oxidase | epoxidation ↓

Dieldrin

NADPH, microsomal oxidase | hydroxylation ↓

Hydroxy-dieldrin
(one of several isomers)

Fig. 11.11 Examples of oxidations, occurring at the endoplasmic reticulum, dependent on NADPH

First, a microsomal system exists in the adrenal cortex which hydroxylates carbon atom 21 of various natural steroids. For example, progesterone is converted to 11-deoxycorticosterone. Similarly, 21-deoxycorticosterone is converted to corticosterone and 21-deoxycortisol is converted to cortisol (Fig. 11.11a). Clearly, the microsomal oxidases play an important role in the metabolism of steroids known to possess hormonal functions.

Secondly, exogenous lipophilic substances which could become harmful if they accumulated within the cell can often be attacked by the microsomal enzyme complex. The object is normally to increase the polarity of these potentially harmful substances, thus making them both more soluble in water and, very often, more readily conjugated with another molecule as an important step in the detoxication mechanism. The insecticide aldrin, notorious for the way it persists in the environment, illustrates this point in an interesting negative way; it is rapidly oxidized by animal liver microsomes to its (still highly toxic) epoxide, dieldrin. The dieldrin is then only *slowly* converted to a variety of less toxic hydroxylated compounds (Fig. 11.11). Oxidation (often followed by conjugation) occurs more or less readily for a wide range of foreign molecules, including medical drugs, pesticides, and other environmental contaminants.

The precise mechanism by which these NADPH-dependent oxidative changes are brought about is still uncertain, although a flavoprotein, a non-haem iron protein and a cytochrome known as cytochrome P_{450} take part in an electron transport system. These carriers are normally closely associated with the membranes of the endoplasmic reticulum (although the mitochondria of cells in some tissues possess similar

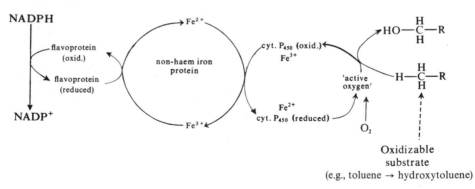

Fig. 11.12   Suggested role of NADPH in microsomal oxidations

oxidative systems). A possible representation of the microsomal oxidative system is shown in Fig. 11.12. Although a marked resemblance to the ordinary NAD-dependent mitochondrial electron transport system exists superficially, the overall form and function of the microsomal system is unique. The substance which donates oxygen to a lipophilic oxidizable substrate, $R—CH_3$, could be an oxygenated form of ferrocytochrome P_{450}—i.e., a compound analogous to oxyhaemoglobin, in that it has oxygen attached to it (p. 372). However, cytochrome P_{450} also resembles mitochondrial cytochrome, in that the iron present in it can assume both the ferrous and the ferric state. The ultimate purpose of the NADPH could well be to reduce the ferric form of cytochrome P_{450} to the ferrous form, ready for re-oxygenation. This concept of the nature and functioning of 'active oxygen' is still tentative, but the existence of an 'active oxygen' which participates in biochemical reactions is no more remarkable than the existence of such substances as 'active phosphate', 'active acetate', 'active carbon dioxide', and 'active formate'.

12
Nitrogen metabolism

12.1 Nitrogen fixation

Atmospheric nitrogen is the ultimate source of the nitrogen present in the proteins, nucleic acids, and other vital nitrogenous components of all living cells.   However, neither animals nor higher plants are able directly to utilize atmospheric nitrogen, but depend instead upon the activities of a relatively few types of bacteria and blue-green algae which are capable of converting atmospheric nitrogen to ammonia.   This process, termed nitrogen fixation, is a fundamental step in the nitrogen cycle, for the ammonium compounds are further transformed to nitrites and nitrates in the soil and into organic nitrogenous compounds in plants.   When plants and animals die and decay, much of this nitrogen returns to the soil or the sea.   Animal excreta also contain nitrogenous materials which are decomposed to simple inorganic nitrogen compounds.   On the other hand a proportion of the nitrogen fixed by nitrifying bacteria finds its way back into the atmosphere as a result of the activities of denitrifying bacteria.

In recent times, man has increasingly supplemented the natural process of nitrogen fixation by such industrial nitrogen fixation reactions as the Haber process.   The ammonium compounds and nitrates obtained by chemical processes are used as fertilizers to replace nitrogenous compounds removed from arable land when crops are harvested.   Nevertheless, microbial nitrogen fixation is very largely responsible for maintaining the level of combined nitrogen in the biosphere.   In view of man's dependence on plant protein, the importance of natural nitrogen fixation cannot be overestimated.   Furthermore, the mechanisms of microbial nitrogen fixation pose some intriguing problems for the biologist and biochemist, for molecular nitrogen is an extremely unreactive substance at ordinary temperatures and pressures, the activation energy (p. 95) of the reaction by which nitrogen is reduced to ammonia being very high. Despite this barrier to reaction, which is still considerable at high temperature unless pressures of several hundred atmospheres are applied, it is a remarkable fact that several types of micro-organisms, quite unrelated to one another, are able to reduce nitrogen to ammonia at a temperature, pressure, and pH compatible with the existence of life.

Some nitrogen-fixing organisms are present in soil, others in root nodules of plants and yet others live in water. Some of the soil bacteria capable of fixing nitrogen are aerobic (e.g., *Azotobacter*) but it may well be that anaerobic species make a quantitatively greater contribution to nitrogen fixation. The latter include species of *Clostridium*, a genus almost universally distributed in soils, and it is upon these that most of the recent investigations into the biochemistry of nitrogen fixation have been performed. Other micro-organisms, belonging to the genus *Rhizobium*, enter into a symbiotic relationship with many leguminous plants (clovers, peas, beans, etc.). Various species of Rhizobium are associated with the nodules on the roots of these plants and are able to convert nitrogen gas into the amino-groups of amino acids, ammonium ions probably being formed as an intermediate. Root nodules are remarkable in that they contain a form of haemoglobin, leghaemoglobin, the presence of which often causes them to have a pink colour. Curiously, the pigment lies outside the bacteria associated with the root nodules, though it is never present in bacteria-free Leguminosae. A third important example of nitrogen fixation is provided by certain blue-green algae, for the production of rice in Asiatic countries relies upon their presence in the water of paddy fields. Eventually the algae die and decompose, liberating ammonia and nitrate ions which are subsequently absorbed by the roots of the growing rice. Several groups of photo-synthetic bacteria also contribute to the nitrogen store in water.

Despite its importance, the biochemical mechanism of nitrogen fixation is still very imperfectly understood, and the account which follows must therefore be regarded as incomplete and, in places, speculative. It is, however, established that three inter-related processes are involved. The first of these produces the energy for the nitrogenase (nitrogen-reducing) reaction; this process, which is ATP-dependent, appears to lead to the formation of the reduced form of *ferredoxin*, a protein containing non-haem iron as well as cysteine and a more labile form of sulphur. This type of electron carrier is of interest in that it structurally resembles, but is certainly not identical with, the non-haem iron proteins which participate in the light reaction of photosynthesis (p. 239), in mitochondrial electron transport (p. 145), and probably in microsomal electron transport.

In the second process, reducing power is transferred from reduced ferredoxin to a special but as yet unspecified, *electron activator*. This activator, X, may be a separate substance but is, according to some authorities, one sort of active site on the surface of nitrogenase. This enzyme is probably a complex rather than a single substance for it has been subdivided into two different proteins with molecular weights of about 100,000 and 50,000 respectively. Whatever the precise nature of X, it has, in its reduced form, the ability to pass electrons on to the third part of the nitrogen fixation process which is responsible for the catalytic reduction of nitrogen to ammonia. This may take place at a nearby site, Y, on the surface of the nitrogenase.

The ATP and reduced ferredoxin needed to drive the subsequent processes are probably produced by different mechanisms in different types of micro-organisms. In the blue-green algae, which are photosynthetic organisms, they may originate directly from photosynthesis by processes analogous to those by which the same two products are formed in the chloroplasts of higher plants (p. 240). Of the nitrogen-fixing bacteria, the most intensively studied is *Clostridium pasteurianum*, and it is of interest that cell-free extracts have been prepared from this organism which are capable of fixing nitrogen. To enable them to do this, the extracts normally require a supply of pyruvic acid, the catabolism of which produces both ATP and reduced ferredoxin. Some part of this process is probably analogous to the oxidative decarboxylation of pyruvate described in section 9.7 (p. 205).

The pyruvic acid is first converted to a thiamine pyrophosphate complex and this then undergoes decarboxylation:

$$CH_3-\overset{\overset{O}{\|}}{C}-COOH \xrightarrow{TPP} CH_3-\overset{\overset{OH}{|}}{\underset{TPP}{C}}+COOH \xrightarrow{CO_2} CH_3-\overset{\overset{OH}{|}}{\underset{TPP}{C}}-H$$

Pyruvic acid

TPP—aldehyde complex

The TPP–aldehyde complex is next converted to acetyl coenzyme A but by a process quite different from the aerobic mechanism illustrated in Fig. 7.4 for it neither involves lipoic acid nor a mitochondrial electron transport system. Instead, in *C. pasteurianum* which, it will be recalled, is an anaerobic organism, oxidation of the aldehyde complex is coupled to the reduction of ferredoxin, a one-electron redox carrier:

$$CH_3-\underset{\underset{TPP}{|}}{\overset{\overset{OH}{|}}{C}}-H + HS.CoA \longrightarrow 2e + 2H^+ \longrightarrow \begin{array}{l} 2\ ferredoxin \\ (reduced\ form) \end{array}$$

$$CH_3-\overset{\overset{O}{\|}}{C}\sim S.CoA + TPP \qquad\qquad \begin{array}{l} 2\ ferredoxin \\ (oxidized\ form) \end{array}$$

The acetyl coenzyme A is not appreciably catabolized by the TCA cycle in this organism but takes part in a two step process which yields ATP. In the first of these, the acetyl coenzyme A reacts with inorganic phosphate to form acetyl phosphate, the energy being conserved since the latter is also a high energy compound.

$$CH_3-\overset{\overset{O}{\|}}{C}\sim S.CoA + P_i \longrightarrow CH_3-\overset{\overset{O}{\|}}{C}\sim \textcircled{P} + HS.CoA$$

$$\text{Acetyl coenzyme A} \qquad \text{Acetyl phosphate}$$

The high energy phosphate is then transferred from the acetyl phosphate to ADP:

$$CH_3-\overset{\overset{O}{\|}}{C}\sim\textcircled{P} + ADP \longrightarrow ATP + CH_3COOH$$

It should be noted that the yield of ATP is very small (one molecule per molecule of acetyl coenzyme A) in contrast to the high efficiency of the aerobic catabolism of acetyl coenzyme A by the TCA cycle (p. 172).

The most poorly understood component of the nitrogen fixation process is the *electron activator*, X. This becomes reduced by reduced ferredoxin to form an extremely powerful reducing agent, $X_{red.}$. The reduction involves a preliminary activation, in the course of which ATP is converted to ADP and inorganic phosphate.

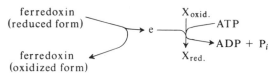

The nature of the substance (or active site) X is very speculative; it has been suggested that it is a molybdenum complex. X_{red} is able, under certain experimental conditions, to reduce hydrogen ions to hydrogen:

$$2H^+ + 2e \rightarrow H_2$$

Normally, however, its function is to reduce nitrogen.

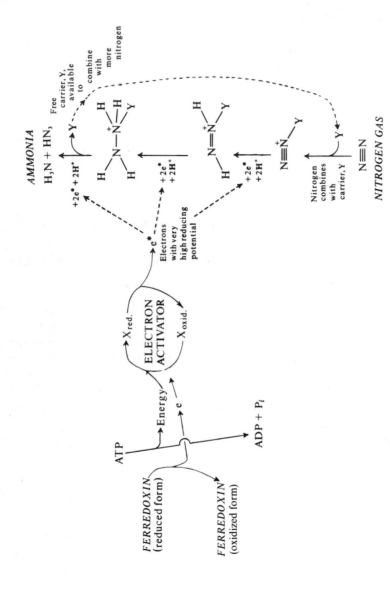

Fig. 12.1   Speculative model for the process of nitrogen fixation

The reduction of a molecule of nitrogen to ammonia requires the participation of 6 electrons:

$$6e + 6H^+ + N_2 \rightarrow 2NH_3$$

Such a process is very unlikely to occur in one step but probably takes place via at least two intermediates:

$$N_2 \xrightarrow[2e]{2H^+} HN{=}NH \xrightarrow[2e]{2H^+} H_2N{-}NH_2 \xrightarrow[2e]{2H^+} 2NH_3$$

di-imide   Hydrazine

(complex)   (complex)

However, no clearly-defined intermediates have been isolated experimentally, probably because they are firmly attached to a carrier (or active site), Y, from the moment that nitrogen combines with the carrier until ammonia is finally released from it. The nitrogen attached to the carrier undergoes a series of reduction processes, possibly in the way indicated in Fig. 12.1. Electrons, released from the special electron activator, having a very high reduction potential, bring about the stepwise reduction of nitrogen. The nature of Y is not established, but it has been suggested that the nitrogen may be complexed to iron at the active site.

12.2 The synthesis of amino groups

Cells of most organisms are able to convert ammonia to the α-amino group of amino acids. This activity is perhaps greatest in plants and micro-organisms since animals rely in large measure on pre-formed dietary amino acids for their protein synthesis. Plant roots readily absorb ammonium or nitrate ions from the soil. Absorbed nitrate is reduced to ammonia by enzymes in plant tissue prior to the formation of amino groups.

When radioactive ammonium salts (e.g., $^{15}NH_4Cl$) are absorbed by the roots of plants, or when micro-organisms such as yeasts are incubated in a medium containing similarly-labelled ammonium compounds, amino acids and other nitrogen compounds are soon found to contain ^{15}N. Quantitatively, the most important initial reaction is the conversion of ammonia to *glutamic acid*. In plant and animal cells the enzyme concerned, *glutamate dehydrogenase*, is situated mainly in the mitochondria, where the α-ketoglutarate, which is the acceptor of the ammonia, is synthesized by the enzymes of the TCA cycle. Glutamate dehydrogenase requires either NADH or NADPH as cofactor for the synthesis of glutamate, the cofactor requirement depending on the sources of the enzyme. The reaction, which is *fully reversible*, probably involves the formation of an imino acid from the α-ketoglutarate and ammonia. This is then reduced to glutamate by electrons donated by NADH or NADPH:

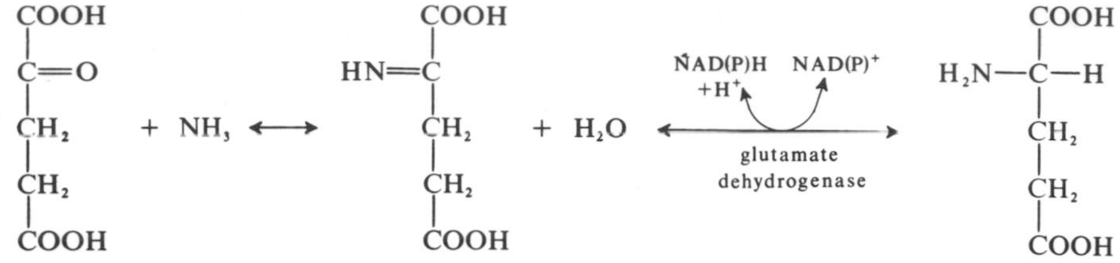

$$
\begin{array}{ccccc}
\text{COOH} & & \text{COOH} & & \text{COOH} \\
| & & | & & | \\
\text{C}=\text{O} & \text{HN}=\text{C} & \text{H}_2\text{N}-\text{C}-\text{H} \\
| & +\ \text{NH}_3 \ \longleftrightarrow & | & +\ \text{H}_2\text{O} & | \\
\text{CH}_2 & & \text{CH}_2 & & \text{CH}_2 \\
| & & | & & | \\
\text{CH}_2 & & \text{CH}_2 & & \text{CH}_2 \\
| & & | & & | \\
\text{COOH} & & \text{COOH} & & \text{COOH}
\end{array}
$$

$\overset{\text{NAD(P)H} \quad \text{NAD(P)}^+}{\underset{\text{glutamate dehydrogenase}}{+\text{H}^+ \ \rightleftharpoons}}$

α-Ketoglutaric acid   Imino acid   L-Glutamic acid

Very soon after radioactive nitrogen derived from $^{15}\text{NH}_3$ appears in glutamic acid, it is found also in the other α-amino acids of proteins. The amino group of glutamic acid is, in fact, *readily transferred* to a variety of α-keto acids which are thereby converted to α-amino acids. The process is known appropriately as *transamination* and is catalysed by enzymes called *transaminases* or *amino transferases*. The typical reaction, which is freely reversible, may be represented in the following way:

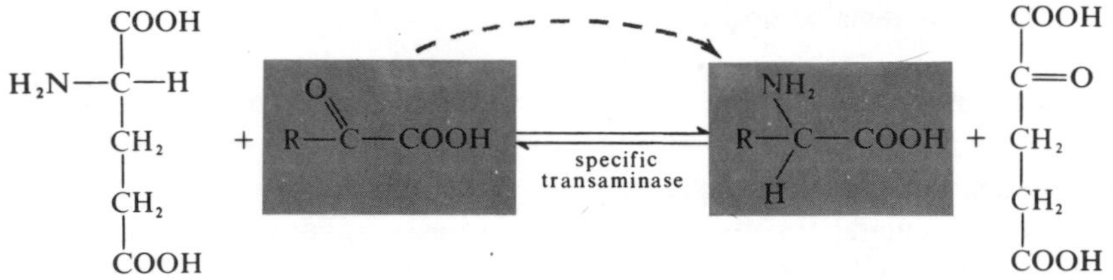

$$
\begin{array}{ccccc}
\text{COOH} & & & & \text{COOH} \\
| & & & & | \\
\text{H}_2\text{N}-\text{C}-\text{H} & & & & \text{C}=\text{O} \\
| & + & \underset{\text{specific transaminase}}{R-\overset{O}{\overset{||}{C}}-\text{COOH} \rightleftharpoons R-\overset{\text{NH}_2}{\underset{H}{\overset{|}{C}}}-\text{COOH}} & + & \text{CH}_2 \\
\text{CH}_2 & & & & | \\
| & & & & \text{CH}_2 \\
\text{CH}_2 & & & & | \\
| & & & & \text{COOH} \\
\text{COOH}
\end{array}
$$

L-Glutamic acid   α-Keto acid   L-α-Amino acid   α-Ketoglutarate

If considered in the *reverse direction* to that shown above, it is clear that transaminases remove the elements of ammonia from L-α-amino acids and pass them to α-ketoglutarate. Transaminases only act on the L-forms of amino acids and a specific transaminase exists for each of these. Pyridoxal phosphate, a derivative of vitamin B_6, is an essential cofactor (section 13.2).

It should be emphasized that the α-ketoglutarate-glutamate inter-conversion (followed by the glutamate dehydrogenase reaction) provides a mechanism which is unique in amino acid metabolism. It enables the majority of amino acids to be deaminated by the donation of their amino

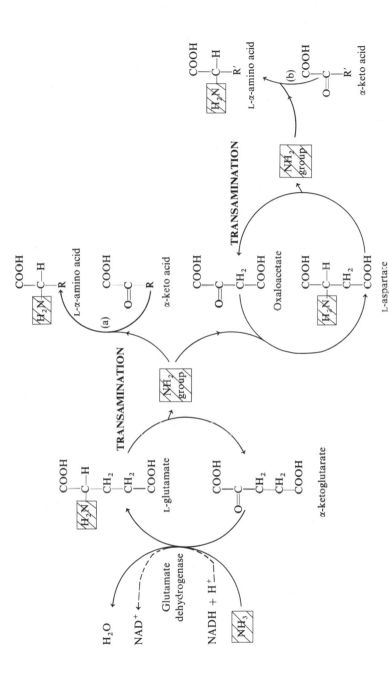

Fig. 12.2 The synthesis of glutamate from α-ketoglutarate and the formation of an α-amino acid from the corresponding α-keto acid. (a) By direct transamination from glutamate (b) By transamination via the oxaloacetate/aspartate system. Each of the reactions is reversible

groups to α-ketoglutarate, the action being catalysed by specific trans-aminases. The glutamic acid so formed is then broken down to produce α-ketoglutarate and ammonia by the reaction catalysed by glutamate dehydrogenase. This ammonia, as will be seen later, is very toxic to cells and if it is not to be excreted at once, it is largely transformed to other useful or excretable nitrogenous compounds (sections 12.5 to 12.8).

Conversely, transamination also occurs during the synthesis of most (if not all) amino acids present in *plant* proteins. The process is more restricted in *animals*, for not all the corresponding α-keto acids occur in the cell—although, if provided artificially in the diet, the keto acids which are *normally absent* can, in most cases, readily accept amino groups by transamination to form 'essential' amino acids.

L-*Aspartate*, like L-glutamate, can act as the donor of amino groups for amino acid synthesis—i.e., specific transaminases enable L-aspartate as well as L-glutamate to be the donor in transamination reactions (Fig. 12.2). Much of the aspartate may itself originate by the transamination of oxaloacetate, L-glutamate being the donor. Oxaloacetate is, of course, readily synthesized by the reactions of the TCA cycle (p. 171).

12.3 General aspects of amino acid metabolism

Plants and autotrophic micro-organisms are able to build up the carbon skeleton of each of the amino acids which occur in their proteins. In most cases, the skeleton is derived from pyruvate or one of the TCA cycle acids. At some stage in the synthesis, the amino groups are added by transamination. In chapter 8 it was possible to give a general account of the catabolism and anabolism of fats, since the metabolism of most of the fatty acids shows a particular pattern of reactions. There are, however, about twenty α-amino acids of major importance and these are metabolized in a variety of ways. In the next chapter, some important aspects of the biochemistry of individual amino acids are discussed. Here we are concerned with the *overall picture* of metabolism under normal and abnormal dietary conditions.

Animals depend directly or indirectly on plants for all their food materials and thus for the amino acids to build proteins. An adult non-pregnant, non-lactating animal requires a certain level of protein in its diet in order to maintain the repair of its tissues and to allow for growth of hair and nails. In the adult human, the daily protein requirement is about 1 g per kg body weight. If the dietary protein level is increased above the minimum amount needed to maintain health, the animal shows no great increase in the overall nitrogen content of its body, but instead, the excess amino acids absorbed from the intestine are catabolized, the nitrogen of the amino groups being excreted—in different types of animals—as ammonia, urea, or uric acid. Thus the nitrogen excretion greatly increases when excess protein is consumed. At the same time the carbon skeletons of amino acids are converted, directly or indirectly, either to oxaloacetate (and hence to glycogen), or to acetyl coenzyme A (which may be oxidized by the TCA cycle or used for the synthesis of lipids). The pathway by which any particular amino acid is degraded will determine whether it is glucogenic or ketogenic (chapter 10, p. 223).

The proteins of an animal's body are constantly being degraded and new ones built up. In other words, there is a state of metabolic flux; a perpetual turn-over is occurring in which tissue proteins are constantly being broken down and replaced. This turn-over may be quite rapid in the case of liver or plasma proteins where the half life is about 7–10 days, but the proteins of muscle and skin often have a half life of several months. The amino acids which are released as the tissue proteins are broken down are largely reincorporated into new proteins, although only after some have been converted to other amino acids. This process is not completely efficient and therefore some of the amino acids are catabolized in the process, their nitrogen being excreted. Thus each animal requires dietary protein to make good this loss. If the animal is given a diet inadequate in protein the problem of inefficiency is aggravated, the animal being obliged to break down relatively dispensable proteins—especially those in the liver—to replace amino acids required for various indispensable enzymes. Under these conditions, the nitrogen excretion decreases to a very low level while the typical symptoms of emaciation appear.

During complete starvation, the animal at first utilizes its reserves of glycogen and lipid to provide energy. Eventually, however, as these reserves become depleted, it has to obtain energy by resorting to the catabolism of its proteins. The more dispensable proteins are utilized first, but eventually even essential proteins are consumed. At this stage of starvation, nitrogen excretion rises rapidly as the amino groups of amino acids are discarded. Death soon results once the essential proteins are appreciably depleted.

Young growing animals and pregnant females require a relatively greater dietary protein intake than normal adults. The results of feeding a low protein diet, or of complete starvation, are consequently much more pronounced and appear at an earlier stage.

12.4 Essential and non-essential amino acids

It will be recalled (chapter 2) that each protein has a definite primary structure, usually comprising residues of most of the twenty or so amino acids. If for some reason a particular amino acid is not made available to the ribosomes, the protein cannot be formed there. If a deficit of certain amino acids temporarily exists, *plants* and many micro-organisms can synthesize them from other amino acids which happen to be present in temporary excess.

In higher *animals*, however, some part of this biosynthetic versatility has been lost, for most of them are unable to make the carbon skeletons of about nine amino acids, so these or their keto acid precursors must be supplied in the diet. The animal is unable to survive unless it is provided with them, and hence they are known as *essential* amino acids. Those amino acids the animal is able to synthesize for itself are termed *non-essential* amino acids.

The amino acids are classified as essential or non-essential in Table 12.1. The list of essential acids could be expanded, for it is found that in young rats, for example, arginine and histidine are *partially essential* since the growth of the animal is so great that although it can synthesize

Table 12.1   Essential and non-essential amino acids

Non-essential	Essential
Alanine	Methionine
Arginine	Threonine
Aspartic acid	Tryptophan
Cysteine	Valine
Glutamic acid	Leucine*
Glycine	Isoleucine†
Histidine	Lysine†
Hydroxyproline	Phenylalanine†
Proline	Tyrosine†
Serine	

* Amino acid is ketogenic
† Amino acid is glucogenic and ketogenic
Amino acids not asterisked are glucogenic only

them it is unable to do so at a sufficiently high rate. Similarly, young chicks require glycine yet the adult is able to synthesize sufficient for its needs. Moreover, cysteine is not essential provided sufficient sulphur in the form of methionine is provided. Animals are able to synthesize tyrosine from phenylalanine, itself an essential amino acid. Therefore, provided the animal's diet contains sufficient phenylalanine to supply the necessary quantity of these two amino acids, tyrosine is not itself essential. It is noteworthy that all the non-essential amino acids are glucogenic; the converse, however, is not true.

In order to test whether a particular amino acid is essential the procedure normally adopted is to feed the animal a special diet which is complete in every way except that the amino acid under consideration is absent. This can be done by feeding no protein and using pure amino acid mixtures to replace it. Under these conditions, if the amino acid is indeed essential, the animal develops a negative nitrogen balance which means that its excretion of nitrogen exceeds the nitrogen intake. This results from the breakdown of the more dispensable proteins in order to release the deficient amino acid for the synthesis of more urgently needed proteins. The excess nitrogen excretion comes from the catabolism of the more common of the non-essential amino acids released from dispensable protein. Thus a diet deficient in an essential amino acid produces symptoms similar to those obtained by feeding a very low protein diet. Long term lack of an essential amino acid prevents growth and produces deficiency diseases and even death—especially in growing animals which need to synthesize a large quantity of protein.

Many plant proteins contain relatively small quantities of the amino acids which are essential to animals. Methionine and lysine, in particular, are often only present in small quantities. This implies that an animal could develop deficiency diseases if fed exclusively on a poorly balanced vegetarian diet.

Many animal proteins, including those in muscle, eggs, and milk, contain all the essential amino acids in suitable quantities for adequate animal nutrition.

Ruminant animals such as cattle and sheep, present a special case. They do not suffer from deficiency of essential amino acids under normal conditions even though their fodder may not contain adequate amounts of them. The reason is that the micro-organisms in their specially adapted stomachs are able to synthesis all the essential amino acids and vitamins, and these are eventually made available to the animal when it subsequently digests the bacteria.

Nitrogen metabolism

12.5 Glutamine, aspartic acid, and carbamyl phosphate

The ammonium ion is very toxic to animals and only small quantities are present in the plasma of animals, the usual level being between 10^{-4} and 10^{-5} g per 100 ml. Nevertheless, the kidney produces large amounts of ammonia whenever it excretes acid urine and ammonia indirectly provides some of the nitrogen atoms present in a variety of complex nitrogen compounds (other than proteins) present in the cell. This ready availability of ammonia, despite its low concentration in plasma and cytoplasm, has a simple explanation: cells of both plants and animals possess what amounts to a nitrogen store from which ammonium ions can readily be formed on demand. In animals, the main store is the δ-amido-group of glutamine; blood contains as much as 10 mg glutamine in 100 ml. In plants, asparagine sometimes partly or completely replaces glutamine as the nitrogen store. *Glutamine* is synthesized from glutamic acid by the action of a widely distributed enzyme, *glutamine synthetase;* ATP is a necessary co-factor:

COOH $^{\delta}$CONH$_2$ ◄---- δ-Amido
| | group
CH$_2$ NH$_3$ H$_2$O $^{\gamma}$CH$_2$
| |
CH$_2$ ⟶ $^{\beta}$CH$_2$
| ATP ADP + P$_i$ |
CHNH$_2$ $^{\alpha}$CHNH$_2$ ◄---- α-Amino
| | group
COOH   COOH

   Glutamic acid   Glutamine

The amide nitrogen of glutamine is readily liberated by the enzyme, glutaminase.

$$\text{Glutamine} + \text{H}_2\text{O} \xrightarrow{\text{glutaminase}} \text{Glutamic acid} + \text{NH}_3$$

This reaction is exergonic, a fact which explains why ATP is necessary in the glutamine synthetase reaction which effectively bypasses the energetically unfavourable direct formation of glutamine from glutamate and ammonia.

In plants, *asparagine* is synthesized in a similar manner from aspartic acid, a dibasic acid with one less methylene group than glutamic acid.

Glutamine is the source of amino groups for a number of reactions. For example, two of the nitrogen atoms of the purine ring system originate from glutamine. The reactions whereby glutamine acts as donor of amino groups superficially resemble transamination. However, the amino group does not arise from the α-amino nitrogen (as it does in true transamination) but from the amide nitrogen:

$$\underset{\text{Acceptor}}{R} \quad + \text{Glutamine} \longrightarrow R{-}NH_2 + \text{Glutamic acid}$$

In addition to transamination and asparagine formation, *aspartic acid* undergoes another reaction which enables nitrogen to enter certain organic compounds. As in the case of transamination, the α-amino nitrogen atom is transferred, but the mechanism of the reaction is quite different. A hydroxyl group of the compound which is to receive the nitrogen reacts with the amino group of aspartic acid. Water is eliminated, to form a double molecule possessing a 'nitrogen bridge'. Energy for this condensation is derived from ATP or from GTP. The 'bridge', having been formed on one side of the nitrogen atom is next cleaved on the other, so that the nitrogen atom effectively changes sides. As the following scheme shows, the carbon skeleton of the aspartate is released as fumarate (which may then be incorporated into the TCA cycle):

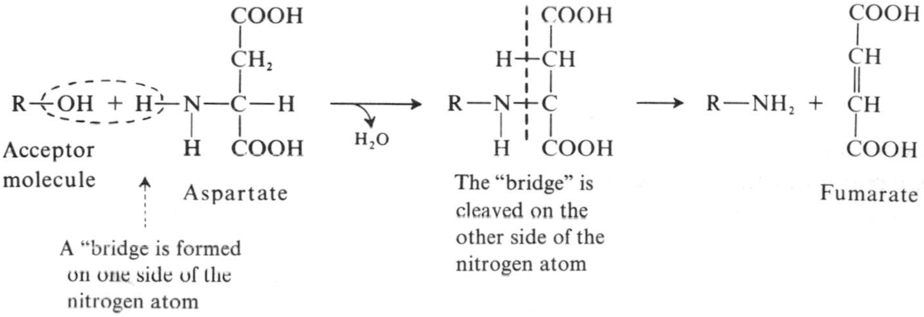

An important reaction of this type occurs in the synthesis of urea (section 12.7) and the amino side-group of the purine base, adenine, of AMP is inserted in this way into its hydroxylated precursor, inosine monophosphate (section 12.9; Fig. 12.5). In section 12.8 it will be shown that one of the nitrogen atoms in the purine ring system also originates from aspartic acid.

In some biosynthetic reactions, a carbon atom and a nitrogen atom are inserted together as a result of the participation of a substance known as *carbamyl phosphate*. This is a high-energy compound, being a 'mixed anhydride' of phosphoric and carbamic acid (compare pyrophosphoric acid or acetic anhydride). Carbamic acid is another name for amino-formic acid, $H_2N{-}COOH$.

Nitrogen metabolism

Carbamyl phosphate is formed from ammonium ions in the mammalian liver by a reaction catalysed by carbamyl phosphate synthase.

$$H_3N \ + \ CO_2 \ + \ 2ATP \ \xrightarrow[\text{synthase}]{\substack{\text{carbamyl} \\ \text{phosphate}}} \ H_2N-C\overset{O}{\underset{O \sim \circled{P}}{\diagup}} \ + \ 2ADP \ + \ P_i$$

It will be observed that two molecules of ATP provide energy for this synthesis, but micro-organisms appear to synthesize carbamyl phosphate by a process which requires the utilization of only one molecule of ATP.

Several biosynthetic functions of carbamyl phosphate will be encountered in later sections. In mammals, it takes part in the synthesis of urea, the major end-product of protein metabolism in these animals. Carbamyl phosphate and aspartate also cooperate in the synthesis of the pyrimidine ring (section 12.9).

12.6 Nitrogen excretion in animals

In the catabolism of amino acids the first reaction is in most cases the removal of the α-amino group by transamination, the product being the corresponding keto acid. The amino group is thereby transferred to either oxaloacetate or α-ketoglutarate so producing aspartate or glutamate respectively.

$$R-CHNH_2-COOH + Oxaloacetate \underset{\text{transaminases}}{\overset{\text{specific}}{\rightleftharpoons}} R-CO.COOH + Aspartate$$

$$R-CHNH_2-COOH + \alpha\text{-Ketoglutarate} \rightleftharpoons R-CO.COOH + Glutamate$$

Aspartate can, in turn, pass on the amino group to α-ketoglutarate. Much of the glutamic acid so formed is then oxidized by the action of the enzyme glutamate dehydrogenase, which, as was mentioned in section 12.2, is located in the mitochondria. The result is that *ammonia* and NADH are produced, and α-ketoglutarate is regenerated. The NADH, upon reoxidation by the mitochondrial electron transport system, yields ATP by oxidative phosphorylation (chapter 7).

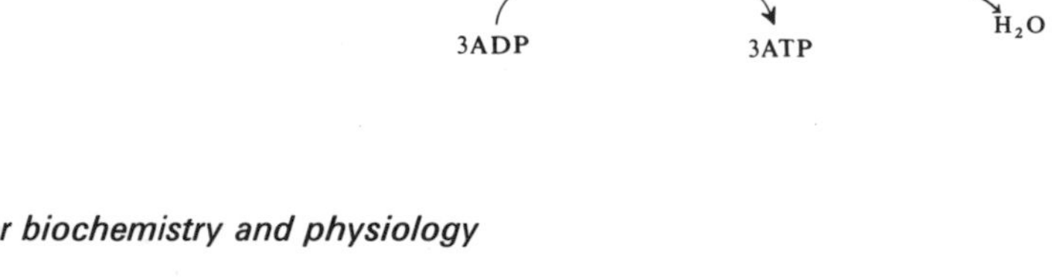

$$Glutamate \xrightarrow{\text{glutamate dehydrogenase}} \alpha\text{-Ketoglutarate} \ + \ NH_3$$

NAD$^+$ NADH + H$^+$

$2e + 2H^+$

(electron transport system)   O$_2$

3ADP   3ATP   H$_2$O

Many aquatic animals excrete the ammonium ion. It is, in particular, the principal nitrogenous substance excreted by aquatic invertebrates, although it is also eliminated by bony fish through their gills. In other animals, several mechanisms have evolved for excreting nitrogen in less toxic forms. Thus cartilaginous fish (e.g., sharks), adult amphibians, and mammals excrete nitrogen, not as ammonia, but as *urea* (Table 12.2), the amide of carbonic acid. Urea is a relatively harmless, soluble

Table 12.2   Nitrogenous excretory products of animals

Compound		Animals excreting the compound
Ammonium ion	$NH_4{}^+$	Most aquatic invertebrates; also excreted by the gills of bony fish
Urea	$H_2N-\overset{\overset{\displaystyle O}{\|}}{C}-NH_2$	Cartilaginous fish, amphibians, and mammals
Trimethylamine oxide	$(H_3C)_3N \rightarrow O$	Excreted by the kidneys of marine bony fish
Uric acid	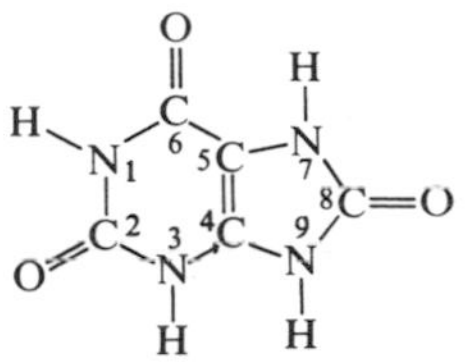	Insects, terrestrial molluscs, terrestrial reptiles, and birds. In many mammals, uric acid is the end-product of purine but not of amino acid metabolism
Guanine	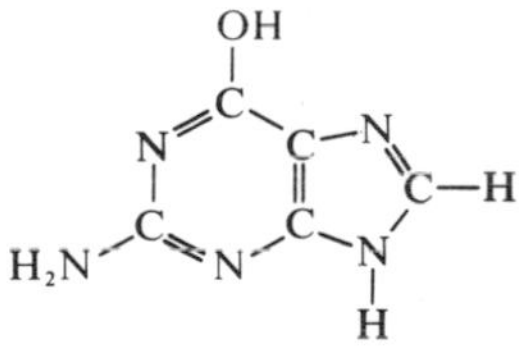	Spiders
Allantoin	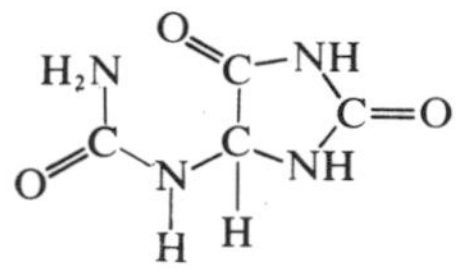	Product of the catabolism of *purines* in **many** mammals (e.g., dog). An enzyme, *uricase,* oxidizes uric acid removing carbon **atom** number 6 to produce allantoin
Creatinine	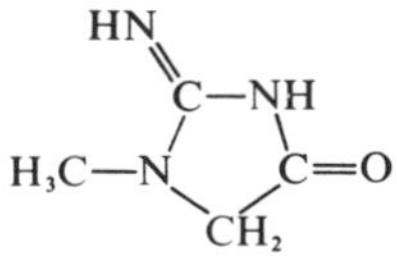	Vertebrates. It is a product of the breakdown of creatine phosphate, the energy store of muscle

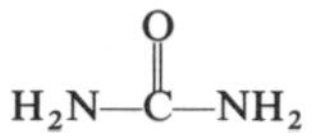

compound so, in animals where it represents the major form of nitrogen excretion, it is possible for it to accumulate to quite high concentrations without deleterious effects. Thus, while an ammonium ion concentration of 0·2 mg per 100 ml in vertebrate blood is abnormally high, the normal level of urea in mammalian blood is about 40 mg per 100 ml. This has an important consequence in amphibia and mammals for it enables them to excrete nitrogen in a concentrated form and so to survive on land where water is a more precious commodity than it is for organisms living in an aqueous environment.

The body fluids of marine vertebrates tend to have a lower osmotic pressure than sea water. Consequently, water would be drawn out of fish by osmosis if they did not have special regulatory mechanisms to oppose the water loss (chapter 15). Like terrestrial animals, therefore, marine fish live in an environment where conservation of body water is vital. Cartilaginous fish have a high blood urea concentration which raises their osmotic pressure to that of sea water, so preventing loss of excess water by osmosis. The urine of many marine bony fish contains trimethylamine oxide which, like urea, is a relatively non-toxic yet highly soluble compound.

From the point of view of conservation of water, birds and insects are the most successful of terrestrial animals, for the bulk of their unwanted nitrogen is excreted in the form of the tri-keto-purine, *uric acid* (Table 12.2). This sparingly soluble substance is excreted as a solid with the result that little water is wasted in eliminating it. Such animals are described as *uricotelic* to contrast them with those which excrete nitrogen predominantly in the form of ammonia or urea—and which are known as *ammoniotelic* and *ureotelic* respectively. Spiders excrete nitrogen mainly as the purine, guanine, which, like the uric acid excreted by insects, is eliminated as a relatively dry mass.

Several nitrogenous compounds are present in mammalian urine. Of these, urea is quantitatively the most important, but small quantities of uric acid, allantoin, ammonia, and creatinine (Table 12.2) are also present. The uric acid, unlike the bulk of the urate produced by uricotelic animals, arises from the catabolism of nucleotide bases of nucleic acids. Some mammals further degrade the uric acid to allantoin. Creatinine arises from creatine, the compound which, as creatine phosphate, acts as a store of high energy phosphate in the muscles (p. 135). The urea, on the other hand, is the main end product of protein catabolism. Thus when mammals are fed a very high protein diet the excretion of urea rather than of other nitrogenous compounds increases (Table 12.3).

Table 12.3 Relative amounts of nitrogenous compounds excreted by mammals on high and low protein diets

	High nitrogen diet	Low nitrogen diet	Difference
Total nitrogen excreted	30	10	20
urea	24	6	18
creatinine	1	1	0
ammonia	1	1	0
uric acid	2	1	1
others	~ 2	~ 1	~ 1

12.7 The formation of urea in mammals and amphibia

The breakdown of urea to carbon dioxide and ammonia is an exergonic reaction catalysed by urease.   Since the reaction is virtually irreversible, the biosynthesis of finite amounts of urea by direct combination of ammonia and carbon dioxide is not possible.   Krebs and his co-workers demonstrated in 1932 that small quantities of arginine or ornithine significantly increased the production of urea by mammalian liver slices. More recently citrulline was shown to have a similar stimulatory effect. It is found that three ATP molecules are required for the formation of one urea molecule—clearly, mammals expend considerable energy simply to render the catabolic product of nitrogen metabolism innocuous.

The complete sequence of reactions by which urea is synthesized from ammonia and carbon dioxide is known as the *ornithine cycle* and can be divided into a preliminary step followed by three others.   All the enzymes necessary for the synthesis are present in the liver, although some of them are present in other organs as well (e.g., in the kidney and the mammary gland).

The preliminary step in the synthesis of urea is the formation of *carbamyl phosphate* from carbon dioxide and ammonia.   It will be recalled (section 12.5) that two molecules of ATP are converted to ADP for each molecule of carbamyl phosphate formed.   The carbamyl phosphate provides the carbon atom and one of the two nitrogen atoms of the urea molecule (Fig. 12.3).

Step 1   In this reaction, the carbamyl phosphate reacts with a diamino acid, *ornithine*.   Ornithine, which gives the cycle of reactions its name, is an acid containing five-carbon atoms in its molecule, and is thus related to *glutaric acid* and *glutamic acid*.   Since ornithine is re-formed

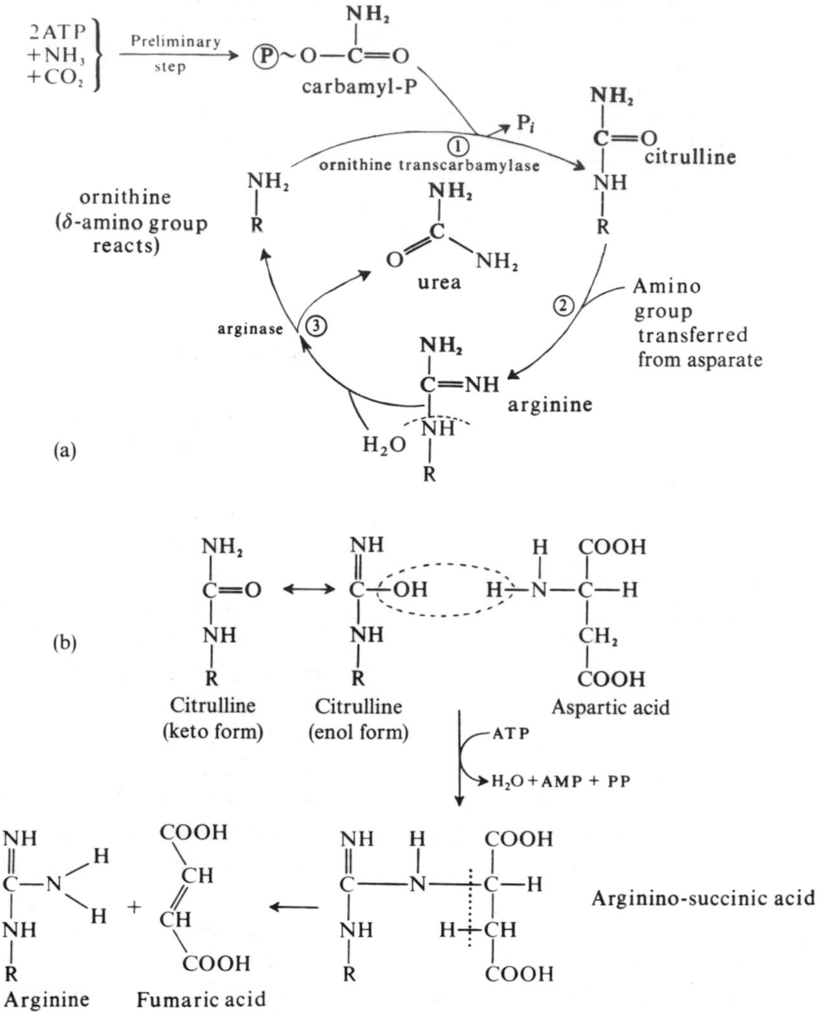

2ATP
+NH$_3$
+CO$_2$ } $\xrightarrow[\text{step}]{\text{Preliminary}}$ ⓟ~O—C=O

NH$_2$

carbamyl-P

P$_i$

ornithine transcarbamylase ①

citrulline
NH$_2$
C=O
NH
R

ornithine
(δ-amino group
reacts)

NH$_2$
R

NH$_2$
C
O   NH$_2$
urea

arginase ③

NH$_2$
C=NH
NH
R
arginine

② Amino
group
transferred
from asparate

H$_2$O

(a)

(b)

NH$_2$
C=O
NH
R
Citrulline
(keto form)

⟷

NH
C—OH
NH
R
Citrulline
(enol form)

H   COOH
H—N—C—H
CH$_2$
COOH
Aspartic acid

ATP

H$_2$O + AMP + PP

NH
C—N
NH   H
R   +
H

COOH
CH
CH
COOH

←

NH   H   COOH
C—N—C—H
NH   H—CH
R   COOH

Arginino-succinic acid

Arginine   Fumaric acid

Fig. 12.3   The ornithine cycle for the synthesis of urea.

(R ≡ —CH$_2$.CH$_2$.CH$_2$.CHNH$_2$.COOH)

(a) Outline of ornithine cycle
(b) Details of step 2, the conversion of citrulline to arginine

in the third and final step of the sequence, it can be regarded as a co-catalyst (Fig. 12.3a).

The carbamyl group is transferred from carbamyl phosphate to ornithine by an enzyme known as *ornithine transcarbamylase*. The amino acid so formed is called citrulline, the other product being inorganic orthophosphate. Citrulline consists of a carbamyl group attached to the terminal or δ-amino group of ornithine:

$$
\underset{\text{Carbamyl phosphate}}{\overset{\displaystyle O \atop \displaystyle \|}{\text{P}-O-C-NH_2}}
\qquad
\overset{\text{ornithine}}{\underset{\text{transcarbamylase}}{\longrightarrow}}
\qquad
\underset{\text{Citrulline}}{\overset{\displaystyle O \atop \displaystyle \|}{C-NH_2}}
$$

H
NH—(CH$_2$)$_3$—CH—COOH
        |
        NH$_2$

Ornithine

P$_i$

C—NH$_2$
NH—(CH$_2$)$_3$—CH—COOH
                 |
                 NH$_2$

Citrulline

Step 2 An amino group is next transferred from aspartic acid to the carbamyl keto group of citrulline, to form the amino acid *arginine*. It should be remembered that the aspartic acid can itself originate from oxaloacetic acid by transamination from glutamic acid and so, via the glutamate dehydrogenase reaction, its nitrogen atom can originate from unwanted ammonium ions (Figs. 12.2 and 12.3). This nitrogen atom eventually becomes the second nitrogen atom of the final urea molecule. The reaction concerned is an example of one in which aspartic acid forms a 'nitrogen bridge' which is later cleaved on the opposite side from that on which it is formed (section 12.5). The intermediate double molecule is called arginino-succinic acid, and the energy for the formation of the 'nitrogen bridge' originates from ATP, the latter being converted to AMP and pyrophosphate. The details are shown in Fig. 12.3b. Upon cleavage of the bridge, the carbon skeleton of the aspartate is released as fumarate. Aspartate can be regenerated from the fumarate by its conversion to oxaloacetic acid by the enzymes of the TCA cycle, followed by transamination of the oxaloacetate so formed.

Step 3 Finally, urea is formed by the hydrolysis of arginine. Ornithine is thereby regenerated and can then participate in the next turn of the cycle. This hydrolytic cleavage of a C—N bond is catalysed by the enzyme *arginase*. The reaction is best visualized as one which leads to the formation of the imino resonance form of urea, followed by intra-molecular rearrangement:

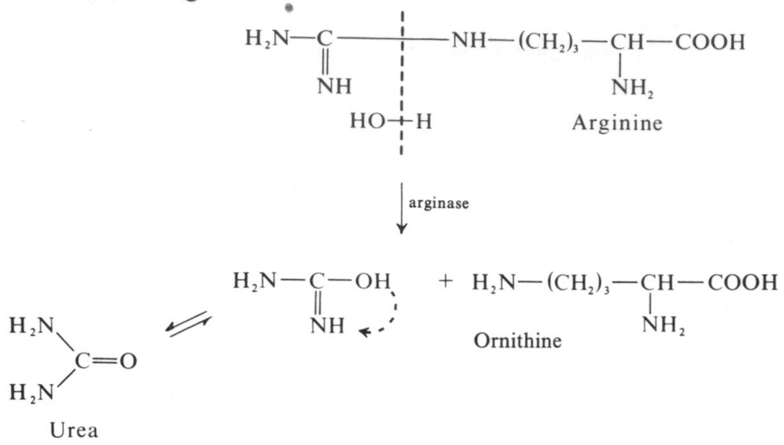

$$H_2N-\underset{\underset{NH}{\|}}{C}\text{------}NH-(CH_2)_3-\underset{\underset{NH_2}{|}}{CH}-COOH$$

$$HO\text{---}H \qquad\qquad \text{Arginine}$$

$$\downarrow \text{arginase}$$

$$\underset{H_2N}{\overset{H_2N}{>}}C=O \qquad \rightleftharpoons \qquad H_2N-\underset{\underset{NH}{\|}}{C}-OH \quad + \quad H_2N-(CH_2)_3-\underset{\underset{NH_2}{|}}{CH}-COOH$$

$$\text{Urea} \qquad\qquad\qquad\qquad\qquad\qquad \text{Ornithine}$$

The urea so produced leaves the liver by the blood stream, which carries it to the kidney.

12.8 Fate of amino-nitrogen in uricotelic animals

The *synthesis of the purine ring system* is essentially the same whether the synthesis is a device leading to excretion in uricotelic animals or whether it is the first step in the anabolism of the nucleotide bases adenine and guanine and hence (in *all* animals and plants) of the nucleic acids. It is thus of *general* importance, as well as being of special interest in relation to excretion. In addition, it is a fascinating example of how cells are able to construct complicated organic molecules, step by step, from relatively simple precursors.

Three nitrogen atoms of the purine ring system originate from ammonia. Of these two are incorporated after the ammonia has been converted to the amide group of *glutamine*; the third molecule of ammonia undergoes prior conversion to *aspartate*. The fourth nitrogen atom of the purine structure, and two of the carbon atoms, originate from a *glycine* molecule. Of the three remaining carbon atoms, one is derived from carbon dioxide, and the other two are incorporated in the form of a compound, which, for the moment, will be called 'active formate'. Free formate cannot be incorporated. The origin of the atoms in the rings can be summarized thus:

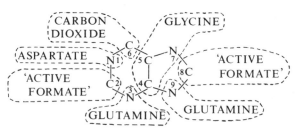

The process leading to purine synthesis is complex and only a simplified account is presented here. It can be followed by constant reference to Fig. 12.4, in which each atom of the ring is numbered in the reaction where it is inserted. The route of synthesis is such that the ribose and phosphate of the prospective nucleotide are inserted first (being shed later in the case of uric acid formation). The *smaller* of the two rings of the purine system is completed before the larger ring.

The purine begins its development by the addition of an amino group to carbon atom number 1 of ribose-5-P. This amino group eventually becomes nitrogen atom number 9 of the completed purine. It is derived from glutamine which is thereby converted to glutamate, ATP providing the necessary energy. Next, a whole glycine molecule is added to the structure in such a way that it provides carbon atoms 4 and 5, and nitrogen atom 7, of the final product. Again, ATP provides energy for the reaction. Now carbon atom number 8 of the system is added in the form of 'active formate'.

In the next step, the five-membered ring is completed by ring closure, and an additional nitrogen atom (number 3) is added by transfer from glutamine, ATP supplying energy for the process. The future carbon atom number 6 is now added but in this case the carbon incorporated is in the form of carbon dioxide (not as 'active formate'). This is followed by the addition of the last nitrogen atom of the purine ring system (atom number 1). It arises from aspartate via an intermediate 'double molecule' with a nitrogen bridge (section 12.5). When cleavage of the bridge takes place, the amino group is transferred to the purine, and the residue is fumarate. ATP provides the energy for the reaction. Finally, carbon atom number 2 of the purine base is added—as 'active formate'—and a complete purine nucleotide, inosine monophosphate (IMP) results as closure takes place to form the larger ring.

In uricotelic animals, uric acid is synthesized by the pathway leading to IMP. The process is completed by the removal of the ribose-P from

Nitrogen metabolism

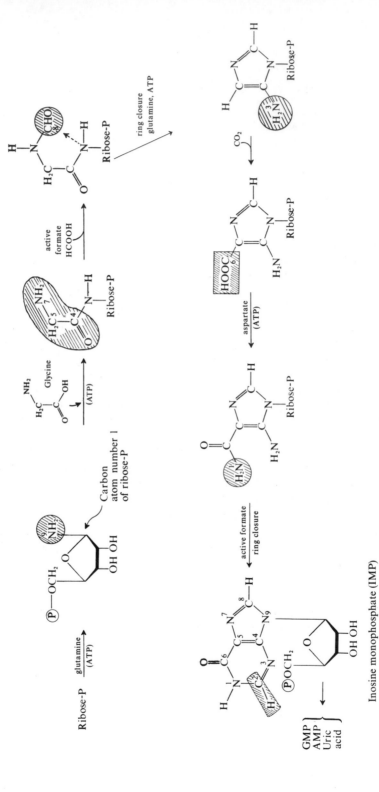

Inosine monophosphate (IMP)

Fig. 12.4 Simplified outline of purine synthesis.   (Numbers refer to atoms in the final purine molecule)

the purine, the latter (hypoxanthine) then being oxidized by flavoprotein enzymes to uric acid:

IMP $\longrightarrow$ Ribose-5-P +

Hypoxanthine
(mono-hydroxy
purine)

oxidation

Xanthine
(di-hydroxy
purine)

oxidation

tautomeric
forms

Uric acid
(triketo purine form)

Uric acid
(trihydroxy purine form)

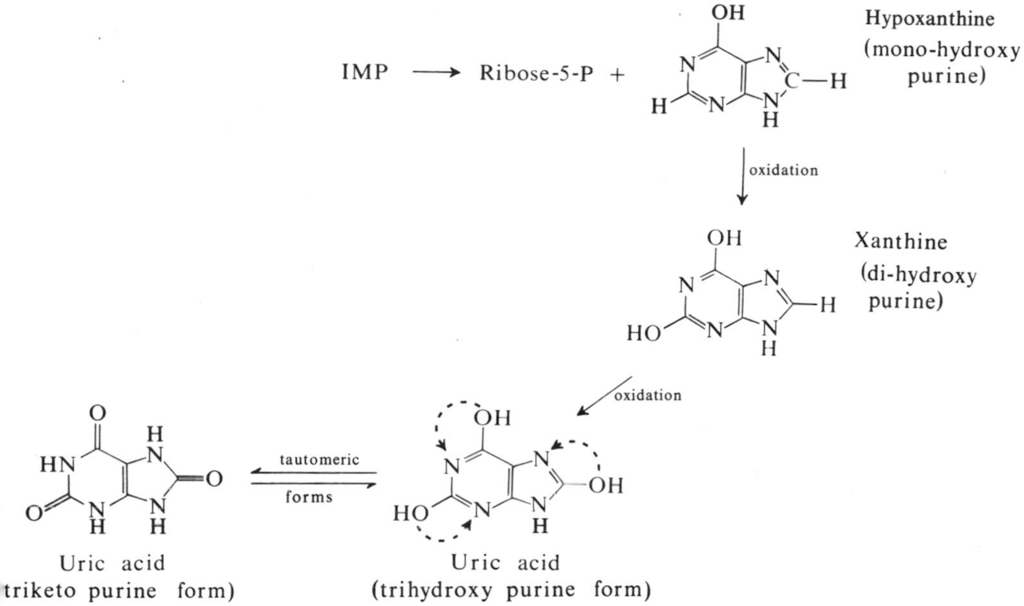

Most non-uricotelic animals, including mammals, excrete some uric acid but this is derived largely from adenine and guanine and is a waste product of nucleic acid metabolism. In this case, the purine bases are hydrolysed from the ribose, deaminated, and then oxidized by flavoprotein enzymes to uric acid.

12.9 The biosynthesis of some important compounds containing nitrogen

(a) The nucleotides AMP and GMP

As we have just seen, the nucleotide inosine monophosphate, IMP, is synthesized from a number of relatively simple compounds. It is, in turn, the precursor of AMP and GMP. IMP is actually *hypoxanthine-ribose-phosphate*, hypoxanthine being a monohydroxy purine (or mono-keto purine since tautomeric change is possible). It is converted to AMP or GMP by the steps shown in the Fig. 12.5; the roles, as nitrogen donors, of aspartate (which forms an intermediate compound with a nitrogen bridge) and of glutamine, are noteworthy.

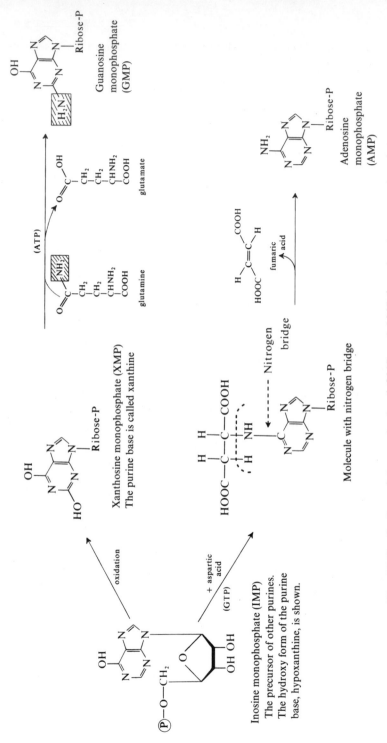

Fig. 12.5 Outline of the synthesis of GMP and AMP from inozine monophosphate

(b) The pyrimidines

As was mentioned earlier, carbamyl phosphate and aspartic acid participate in pyrimidine biosynthesis. They first unite to give carbamyl aspartate, inorganic phosphate being eliminated. The former then loses a molecule of water in the manner indicated in Fig. 12.6 to produce a heterocyclic ring containing two nitrogen atoms. The product, dihydro-orotic acid is the precursor of all the pyrimidines of biological importance, for, in a series of steps, it is oxidized, decarboxylated, and finally (in contrast to the purines) attached to ribose phosphate to give one or other of the common pyrimidine nucleotides (p. 254).

(c) The porphyrins

The *porphyrin* ring system occurs in the cytochromes (p. 144) and in haemoglobin. It contains four pyrrole rings united by methine ($-CH=$) bridges. An example, protohaem, is depicted in Fig. 6.6.

The synthesis begins with a reaction in which the TCA cycle intermediate, succinyl coenzyme A, reacts with glycine. Coenzyme A and the carboxyl group of the glycine molecule are eliminated, producing δ-aminolaevulinic acid. This acid contains five carbon atoms and could also be called γ-keto,-δ-amino-valeric acid:

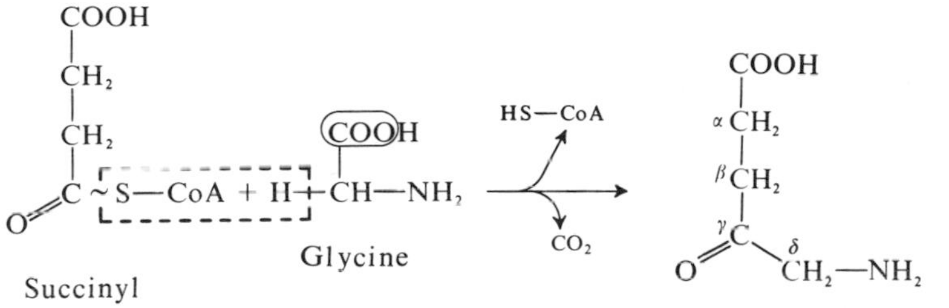

Succinyl coenzyme A   Glycine δ-Aminolaevulinic acid

The five-membered heterocyclic pyrrole ring is then formed by the condensation of two molecules of δ-aminolaevulinic acid, two molecules of water being eliminated, the oxygen atom in each case being provided by the γ-carbonyl groups. The product is known as *porphobilinogen*; two of its carbon atoms and both nitrogen atoms are derived from the two original glycine molecules (Fig. 12.7).

The porphyrin structure is derived from four porphobilinogen molecules which link together with the elimination of the side-chain amino groups. The mechanism is complicated and only partially understood. One curious feature is that an isomerization occurs in the pyrrole ring number

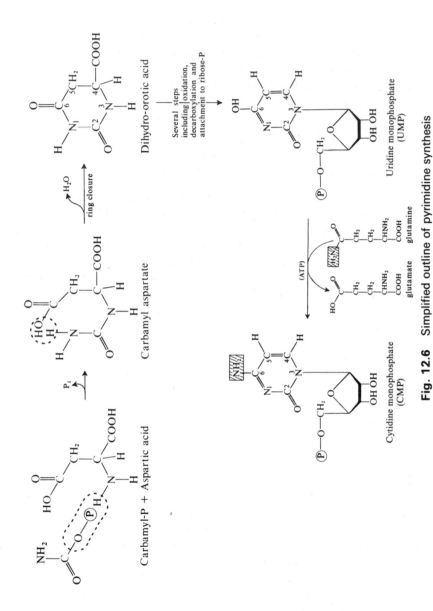

Fig. 12.6 Simplified outline of pyrimidine synthesis

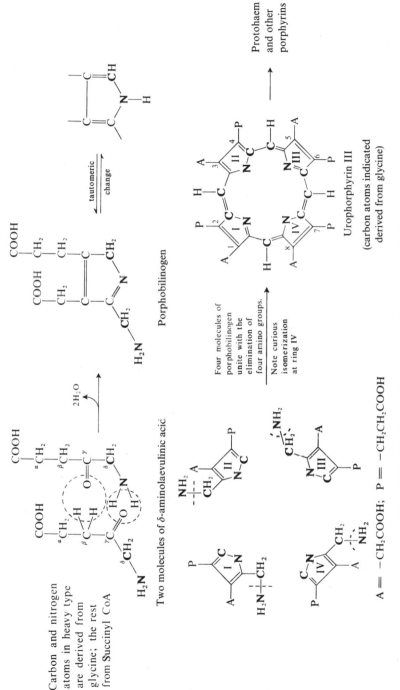

Carbon and nitrogen
atoms in heavy type
are derived from
glycine; the rest
from Succinyl CoA

Two molecules of δ-aminolaevulinic acid

Porphobilinogen

Four molecules of
porphobilinogen
unite with the
elimination of
four amino groups.

Note curious
isomerization
at ring IV

Uroporphyrin III

(carbon atoms indicated
derived from glycine)

Protohaem
and other
porphyrins

tautomeric
change

A = —CH₂COOH; P = —CH₂CH₂COOH

Fig. 12.7 Outline of the synthesis of the porphyrin ring system

IV (Fig. 12.7). Thus side groups 1, 3, and 5 of the first identifiable porphyrin, uroporphyrin III, are —$CH_2 \cdot COOH$ groups but it is side group 8, and not 7, which has this structure. The figure shows that each of the methine groups linking the pyrrole rings is derived from the original glycine precursor, as, of course, are the nitrogen atoms of the pyrrole rings. The only other compound contributing to the porphyrin structure is succinyl coenzyme A.

Most other porphyrins are derived from uroporphyrin III by modification of the side groups. For example, in protohaem (Fig. 6.6, p. 144) each of the side groups of the type —$CH_2 \cdot COOH$ has been decarboxylated to a methyl group, while the side group —$CH_2 \cdot CH_2 COOH$ is unmodified at positions 6 and 7 but has been converted to —$CH = CH_2$ at positions 2 and 4.

Metabolism of the carbon skeletons of the amino acids

The L-amino acids occupy a central position in cellular metabolism since almost all biochemical reactions are catalysed by enzymes comprising L-amino acid residues. Moreover, the routes of synthesis and catabolism of amino acids link up with the biochemical pathways followed by carbohydrates and fatty acids, so amino acid biochemistry cannot be isolated from that of these other two groups of biologically important compounds.

Green plants and many bacteria can synthesize each of the L-amino acids from relatively simple carbon compounds such as triose phosphate or acids of the TCA cycle and these, in turn, are the products of lipid or of carbohydrate metabolism (chapters 7, 8, and 9). Oxaloacetic acid, for example, can act as the direct precursor of the carbon skeletons of some five amino acids. Animals, however, have lost the ability to synthesize about nine amino acids and they must, therefore, obtain these 'essential' amino acids from their diet (p. 272).

The catabolism of amino acids leads to the formation of ammonia and other nitrogen compounds. The carbon skeletons which remain are mostly converted to acids of the TCA cycle, to pyruvate, propionate, or to acetyl coenzyme A. As discussed in chapter 10, several of these nitrogen-free acids can act as the precursors of polysaccharides although, in animals, acetyl coenzyme A can only be oxidized by the TCA cycle or used for the synthesis of lipids. In addition, several amino acids provide the raw material for the synthesis of other types of important compounds; the formation of purines and porphyrins was mentioned in the previous chapter and other examples are discussed in section 13.10. The present chapter deals mainly with the synthesis of amino acids and with a number of important conversions they undergo. Some of these metabolic

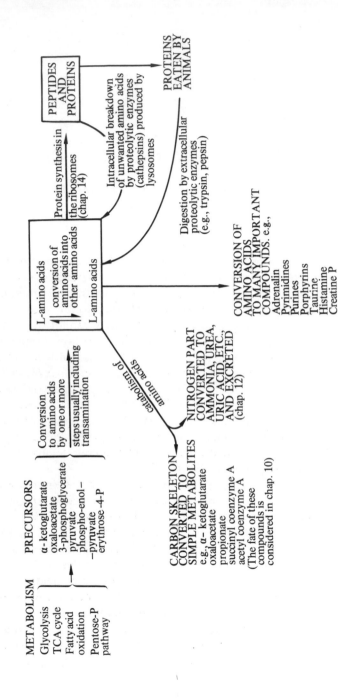

Fig. 13.1 Outline of the major metabolic reactions involving amino acids

changes are summarized in Fig. 13.1.   Protein synthesis, being a somewhat special topic, is described in chapter 14.

13.1   The transfer of units containing only one carbon atom

Attention must now be drawn to two biochemical systems which transfer one-carbon fractions—methyl, formyl, formimino, and formaldehyde—from one molecule to another, for such systems are frequently involved in the metabolism of amino acids.

It will be recalled (chapter 8) that two-carbon (2-C) acetate units seldom participate directly in cellular biochemistry but in the form of 'active acetate'.   As has been seen, 'active acetate' owes its activity to the fact that the acetate group is actually attached to a carrier or coenzyme by a high-energy bond, being, in fact, in the form of acetyl coenzyme A.   The activity of 1-C fragments is somewhat similar, though the carriers are different.   One carrier functions as a carrier of active methyl groups, and a second as a carrier of 'active formate' and of the two similar fragments containing one carbon atom mentioned above.

Methionine is not only an essential amino acid but acts as a donor of methyl groups.   In order that it can fulfil this function, its sulphur atom has to assume the charged *sulphonium* form and this is achieved by reaction with ATP.   'Active methyl' is thus usually 'active methionine' and the latter is *S-adenosyl methionine* (the S- indicating that the adenosine is attached to the sulphur atom).   The structure of this compound is shown in Fig. 13.2.   Following heterolytic cleavage, the positively-charged methyl group can attach itself to several types of atoms possessing a suitably-placed lone pair of electrons.   After donating its methyl group, the methionine is liberated as a sulphydryl amino acid; this is known as homocysteine because it contains one methylene group more than cysteine:

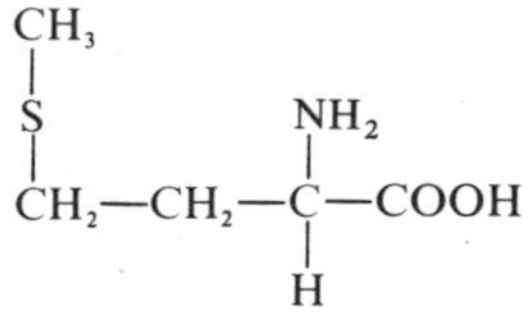

$$CH_3$$
$$|$$
$$S \qquad\qquad NH_2$$
$$| \qquad\qquad\quad |$$
$$CH_2-CH_2-C-COOH$$
$$|$$
$$H$$

Methionine

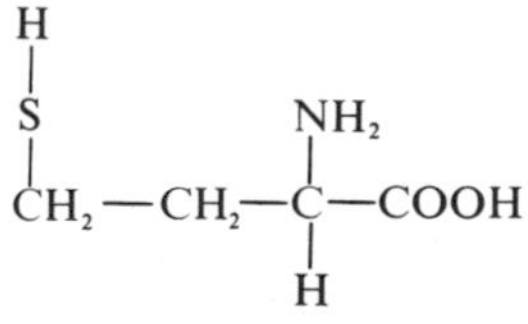

$$H$$
$$|$$
$$S \qquad\qquad NH_2$$
$$| \qquad\qquad\quad |$$
$$CH_2-CH_2-C-COOH$$
$$|$$
$$H$$

Homocysteine

$$H$$
$$|$$
$$S \qquad\quad NH_2$$
$$| \qquad\qquad |$$
$$CH_2-C-COOH$$
$$|$$
$$H$$

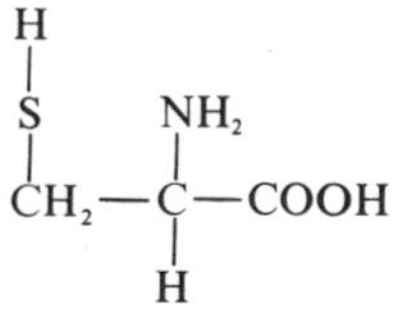

Cysteine

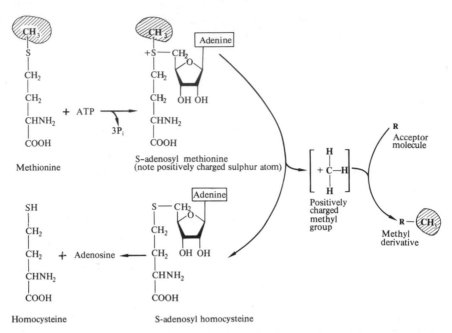

Fig. 13.2 The structure of S-adenosyl methionine ('active methionine') and the mechanism of transmethylation

In its simplest form, and without introducing the appropriate cofactors and enzymes, the transfer of a methyl group can be represented as follows:

$$H_3C \cdots S—(CH_2)_2.CHNH_2.COOH \qquad\qquad H—S—(CH_2)_2.CHNH_2.COOH$$

Methionine (donor)   Homocysteine

$$HO—CH_2CH_2—N\begin{smallmatrix}CH_3\\CH_3\end{smallmatrix} \quad + \quad \begin{smallmatrix}H\\-\!-\!\vert\!-\!-\\OH\end{smallmatrix} \quad\longrightarrow\quad HO—CH_2CH_2—\overset{+}{N}\begin{smallmatrix}CH_3\\CH_3\\CH_3\end{smallmatrix} \quad + \quad OH^-$$

Dimethylethanolamine (acceptor)   Choline

The transfer of formyl and similar fractions is achieved by the mediation of a carrier derived from the vitamin *folic acid*. It is known both as *tetrahydrofolate* and as *coenzyme F*. We shall adopt the second name, to bring it into line with that of coenzyme A, the carrier for 2-C units. It has the somewhat complex structure shown in Fig. 13.3. The presence of a p-aminobenzoic acid residue is of interest for it is the structural

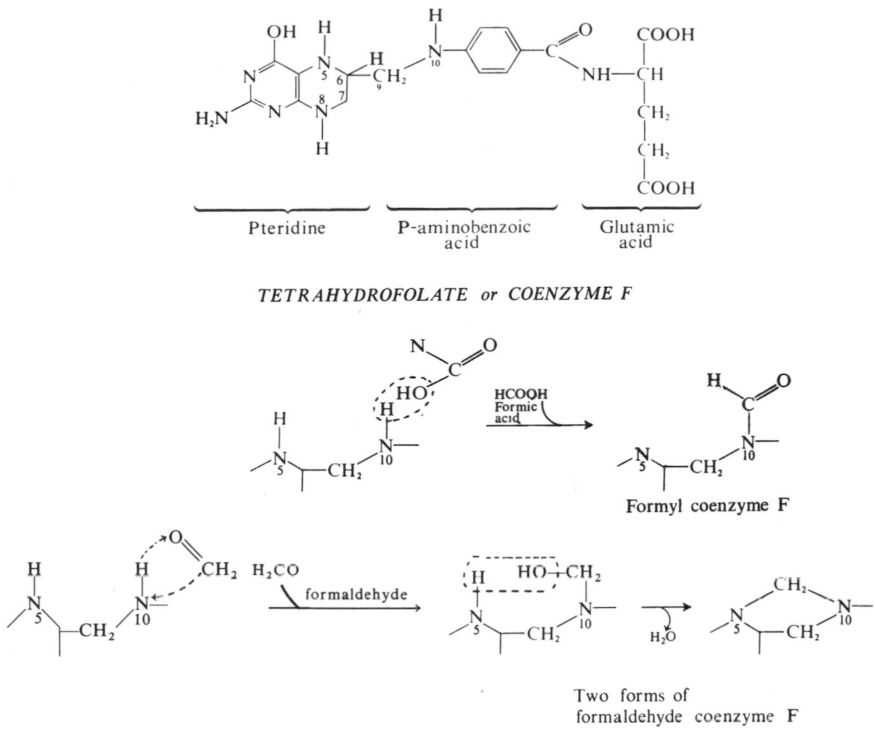

OH H

H_2N

Pteridine   P-aminobenzoic acid   Glutamic acid

COOH
NH—CH
CH₂
CH₂
COOH

TETRAHYDROFOLATE or COENZYME F

N
C O
HO
H
H
N—
5
CH₂
10

HCOOH
Formic acid

H
C O
N
5
CH₂
N—
10

Formyl coenzyme F

H
N
5
CH₂
H
N
10
O
CH₂

H₂CO
formaldehyde

H
N
5
CH₂
HO CH₂
N—
10

H₂O

N
5
CH₂
CH₂
N—
10

Two forms of
formaldehyde coenzyme F

Fig. 13.3 The structure of coenzyme F and examples of 'one-carbon' (1-C) compounds

similarity of the sulphonamide drugs to this compound which accounts for their bacteriostatic and bactericidal action. The functional part of coenzyme F is that part of the molecule involving the nitrogen atoms marked 5 and 10, and the figure shows how various labile 'one-carbon' (1-C) fractions are attached (but not the precise mechanism leading to their attachment). Note that the nature of the attachment is such that formyl coenzyme F, produced from formate, looks superficially like an aldehyde, and that 'formaldehyde coenzyme F', in one of its two forms, bears a superficial resemblance to an alcohol.

The use of coenzyme F derivatives as donors of 1-C fractions will be illustrated in relation to serine, for this amino acid very commonly donates 1-C units to coenzyme F. It was mentioned in the previous chapter that 'active formate' participates in the synthesis of purines; it is also necessary for the synthesis of such important compounds as the porphyrins, histidine,

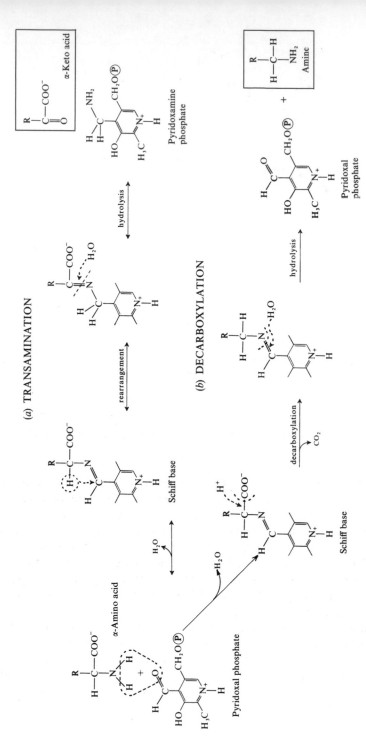

Fig. 13.4 The role of derivatives of vitamin B₆ in the mechanism of (a) transaminases and (b) amino acid decarboxylases

and tryptophan. Furthermore, there is an incompletely-understood mechanism, but one which involves vitamin B_{12} (the vitamin that contains cobalt), which allows some form of reversible cross-connection between the 'active formate' and 'active methyl' systems. Indirectly, and by means of this bridge, coenzyme F derivatives can be used to synthesize methionine and choline.

13.2   The role of derivatives of vitamin B_6 in amino acid metabolism

Many enzymes which catalyse reactions involving amino acids contain derivatives of vitamin B_6. This vitamin occurs in forms known as *pyridoxal*, which is an aldehyde, and *pyridoxamine*, an amine. However, the vitamin does not take part directly in metabolism, but is first phosphorylated at the expense of ATP to form *pyridoxal phosphate* or *pyridoxamine phosphate*, the formulae of which are shown in Fig. 13.4. These are essential cofactors for the transaminases, enzymes which are also known as *amino transferases*. Transamination was discussed in the previous chapter and it is now appropriate to describe the mechanism by which it occurs.

Pyridoxal phosphate is attached to amino transferases by links which probably involve the phosphate group. Acting as the prosthetic group of the enzyme, pyridoxal phosphate reacts with an α-amino acid in such a way that the aldehyde group condenses with the α-amino group of the acid with the elimination of water, the product being a 'Schiff base' type of compound$^\#$ (Fig. 13.4a). There are several possible resonance forms of the Schiff base; the one illustrated is readily hydrolysed to form pyridoxamine phosphate, still attached to the enzyme, and a free α-keto acid corresponding to the original α-amino acid. The reactions illustrated are reversible. Thus with the enzyme alanine:α-ketoglutarate amino transferase, the α-amino group of alanine may be transferred to enzyme-bound pyridoxal phosphate to form pyruvate and pyridoxamine phosphate. The amino group is transferred from the latter to α-*ketoglutarate* to form glutamic acid and thus the prosthetic group is converted back to the pyridoxal phosphate form:

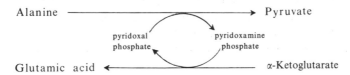

Thus, during transamination, the aldehyde and amino forms of the cofactor are repeatedly undergoing interconversion. The importance of the cofactor is shown by the fact that the tissues of animals which are deficient in vitamin B_6 have reduced transaminase activity, the activity being restored to normal when the vitamin is supplied in the diet.

The *amino acid decarboxylases* represent a second group of enzymes for which pyridoxal phosphate is the cofactor. These enzymes catalyse the general reaction where the α-carboxyl group of an amino acid is removed as carbon dioxide, leaving the corresponding primary amine:

$$R—CHNH_2—COOH \rightarrow CO_2 + RCH_2NH_2$$

The reaction is brought about by the amino acid forming a Schiff base with enzyme-attached pyridoxal phosphate (Fig. 13.4b). The carboxyl group of the amino acid is thereby made labile so that it breaks away as carbon dioxide. Finally the remaining part of the original amino acid, the amine, is hydrolysed from the pyridoxal phosphate. Several examples of amino acid decarboxylases are described in section 13.10.

Other instances where derivatives of vitamin B_6 are involved as cofactors for enzymes which catalyse various aspects of amino acid metabolism are considered later in this chapter. In view of their number and variety it is not surprising that deficiency of the vitamin produces severe physiological disturbance or even death in animals.

13.3 The general pattern of amino acid synthesis and catabolism

In *plants,* the carbon skeleton of each of the twenty or so amino acids is derived from simple precursors. Thus two acids of the TCA cycle, namely, *oxaloacetate* and *α-ketoglutarate,* each give rise to some five amino acids. An intermediate compound of glycolysis, *3-phosphoglyceric acid,* is the precursor of four amino acids, while three more arise from *pyruvate.* D-*Erythrose-4-P,* synthesized in the pentose-phosphate pathway (chapter 11), and *phosphoenol pyruvate* together provide carbon atoms for phenylalanine, tyrosine, and tryptophan. The origin of the carbon skeletons of amino acids is summarized in Table 13.1. Two amino acids, lysine and histidine, do not readily fit into the scheme. The synthesis of lysine differs in various groups of organisms—it is an essential amino acid in animals. The purine ring of ATP provides most of the atoms for histidine synthesis.

Table 13.1 Origin and major catabolic fate of carbon skeletons of amino acids

Precursor	Amino acid	Major product of catabolism
α-Ketoglutarate	glutamic acid, glutamine, arginine, proline, hydroxyproline	α-ketoglutarate
	lysine	various products
Oxaloacetate	aspartic acid, asparagine	oxaloacetate
	methionine, threonine	propionate
	isoleucine	propionyl coenzyme A + acetoacetyl coenzyme A*
3-Phosphoglycerate	glycine, serine, cysteine, cystine	pyruvate
	alanine	
Pyruvate	valine	succinyl coenzyme A
	leucine	acetoacetate* + acetyl coenzyme A*
Erythrose-4-P + phosphoenolpyruvate	phenylalanine ↓ animals, tyrosine	acetoacetic acid* + fumarate
	tryptophan	various products
Purine ring of ATP	histidine	α-ketoglutarate

Dashed-arrow annotations: *fungi*; *green plants and some bacteria*; *plants*; *plants*; *plants*

* Indicates the product is potentially ketogenic

(Dashed arrows indicate that the pathways are largely absent in animals)

Considerable variability in the pattern of amino acid metabolism is to be found in nature, and only the major pathways in the majority of animals and some micro-organisms are described here.   The pathways indicated by dashed lines in Table 13.1 only occur in green plants or in certain micro-organisms; they are probably absent in animals—with the result that the amino acids are 'essential' and must be supplied in the diet (section 12.4).   It is often convenient to consider amino acids in groups. For example, the α-ketoglutarate family of amino acids share a common precursor, have similar fates, and considerable interconversion occurs between amino acids within the group.

Table 13.1 also shows the major products of catabolism of the carbon skeletons of the amino acids. Most of these can be converted to carbohydrate as described in chapter 10, or undergo complete oxidation in the TCA cycle.

13.4 The α-ketoglutarate family of amino acids (glutamic acid, glutamine, arginine, proline, hydroxyproline)

From α-ketoglutarate, a member of the TCA cycle, most organisms, including mammals, are able to synthesize at least five amino acids (Fig. 13.5). The first step in all cases is the formation of L-glutamic acid. This can arise either by transamination (p. 268) or by the action of glutamate dehydrogenase. Glutamate formed in this way, as well as dietary glutamate, may be incorporated into proteins directly or undergo conversion to other amino acids. It can also be readily converted to glutamine by the enzyme glutamine synthetase (section 12.5). The role in general nitrogen metabolism of glutamate and glutamine was discussed in the previous chapter.

Most organisms are able to reduce glutamic acid to a compound known as *glutamate semi-aldehyde*. This substance resembles glutamic acid except that the δ-carbon atom is part of an aldehyde instead of a carboxylic

$$\text{COOH}$$
$$\text{C=O}$$
$$\text{CH}_2$$
$$\text{CH}_2$$
$$\text{COOH}$$
α-Ketoglutarate

transaminase
or
glutamate dehydrogenase

$$\text{COOH}$$
$$\text{H}_2\text{N.C.H} \quad \alpha$$
$$\text{CH}_2 \quad \beta$$
$$\text{CH}_2 \quad \gamma$$
$$\text{COOH} \quad \delta$$
Glutamic acid

$\text{NH}_3 + \text{ATP} \quad \text{ADP} + \text{P}$
glutamine synthase

$\text{NH}_3 \quad \text{H}_2\text{O}$
glutaminase

$$\text{COOH}$$
$$\text{H}_2\text{N.C.H}$$
$$\text{CH}_2$$
$$\text{CH}_2$$
$$\text{C}$$
$$\text{H}_2\text{N} \quad \text{O}$$
Glutamine

$\text{NADH} + \text{H}^+$

NAD^+

glutamate -δ-semi-aldehyde dehydrogenase

To glutamate-δ-semi-aldehyde

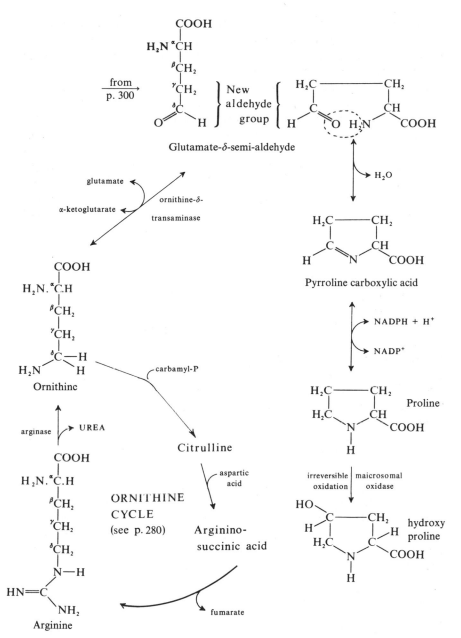

Fig. 13.5 Outline of the synthesis of amino acids of the α-ketoglutarate family

acid group. It is called a *semi*-aldehyde since only one of the two carboxylic acid groups of glutamate is in the reduced aldehydic form. The reaction is catalysed by glutamate-δ-semi-aldehyde dehydrogenase, an enzyme which is probably NAD-linked. Glutamate semi-aldehyde is the precursor of arginine and of the two imino acids present in proteins.

The ornithine cycle not only takes part in the synthesis of urea (section 12.7) but also serves as the means by which *arginine* is synthesized. Arginine is an amino acid commonly found in proteins, whereas proteins only rarely contain the di-amino acid, ornithine. Ornithine is formed from glutamate semi-aldehyde by a reaction in which an amino group is transferred to it from glutamic acid. It is noteworthy that this reaction is somewhat atypical for although it superficially resembles transamination of the conventional sort (the enzyme concerned is called ornithine-δ-transaminase) it clearly differs from ordinary transamination in that the δ-position and not an α-keto (or α-amino) group is involved in the reaction.

Proline is synthesized from glutamate semi-aldehyde by two steps. The first of these may occur spontaneously and involves elimination of water to form the cyclic compound, pyrroline carboxylic acid. This is then reduced and—as so often occurs in synthetic processes—it is NADPH and not NADH which supplies the electrons (Fig. 13.5).

Hydroxyproline is synthesized from proline but the details are somewhat obscure. In animals, collagen is particularly rich in hydroxyproline residues and there is evidence that the hydroxylation actually occurs while proline is being incorporated into the protein. It takes place at the endoplasmic reticulum of cells. The enzyme system responsible may be the same microsomal oxidase system that takes part in the oxidation of foreign molecules (p. 258) and in some steps in the synthesis of complex lipids.

Most of the reactions indicated in Fig. 13.5 are reversible. Proline and glutamic acid are readily converted to α-ketoglutarate and so these amino acids are potentially glucogenic. Arginine is catabolized by first being converted to ornithine by the elimination of urea (p. 280). However, glutamine is not converted to glutamate by a reversal of the glutamine synthase reaction (which would be endergonic and linked with the synthesis of ATP from ADP) but instead a second enzyme, *glutaminase*, catalyses a straightforward hydrolytic deamination of the δ-amide group (p. 274).

The microsomal oxidase reaction by which hydroxyproline is converted to proline is *not reversible*. During the breakdown of the collagen in bone, hydroxyproline is excreted in the urine; it thus seems probable that hydroxyproline is not extensively catabolized. With this probable exception, each of the amino acids shown in Fig. 13.5 is readily converted to each of the others. Arginine, for example, can give rise to both proline and glutamine.

13.5 The 3-phosphoglycerate family of amino acids (glycine, serine, cysteine, cystine)

The carbon skeletons of serine, cysteine, and glycine arise from 3-phosphoglycerate, an intermediate of the glycolytic pathway. First, 3-phospho-hydroxypyruvate is formed from this intermediate by a NAD-linked dehydrogenase and then phosphoserine is produced by transamination (Fig. 13.6). Serine is readily formed from phosphoserine by a phosphatase which removes the phosphate group as inorganic orthophosphate. This dephosphorylation is almost irreversible, so the catabolism of serine and amino acids derived from it does not take place by a reversal of the anabolic route.

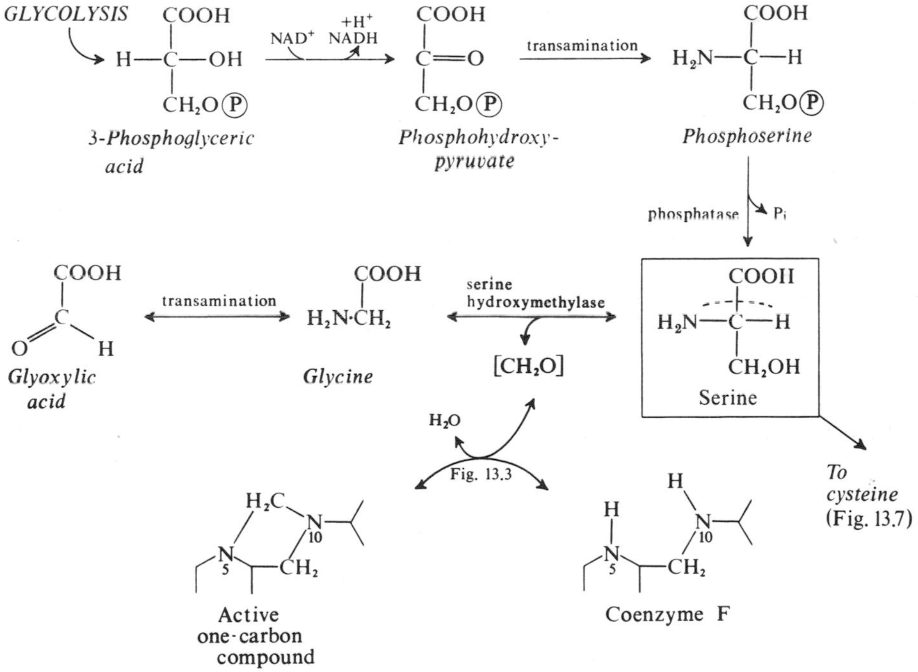

Fig. 13.6 The synthesis and interconversion of serine and glycine (3-phosphoglycerate family)

Metabolism of the carbon skeletons of the amino acids

Serine is converted to glycine by *serine hydroxymethylase,* an enzyme which has pyridoxal phosphate and coenzyme F as cofactors. The hydroxymethyl group is removed from the serine, thereby producing glycine. By forming a methylene group attached to the nitrogen atoms numbers 5 and 10 of coenzyme F, it thus gives rise to 'active formaldehyde'. This is indicated in Fig. 13.6 where the structure of coenzyme F is simplified. Since mechanisms appear to exist which permit ready interconversion of 'active formaldehyde' and 'active formate', the one-carbon fragment which is thus removed from serine can eventually act as 'active formate' in purine synthesis (p. 282) and for a number of other important anabolic processes. The conversion of serine to glycine is reversible so that serine can be synthesized from glycine if 'active formaldehyde' is available.

Cysteine is readily formed from serine by incorporation of the sulphur atom of methionine (homocysteine being an intermediate). Therefore, although methionine is an essential amino acid for animals, cysteine is not essential, provided an adequate amount of dietary methionine is available to supply the sulphur. When the methyl group of methionine is removed for transfer to other compounds (as it is, for example, in the synthesis of choline) a second product is homocysteine (Section 13.1). It is this compound which directly provides the sulphur for cysteine synthesis. An enzyme which is a dehydratase catalyses the formation, with the elimination of water, of a 'sulphur bridge' between serine and homocysteine. The 'bridged' molecule so formed is called *cystathionine*

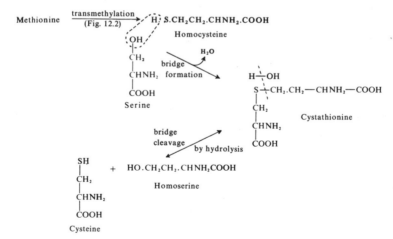

Fig. 13.7 Conversion of serine to cysteine (3-phosphoglycerate family)

(Fig. 13.7) and should be compared to the somewhat similar molecules containing 'nitrogen bridges' (pp. 275, 280, 286). Cysteine is formed from cystathionine by the action of a hydratase which splits the molecule to produce homoserine.

Cystine is formed from cysteine by oxidation. In protein synthesis, it is usual for the cysteine residues to be oxidized to cystine after they have been incorporated into the protein. The disulphide bridge of cystine is essential for maintaining the secondary structure of many proteins (p. 43). Cystine is perhaps better considered as a derivative of cysteine than as a separate amino acid in its own right.

The carbon skeletons of serine and cysteine are readily converted to glycogen and therefore both of these amino acids are glucogenic. Although transamination may occur, this is not the usual metabolic route. Instead, an enzyme called *serine dehydratase* removes water from serine to form an imino compound which is then hydrolysed to give pyruvate and ammonia. Cysteine undergoes a similar reaction, but hydrogen sulphide is removed initially instead of water (compare: H_2S; H_2O). The enzyme in this case is *cysteine desulphydratase*.

$$HO\overset{H}{\underset{\overset{|}{CH_2}}{\text{---}}}\overset{}{\underset{COOH}{C}}\text{---}NH_2 \qquad \xrightarrow[\text{serine hydratase}]{H_2O}$$

Serine

$$H_2C=C\overset{NH_2}{\underset{COOH}{}} \quad \rightleftharpoons \quad H_3C\text{---}\overset{NH}{\overset{||}{C}}\text{---}COOH \quad \xrightarrow{H_2O \quad NH_3} \quad H_3C\text{---}\overset{O}{\overset{||}{C}}\text{---}COOH$$

Imino compound (2 forms) $\qquad\qquad\qquad\qquad\qquad$ Pyruvate

$$HS\overset{H}{\underset{\overset{|}{CH_2}}{\text{---}}}\overset{}{\underset{COOH}{C}}\text{---}NH_2 \qquad \xrightarrow[\text{cysteine desulphydratase}]{H_2S}$$

Cysteine

$$H_2C=C\overset{NH_2}{\underset{COOH}{}} \quad \rightleftharpoons \quad H_3C\text{---}\overset{NH}{\overset{||}{C}}\text{---}COOH \quad \xrightarrow{H_2O \quad NH_3} \quad H_3C\text{---}\overset{O}{\overset{||}{C}}\text{---}COOH$$

Imino compound (2 forms) $\qquad\qquad\qquad\qquad\qquad$ Pyruvate

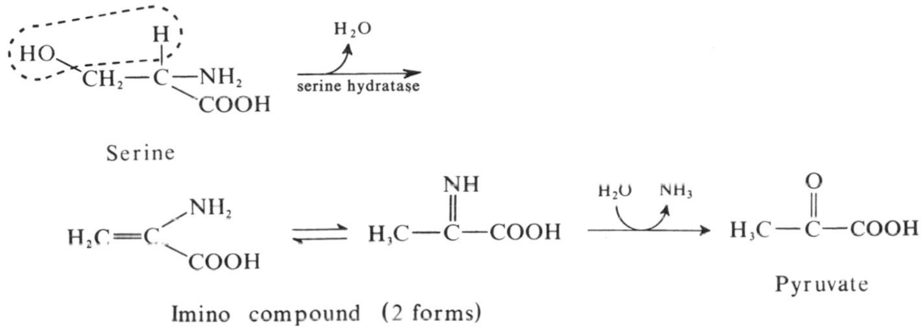

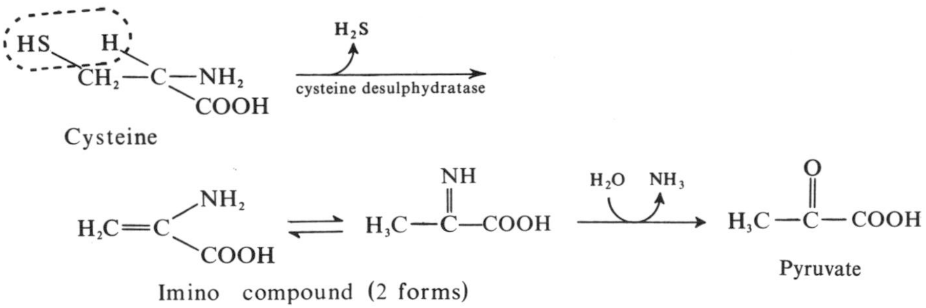

The enzymes which catalyse these reactions have pyridoxal phosphate as cofactor.   The pyruvate so formed may be converted to glycogen by the pathway described in chapter 10.   In animals, much of the hydrogen sulphide formed by the action of desulphydratase is oxidized to sulphate which is excreted by the kidneys.

Glycine is a glucogenic amino acid which can be metabolized in several ways.   It is readily converted by serine hydroxymethylase to the glucogenic amino acid, serine.   Glycine can also be deaminated by transaminases with the formation of the corresponding α-keto acid, glyoxylate.

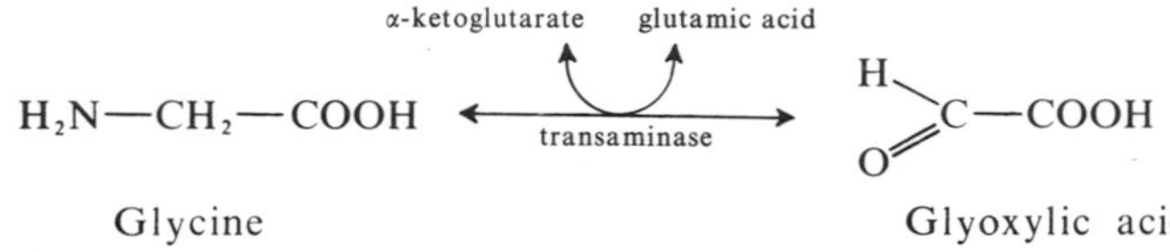

α-ketoglutarate   glutamic acid

$H_2N—CH_2—COOH \underset{\text{transaminase}}{\longleftrightarrow} \overset{H}{\underset{O}{}}C—COOH$

Glycine   Glyoxylic acid

In plants and micro-organisms, glyoxylate can be converted to oxaloacetate by the reactions of the glyoxylate shunt (p. 225).   In these organisms, glyoxylate may be the main precursor of glycine.

13.6   The oxaloacetate family of amino acids (aspartate, asparagine, methionine, threonine, isoleucine, lysine)

In *green plants and bacteria* oxaloacetate is the starting material for the synthesis of the carbon skeletons of six amino acids (Table 13.1).   Animals, however, have largely lost this capacity for synthesis for they are only able to form two amino acids of the group—aspartate and asparagine; the

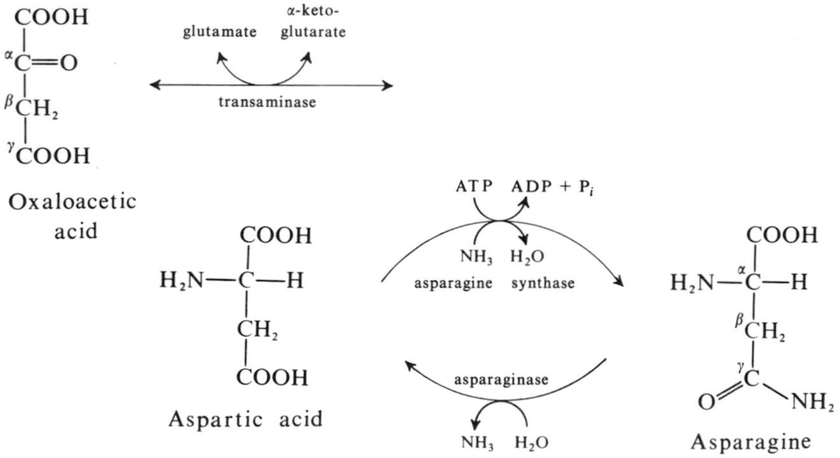

COOH
|
$^{\alpha}$C=O
|
$^{\beta}$CH$_2$
|
$^{\gamma}$COOH

Oxaloacetic acid

glutamate   α-keto-glutarate   transaminase

ATP   ADP + P$_i$

NH$_3$ H$_2$O   asparagine synthase

asparaginase   NH$_3$ H$_2$O

COOH
|
$H_2N—C—H$
|
CH$_2$
|
COOH

Aspartic acid

COOH
|
$H_2N—^{\alpha}C—H$
|
$^{\beta}$CH$_2$
|
$^{\gamma}$C
O $\diagdown$ NH$_2$

Asparagine

others (methionine, threonine, isoleucine, and lysine) are therefore essential dietary factors.

The biosynthesis of *aspartic acid* and *asparagine* from oxaloacetate is very similar to the synthesis of glutamate and glutamine from α-ketoglutarate, the other α-keto acid of the TCA cycle (Fig. 13.5).

The general roles of aspartate and asparagine in nitrogen metabolism were considered in the previous chapter. Both amino acids are glucogenic. The metabolism of asparagine to form aspartate is catalysed by *asparaginase*; this enzyme removes the γ-amino group by hydrolysis, ammonium ions being released into solution. The reaction resembles that of glutaminase, and, for the same energetic reasons (p. 274), the route leading to the synthesis of asparagine does not involve the participation of asparaginase. Aspartic acid is readily converted to oxaloacetate by transaminases and the oxaloacetate formed in this manner (or indirectly from asparagine) may give rise to sugars by means of the reactions discussed in chapter 10.

The catabolism of the three essential amino acids of the oxaloacetate family—methionine, threonine, and isoleucine—produces propionate or propionyl coenzyme A and for this reason these amino acids are glucogenic (p. 225). *Threonine* is often broken down by a route initiated by *threonine dehydratase*, an enzyme which brings about a change analogous to that of serine dehydratase (section 13.5). Threonine dehydratase, like its serine counterpart, contains pyridoxal phosphate; it removes the amino group of threonine as ammonia, the product being α-ketobutyrate.

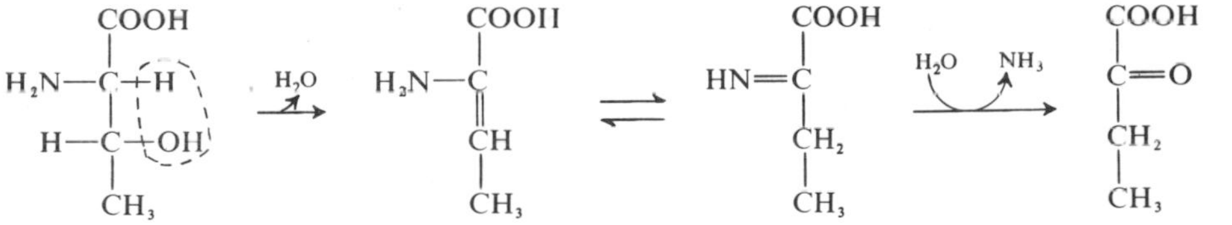

| Threonine | Imino acid (2 forms) | α-Ketobutyrate |

A major catabolic route for *methionine* results in the conversion of its carbon skeleton to homoserine (p. 304). Homoserine is then converted to α-ketobutyrate, though some details of the route are uncertain.

The *α-ketobutyrate*, produced when threonine and methionine are metabolized, is readily decarboxylated:

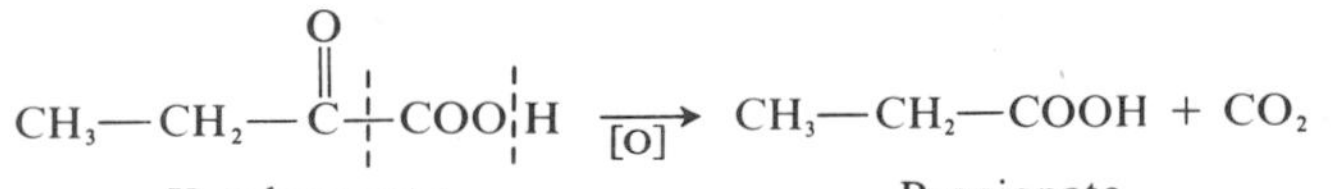

α-Ketobutyrate   Propionate

The propionate so formed can be converted to polysaccharides by way of the reactions discussed in chapter 10.

Isoleucine is both ketogenic and glucogenic, for the products of its catabolism are propionyl coenzyme A (glucogenic) and acetyl coenzyme A (ketogenic).

13.7   The pyruvate family of amino acids (alanine, valine, leucine)

Of the three amino acids in this group, only *alanine* can be synthesized by animals.   The transamination of pyruvate produces alanine and the reaction is readily reversible.   Alanine is, for this reason, a glucogenic amino acid.

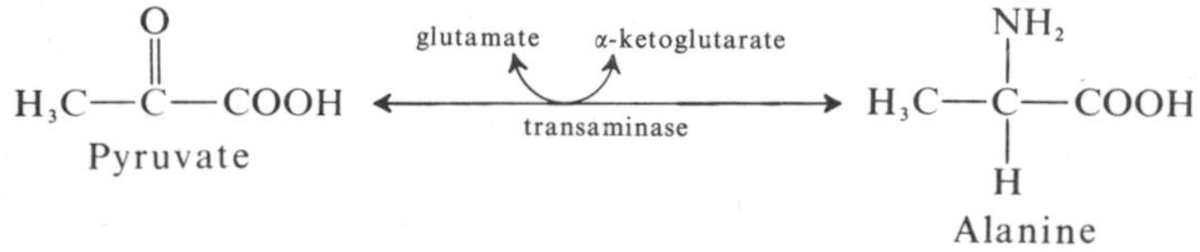

$$\underset{\text{Pyruvate}}{H_3C-\overset{\displaystyle O}{\overset{\|}{C}}-COOH} \quad\underset{\text{transaminase}}{\overset{\text{glutamate}\quad\alpha\text{-ketoglutarate}}{\rightleftharpoons}}\quad \underset{\underset{H}{|}}{\overset{\overset{NH_2}{|}}{H_3C-\underset{|}{\overset{|}{C}}-COOH}} \;\;\text{Alanine}$$

Valine and *leucine* are synthesized from pyruvate by plants and micro-organisms but they are essential amino acids in higher animals.   The initial stages of their catabolism resemble those of the other essential branched-chain aliphatic amino acid, isoleucine, which is a member of the oxaloacetate group of amino acids (Table 13.1).

The first step in the catabolism is transamination.   This is reversible so animals could synthesize these three amino acids if provided with the appropriate α-keto acids.   The next step is oxidative decarboxylation of the carboxylic acid group by a reaction which somewhat resembles the decarboxylation of pyruvate.   As for that substance, a coenzyme A derivative is first formed:

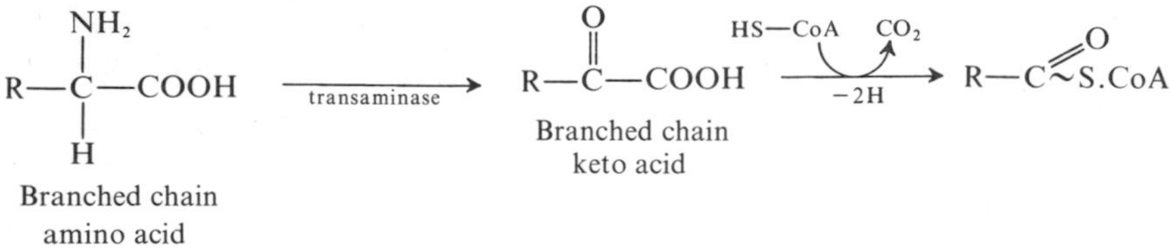

$$\underset{\substack{\text{Branched chain}\\\text{amino acid}}}{\overset{\overset{NH_2}{|}}{R-\underset{\underset{H}{|}}{\overset{|}{C}}-COOH}} \;\xrightarrow{\text{transaminase}}\; \underset{\substack{\text{Branched chain}\\\text{keto acid}}}{R-\overset{\displaystyle O}{\overset{\|}{C}}-COOH} \;\xrightarrow[-2H]{\text{HS}-\text{CoA}\quad CO_2}\; R-C\overset{\displaystyle O}{\underset{\displaystyle}{\sim}}S.CoA$$

Thereafter, the catabolism of the carbon skeletons of the three amino acids differs in each case.   Valine gives rise to succinyl coenzyme A which is easily converted to oxaloacetate by the TCA cycle and so it is potentially glucogenic. Leucine has the distinction of being the only strongly ketogenic amino acid. The reason is that it is metabolized to the 'ketone body', acetoacetic acid, and to acetyl coenzyme A, from which ketone bodies can arise (p. 228). Normally, however, both products are oxidized by the TCA cycle or are converted to lipids.

13.8 The cyclic amino acids (phenylalanine, tyrosine, tryptophan)

The carbon skeletons of phenylalanine, tyrosine, and tryptophan all contain an aromatic (benzene) ring, tryptophan containing a five-membered heterocyclic ring as well. Animals cannot synthesize the benzene ring and therefore these amino acids are essential dietary factors. Plants and micro-organisms, on the other hand, are able to synthesize the benzene ring from D-erythrose-4-P and phosphoenol pyruvate. Since, however, animals can insert a para-hydroxyl group into phenylalanine to form tyrosine, the latter is not an essential amino acid so long as the diet contains adequate amounts of phenylalanine.

Phenylalanine is converted to tyrosine by a microsomal enzyme called *phenylalanine hydroxylase*, which is widely distributed in both plants and animals. As with so many hydroxylating and oxidizing systems associated with microsomes, NADPH is an essential cofactor, but a peculiarity of the phenylalanine hydroxylating system appears to be its additional dependence on *reduced pteridine*. Pteridine is part of the molecule of coenzyme F (Fig. 13.3). Under normal physiological conditions the hydroxylation is irreversible; consequently, while phenylalanine can act as a precursor of tyrosine, if tyrosine is fed to animals they cannot convert it to phenylalanine.

The degradation of phenylalanine and tyrosine leads normally to fumarate, which is glucogenic, and to acetoacetate, which is ketogenic. While the main steps of the degradation are known, some of the mechanisms are not, and the process will not be considered exhaustively.

Phenylalanine is first hydroxylated and then the tyrosine so formed undergoes transamination to give the corresponding keto acid, p-hydroxyphenyl pyruvic acid. A little-understood and multi-step reaction, catalysed by one enzyme, leads to homogentisic acid. Another enzyme then causes the ring to open in the position shown in Fig. 13.8 to give maleylacetoacetic acid; the latter compound is so named because it can

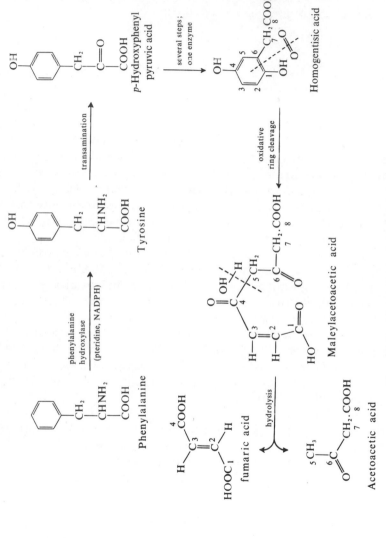

Fig. 13.8 Catabolism of phenylalanine and tyrosine

be regarded as being theoretically derived from maleic acid and aceto-acetic acid:

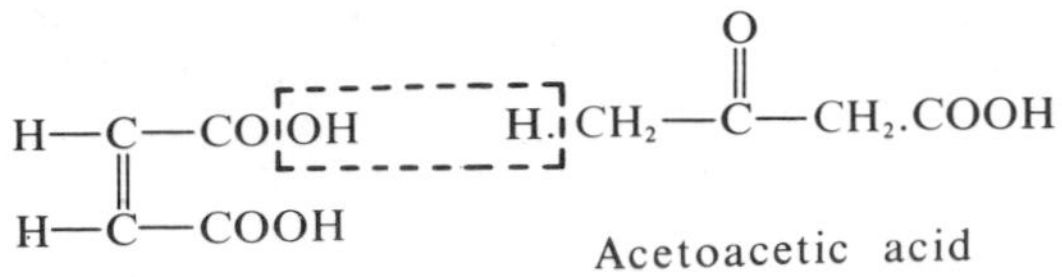

$$H-\underset{\underset{\displaystyle H-C-COOH}{\displaystyle \|}}{C}-CO\overline{|OH} \qquad H.\overline{|CH_2}-\underset{\displaystyle \|}{\overset{\displaystyle O}{C}}-CH_2.COOH$$

Maleic acid   Acetoacetic acid

Maleylacetoacetate is converted to its isomer, fumarylacetoacetate, which then splits to give fumaric acid (a TCA cycle intermediate) and the ketogenic substance, acetoacetic acid.

Tryptophan is incompletely metabolized by mammals, one of the products, anthranilic acid (o-aminobenzoic acid) being excreted in the urine. Its metabolism is of interest in that oxidative cleavage of the heterocyclic ring precedes the loss of nitrogen (Fig. 13.9). The formyl kinurenine so formed presumably loses its labile formyl group to coenzyme F, and certainly gives rise to kynurenine. Kynurenine undergoes hydrolytic cleavage in the position shown in the figure, the products being

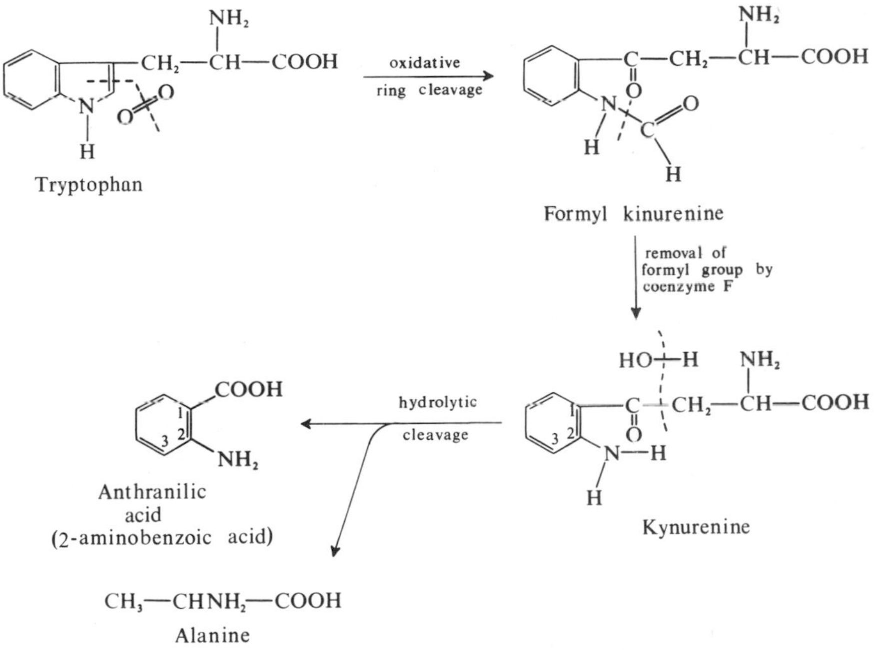

Tryptophan

oxidative ring cleavage →

Formyl kinurenine

removal of formyl group by coenzyme F

Anthranilic acid (2-aminobenzoic acid)

hydrolytic cleavage ←

Kynurenine

CH_3-CHNH_2-COOH

Alanine

Fig. 13.9 Catabolism of tryptophan

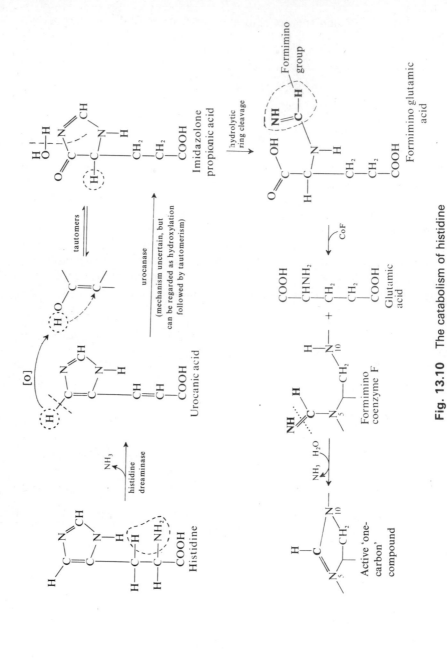

Fig. 13.10 The catabolism of histidine

anthranilic acid and alanine. The latter is, of course, glucogenic; the anthranilic acid is excreted if not required for anabolic processes. Kynurenine can also be degraded by other routes, a common end-product being hydroxyanthranilic acid (Fig. 13.14).

13.9 The metabolism of histidine

Plants and micro-organisms synthesize histidine largely from the purine ring system of ATP. Studies of the amino acid requirement of the rat suggest that histidine is an essential dietary amino acid. Man, however, does not appear to require a supply of histidine so there may be species variation in the extent to which different mammals are able to synthesize it. Such examples as this are salutary reminders that the results of experiments on one species of animal cannot always be transposed to other species without experimental corroberation.

The principal route for the catabolism of histidine in animal tissues leads to the formation of glutamic acid, a glucogenic amino acid. The main steps are indicated in Fig. 13.10. The first reaction is not transamination but *deamination*. It produces an unsaturated carboxylic acid, *urocanic acid*, so named because it was first observed in the urine of a dog. Urocanic acid is hydrolysed by urocanase in a reaction which involves considerable molecular rearrangement. The imidazolone ring of the product is next cleaved by hydrolysis to give formimino glutamic acid. The latter is excreted in the urine of animals deficient in folic acid because coenzyme F is involved in the next reaction—the removal of the one-carbon unit, *formimino*.

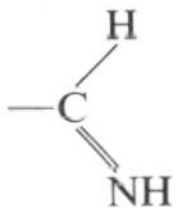

$$-C\begin{smallmatrix} H \\ \\ NH \end{smallmatrix}$$

The transfer of this unit is catalysed by a specific transferase. The reaction leads to the formation of glutamic acid and of a coenzyme F derivative in which the formimino group is attached to nitrogen atom number five. After the removal of the nitrogen from the formimino group, the one-carbon unit is available for numerous anabolic reactions.

13.10 Some important compounds synthesized from amino acids

Numerous substances of biochemical or physiological importance are synthesized from the amino acids. As was seen in chapter 12, these include the purines, pyrimidines, and porphyrins. A few of the many other compounds which are derived from amino acids are considered briefly in this section.

(a) Compounds resulting from the decarboxylation of amino acids

A number of specific amino acid decarboxylases catalyse the removal of carbon dioxide from the carboxylic acid group of amino acids and lead to the formation of primary amines. Pyridoxal phosphate is the cofactor for these enzymes (section 13.2).

Many micro-organisms possess decarboxylases for almost every α-amino acid. This is why primary amines are characteristic products of the bacterial decay of the proteins of dead flesh. For example, two compounds—putrescine and cadaverine—are produced by the decarboxylation of ornithine and lysine.

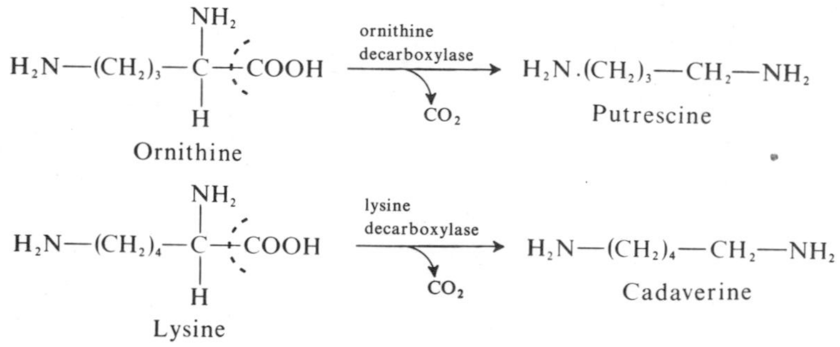

$$H_2N-(CH_2)_3-\underset{\underset{\displaystyle H}{|}}{\overset{\overset{\displaystyle NH_2}{|}}{C}}\!\!\not{-}COOH \xrightarrow[CO_2]{\text{ornithine} \atop \text{decarboxylase}} H_2N.(CH_2)_3-CH_2-NH_2$$

Ornithine — Putrescine

$$H_2N-(CH_2)_4-\underset{\underset{\displaystyle H}{|}}{\overset{\overset{\displaystyle NH_2}{|}}{C}}\!\!\not{-}COOH \xrightarrow[CO_2]{\text{lysine} \atop \text{decarboxylase}} H_2N-(CH_2)_4-CH_2-NH_2$$

Lysine — Cadaverine

Some amino acid decarboxylases occur in almost all organisms so far investigated. Particularly common are those which catalyse the decarboxylation of serine, aspartic acid, and cysteine. The products are of great importance—ethanolamine is a precursor of choline (p. 294),

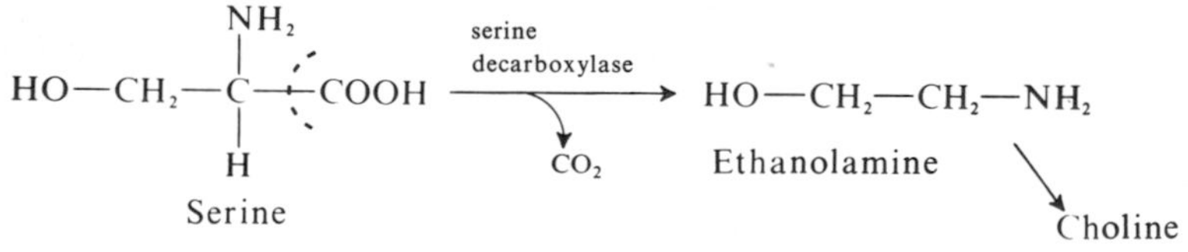

$$HO-CH_2-\underset{\underset{\displaystyle H}{|}}{\overset{\overset{\displaystyle NH_2}{|}}{C}}\!\!\not{-}COOH \xrightarrow[CO_2]{\text{serine} \atop \text{decarboxylase}} HO-CH_2-CH_2-NH_2$$

Serine — Ethanolamine — Choline

and both β-thioethylamine and β-alanine contribute to the structure of coenzyme A (see (d) below).

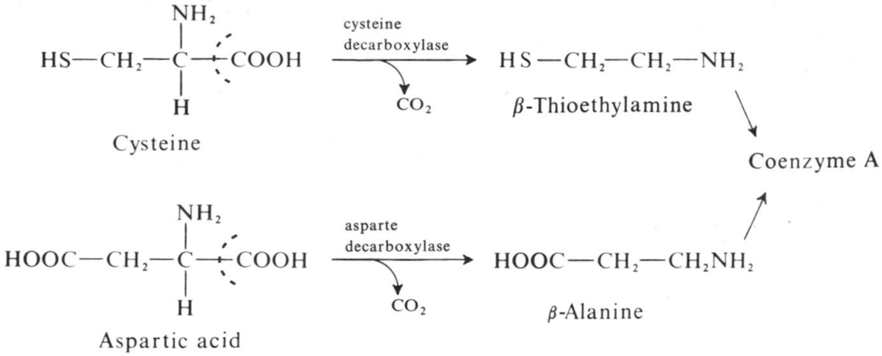

$$HS-CH_2-\underset{\underset{H}{|}}{\overset{\overset{NH_2}{|}}{C}}-COOH \quad \xrightarrow[\substack{\downarrow \\ CO_2}]{\substack{\text{cysteine} \\ \text{decarboxylase}}} \quad HS-CH_2-CH_2-NH_2$$

Cysteine $\qquad\qquad\qquad$ β-Thioethylamine

Coenzyme A

$$HOOC-CH_2-\underset{\underset{H}{|}}{\overset{\overset{NH_2}{|}}{C}}-COOH \quad \xrightarrow[\substack{\downarrow \\ CO_2}]{\substack{\text{asparte} \\ \text{decarboxylase}}} \quad HOOC-CH_2-CH_2NH_2$$

Aspartic acid $\qquad\qquad\qquad$ β-Alanine

Many vertebrate tissues contain histidine decarboxylase, an enzyme which catalyses the production of the local hormone, *histamine*.

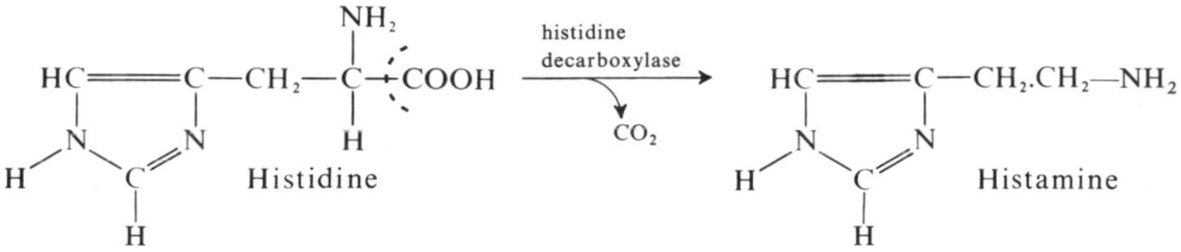

$$HC{=\!=\!=}C-CH_2-\underset{\underset{H}{|}}{\overset{\overset{NH_2}{|}}{C}}-COOH \quad \xrightarrow[\substack{\downarrow \\ CO_2}]{\substack{\text{histidine} \\ \text{decarboxylase}}} \quad HC{=\!=\!=}C-CH_2.CH_2-NH_2$$

Histidine $\qquad\qquad\qquad\qquad\qquad\qquad$ Histamine

It will be shown later that decarboxylases are also important in the synthesis of adrenalin and taurine.

(b) The synthesis of creatine and its derivatives

The importance of creatine phosphate as an energy store in muscle was mentioned in chapter 5 and is considered again in chapter 16 (p. 402). It will be recalled that the Lohmann reaction enables ATP to be formed as required from creatine phosphate and ADP.

Creatine is synthesized from glycine and arginine. Arginine, itself synthesized by the ornithine cycle (p. 280), is converted to ornithine by the transfer of the amidino group to glycine, the other product being *guanidine acetic acid* (also known as glycocyamine). The transfer is catalysed by an enzyme called glycine transamidinase (Fig. 13.11). Finally a methyl group transferred from active methionine converts the guanidine acetic acid to creatine.

In muscle, creatine is phosphorylated *reversibly* to form *creatine phosphate*. This, the Lohmann reaction, is catalysed by creatine kinase (p. 134). As shown in Fig. 13.11, the phosphorus atom of the phosphate group is linked directly to a nitrogen atom in creatine phosphate—so

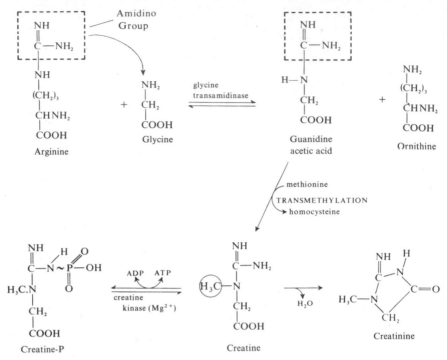

Fig. 13.11 The synthesis of creatine, creatine phosphate, and of creatinine

preserving the anhydride nature of the phosphate and, with it, its high energy content.

Excess creatine is excreted in vertebrates as the ring compound, creatinine (p. 277). This is formed from creatine by the elimination of water, a reaction which produces an intramolecular bond between the carboxylic acid group and a terminal amino group.

(c) The contribution of glycine and cysteine to the formation of the bile acids

The bile acids facilitate the hydrolysis of triglycerides in the intestine (p. 366). The structure of cholic acid, a steroid with a carboxylic acid group on its side chain, was considered in chapter 1 (p. 15). Normally, the carboxylic acid group is conjugated either to glycine or to *taurine*, a product of cysteine metabolism.

Synthesis of taurine involves a multi-step oxidation which converts the sulphydryl (2-valent) sulphur atom to a substituted sulphonic acid in

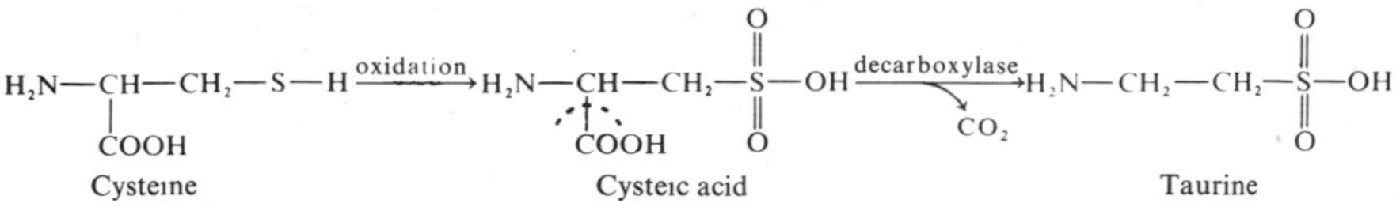

(a) Synthesis of taurine from cysteine

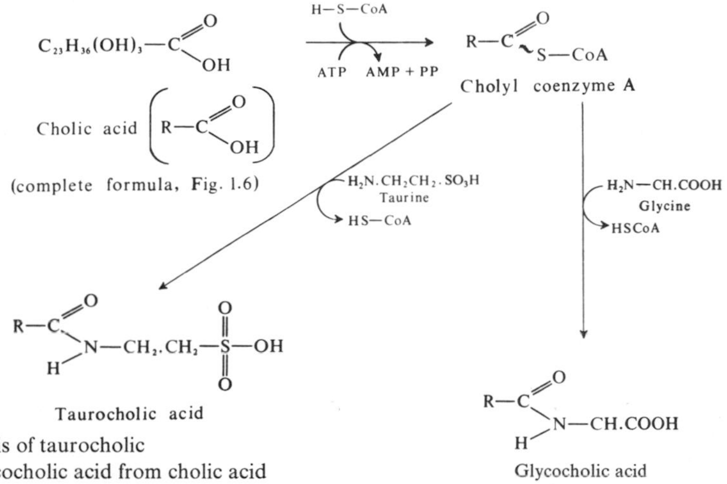

(b) Synthesis of taurocholic
    and glycocholic acid from cholic acid

Fig. 13.12 The synthesis of taurine and of taurocholic and glycocholic acids

which the sulphur is apparently 6-valent. The final step is a straight-forward decarboxylation catalysed by a specific decarboxylase (Fig. 13.12).

Glycine and taurine cannot be directly attached to cholic acid. As is so often the case in reactions involving carboxylic acid groups, the cholic acid must first be activated by union with coenzyme A. As in other examples we have encountered, the formation of cholyl coenzyme A involves the conversion of ATP to AMP (compare, for instance, with the priming reaction in fatty acid catabolism (p. 179)). Taurine or glycine are conjugated to cholic acid by reaction with cholyl coenzyme A, the products being taurocholic and glycocholic acids.

(d) The synthesis of coenzyme A

Coenzyme A is a remarkable example of a complex molecule built largely from the carbon skeletons of amino acids. The first step is the synthesis of *pantothenic acid*. For higher animals, this is an essential dietary factor

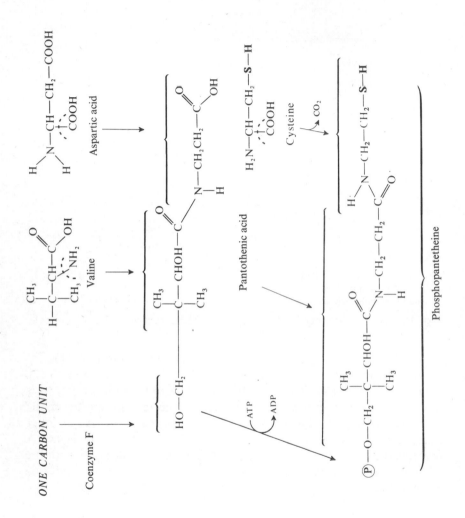

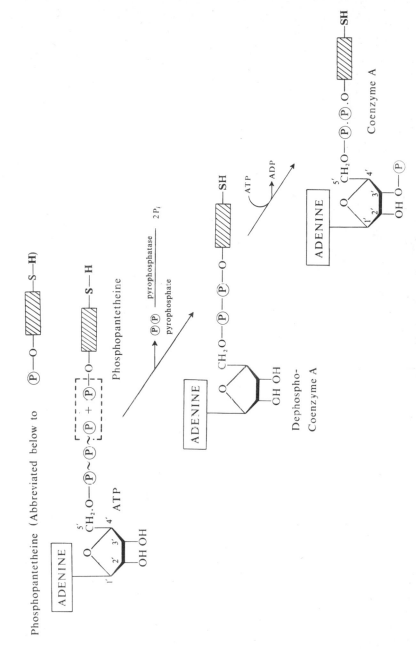

Fig. 13.13 Outline of the synthesis of coenzyme A

Metabolism of the carbon skeletons of the amino acids

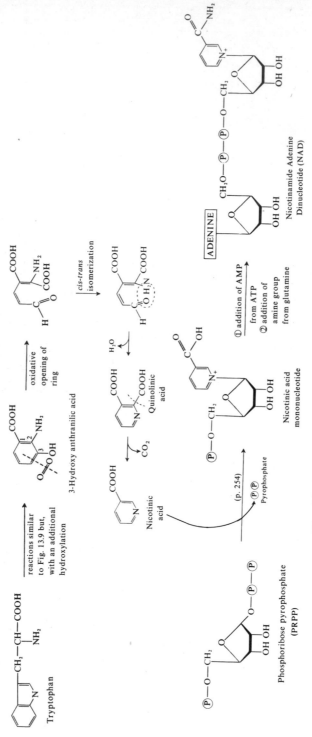

Fig. 13.14 Outline of the synthesis of NAD from tryptophan

since they apparently lack the enzymes for its synthesis. Plants and micro-organisms synthesize pantothenic acid from aspartic acid, which is first decarboxylated to form β-alanine (p. 315), and from valine which has lost its amino group by transamination. A 1-C unit (HO.CH$_2$—), probably transferred from coenzyme F, also contributes to the structure of pantothenic acid (Fig. 13.13).

Pantothenic acid is next converted to *phosphopantetheine* by the addition of β-thioethylamine to the carboxyl end of the molecule and of a phosphate group to the hydroxyl group at the other end. The phosphate is derived from ATP. β-Thioethylamine (see (a) above) is the product of the decarboxylation of cysteine. Phosphopantetheine is the prosthetic group of acyl carrier protein (ACP) involved in fatty acid biosynthesis (p. 186).

Coenzyme A is formed from phosphopantetheine in two steps. First, ATP reacts with phosphopantetheine with the elimination of pyrophosphate to produce a dinucleotide-type structure, *dephospho-coenzyme A*. This is then phosphorylated at carbon atom number 3′ of the ribose moiety to form coenzyme A (Fig. 13.13). Thus aspartic acid, valine, cysteine, and a 1-C unit contribute to the structure of coenzyme A. It will be recalled, moreover, that the purine part of the molecule is itself synthesized from glycine, glutamine, and aspartate (p. 283).

(e) Formation of nicotinic acid, NAD(P), and indole acetic acid from tryptophan

Tryptophan is the precursor of the nicotinamide moiety of NAD and NADP. Several pathways are known for the synthesis of nicotinic acid and these two important dinucleotides, but only the outline of one route of synthesis will be described.

The initial reactions by which tryptophan is converted to nicotinic acid closely resemble those shown in Fig. 13.9. Tryptophan is first converted to kynurenine but, before cleavage of the side chain, hydroxylation occurs at the 3-position. The product, 3-hydroxy anthranilic acid, then undergoes the reactions depicted in Fig. 13.14. Oxidative cleavage of the homocyclic ring produces a compound which undergoes a *cis–trans* isomerization to bring the amino group into close proximity to the carbonyl group. Elimination of water then results in the formation of a heterocyclic compound, quinolinic acid, which is converted to nicotinic acid by decarboxylation. This sequence of reactions is remarkable because, in effect, one carbon atom is removed from the six-membered ring of tryptophan and is replaced by the nitrogen atom of the original five-membered ring.

Nicotinic acid is converted to nicotinic acid mononucleotide by the reaction which typically links ribose to nitrogenous bases (p. 254). Thus nicotinic acid reacts with *phosphoribose pyrophosphate* (PRPP) with the elimination of pyrophosphate (the hydrolysis of which, in the presence of pyrophosphatase, draws the reaction towards completion). The nicotinic acid mononucleotide so formed next reacts with ATP to form a dinucleotide resembling NAD except that it lacks the amide group.

$$\text{ATP} + \text{nicotinic acid nucleotide} \rightarrow \text{pyrophosphate} + \text{desamido-NAD}$$

Again the reaction tends to be drawn to completion by enzymic hydrolysis of the pyrophosphate. Finally, NAD is formed by reaction with glutamine

$$\text{Desamido-NAD} + \text{glutamine} \rightarrow \text{NAD}^+ + \text{glutamate}$$

Some NAD is phosphorylated by a specific kinase to form NADP

$$\text{NAD}^+ + \text{ATP} \rightarrow \text{NADP}^+ + \text{ADP}$$

Growth regulators, or auxins, control the responses of plants to light and gravity. They are produced in meristematic tissue and alter the rate of growth of individual cells. An important auxin is *β-indolyl acetic acid*:

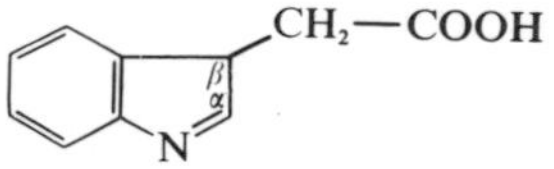

β-Indolyl acetic acid

It is known to be formed from tryptophan. The simplest (but not fully substantiated) pathway for its synthesis involves transamination to form indolyl pyruvic acid, followed by decarboxylation to give indolyl acetaldehyde which is then oxidized to indolyl acetic acid.

(f) Formation of adrenalin and nor-adrenalin from phenylalanine

The hormone adrenalin (epinephrin) is secreted by the medulla of the adrenal gland. In addition to increasing the level of glucose in the blood (p. 232), it produces a remarkable set of physiological reactions. It raises the rate of heart beat, increases the blood pressure and causes dilatation of blood vessels in the skeletal muscles. These effects enable more oxygen and glucose to reach the muscle cells and so prolong their aerobic metabolism and enhance their ability to contract. Adrenalin is secreted in response to fear and its effect is to enable the animal to make a rapid escape from danger. Nor-adrenalin, though chemically very similar, has several physiological effects different from those of

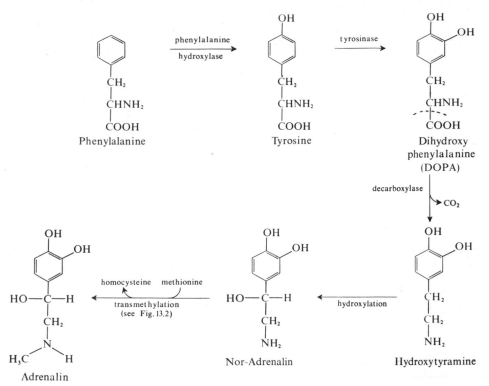

Fig. 13.15 Biosynthesis of adrenalin and nor-adrenalin from phenylalanine

adrenalin—it has, for example, only a small effect on the blood glucose level. Its main function is as a local chemical transmitter which mediates the passage of signals across neuro-motor junctions (p. 394).

In the presence of NADPH, phenylalanine is oxidized to tyrosine by the hydroxylase system already mentioned (p. 309). Tyrosinase inserts a second hydroxyl group; this enzyme is a flavoprotein which requires copper ions as co-factor. Decarboxylation then occurs to give hydroxy-tyramine (Fig. 13.15). This is next hydroxylated in the side chain on the carbon atom nearest the ring. The product is nor-adrenalin; two optical isomers of this substance can be synthesized in the laboratory, but as so often occurs in nature, asymmetric synthesis facilitated by enzymes leads only to the production of the active (laevo) isomer.

L-Nor-adrenalin is converted to L-adrenalin by the addition of a methyl group. The latter is usually derived from methionine by the mechanism shown in Fig. 13.2, the acceptor molecule in this case being nor-adrenalin.

Polynucleotides and protein synthesis

14.1   The location and structure of the nucleic acids

Nucleoproteins are conjugated proteins (p. 35) comprising a basic protein united to a non-protein compound called a *nucleic acid*.   The acid is loosely bound to the protein by electrovalent acid-base forces.   Nucleic acids are large molecules formed by the union of tens, hundreds, or even thousands of mononucleotides (section 3.8), the phosphate group of one nucleotide being attached to the sugar unit of the next.

Two types of nucleic acid exist, distinguished by the nature of the sugar units they contain and by a minor difference in the structure of one of the four nucleotide bases characteristically present.   In addition the two nucleic acids are differently located within the cell, DNA being largely in the nucleus whereas RNA, although present in the nucleus, is for the most part outside it.   It should be added that extra-nuclear DNA also exists—it occurs, for example, in mitochondria; its structure, however, is probably different from that of nuclear DNA.

Although unusual nucleotides are occasionally present, each of the two nucleic acids typically hydrolyses to a mixture of four sorts of nucleotides, two of which contain purine bases and two pyrimidines.   In both nucleic acids, the purine bases are adenine and guanine but whereas RNA contains the pyrimidine bases cytosine and *uracil*, DNA contains cytosine and *thymine*.   The formulae of all five bases are given in Fig. 3.9; it will be observed that thymine is, in fact, 5-methyl uracil.

The second structural difference between DNA and RNA relates to the nature of the 'sugar' part of the mononucleotides from which they are

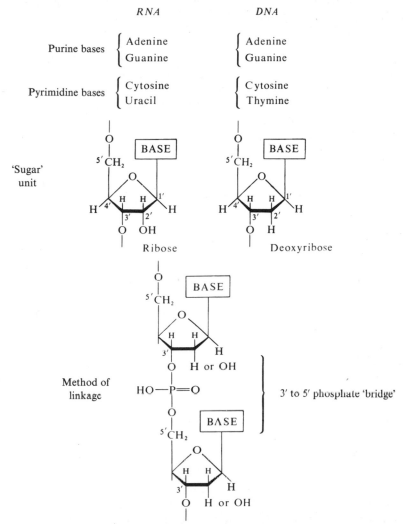

RNA   DNA

Purine bases { Adenine   { Adenine
              { Guanine   { Guanine

Pyrimidine bases { Cytosine   { Cytosine
                 { Uracil   { Thymine

'Sugar' unit

Ribose   Deoxyribose

Method of linkage   HO—P=O   3' to 5' phosphate 'bridge'

Fig. 14.1 Primary structure of RNA and DNA. (Bases other than those named are sometimes present)

constructed, for the mononucleotide units of RNA (*ribo*nucleic acid) contain the furanose form of ribose, whereas those of DNA (*deoxy*ribonucleic acid) contain deoxyribose (p. 81). The latter possesses two hydrogen atoms, and has no hydroxyl group, on carbon atom 2' of the furanose ring (Fig. 14.1). In addition, as will shortly be evident, structural differences of a secondary nature also exist, for whereas RNA is often single-stranded, DNA consists typically of two chains of DNA linked together to form a double helix.

It will be recalled that the dinucleotide NAD comprises two mono-nucleotide residues united by a pyrophosphate linkage (p. 84). Clearly, such a method of union cannot result in the formation of a molecule larger than a dinucleotide—the 'monomeric' units of a polynucleotide, like those of a polysaccharide, must unite in such a way that an indefinite repetition of the linkage is possible. In polynucleotides, the phosphate of one mononucleotide unites *with the 3' hydroxyl group* of the sugar unit of the next mononucleotide in the sequence. The result is a phosphate 'bridge' leading from carbon atom 5' of one nucleotide to carbon atom 3' of the next (Fig. 14.1; nomenclature is discussed in section 3.8).

Finally, it should be mentioned that, while adenine, guanine, cytosine, and thymine are by far the most common bases present in DNA, a small proportion of certain others may also be present. Perhaps the most important of these is 5-methyl cytosine, which may account for 20% of the nucleotide bases of plant DNA. Traces of this base are occasionally present in animal DNA. 5-Hydroxymethyl cytosine is present in a virus, the T-2 bacteriophage, which attacks *E. coli*. Halogenated uracil derivatives can also be introduced instead of thymine into the DNA of certain mutants of *E. coli*.

14.2 The double helix of DNA

Much painstaking analysis, including X-ray crystallographic studies, paved the way which led to the now famous concept of the 'double helix'. Astbury, in 1947, demonstrated that X-ray patterns showed a repetition of structure every 0.34 mμ and this, with other data, led Pauling to suggest early in 1953 that a helical structure might exist. Meanwhile, analysis of the base composition of DNA isolated from a variety of micro-organisms indicated that, although the guanine content of different DNAs varied about three-fold, in all cases guanine was always matched by an equimolar amount of cytosine. That is to say $G = C$ (Fig. 14.3). Similarly, equimolar amounts of adenine and thymine were characteristically present; i.e., $A = T$. Another remarkable fact revealed by this early analytical approach was that the total number of molecules of pyrimidine bases present was always the same as the total number of molecules of purines; i.e., $A + G = T + C$.

Armed with this information and with the results of their own X-ray investigations into the structure of purified preparations of DNA, Watson and Crick, late in 1953, proposed a double helical structure which explained all the facts available at that time. They suggested that the DNA molecule might consist of two nucleotide chains facing in opposite directions and wound around a common axis in such a way as to form two right-handed helices. The 'direction' of each chain is determined by the nature of the phosphate bridge between nucleotides, one direction being determined by the order $C_{3'}$—P—$C_{5'}$ and the other by the reverse, namely, $C_{5'}$—P—$C_{3'}$. The outside 'railing' of the 'spiral staircase' is thus provided by a continuous alternation of deoxyribose and phosphate groups, the mononucleotide bases projecting inwards at right angles to the axis of the helix. This means that, if the double helix were to be imagined as untwisted, each base would help to form the step of a ladder (Fig. 14.2a).

Each base 'helps' to form the step of the ladder, or spiral staircase, in the sense that each step is actually formed from two bases, one from each of the helices in the double structure. Furthermore, for geometric reasons, a 'step' of constant width implies the association of a pyrimidine from one spiral with a purine from the other, since the purine ring structure is larger than that of the pyrimidine. This type of structure thus explains why earlier analyses had always indicated equal numbers of purine and pyrimidine bases in the molecule of DNA.

If, however, each 'step' of the 'spiral staircase' consists of two bases associated together, it is next necessary to enquire into the nature of the chemical forces responsible for that association. It is not only essential that the mechanism of association should be *chemically feasible*, but also that the *strength of the bonding* can be assessed, for it will be recalled (chapter 1) that the biological function of DNA is to carry information. The central position of DNA in relation to chromosomal structure, the 'gene' being some part of the length of the helix, necessitates that a mechanism of self-replication of DNA must exist in a cell which is undergoing rapid mitosis, and it is now accepted that this requires a *temporary separation* of the bases paired together to form each 'step' of the double helix.

It follows that the forces between the associated bases must be rather weak, and the accepted interpretation—that the bases are held together by hydrogen bonding—is not only consistent with geometric considerations, but at the same time elegantly explains another puzzling feature of the earlier analytical work, namely, that DNA always contains as many

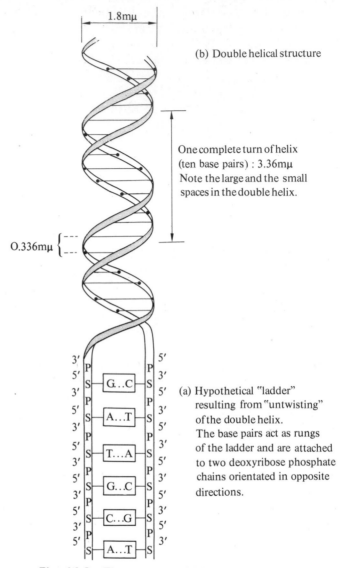

1.8mμ

(b) Double helical structure

One complete turn of helix
(ten base pairs) : 3.36mμ
Note the large and the small
spaces in the double helix.

0.336mμ

3′   5′
5′ P   P 3′
3′ S—G...C—S 5′
5′ P   P 3′
3′ S—A...T—S 5′
5′ P   P 3′
3′ S—T...A—S 5′
5′ P   P 3′
3′ S—G...C—S 5′
5′ P   P 3′
3′ S—C...G—S 5′
5′ P   P 3′
  S—A...T—S

(a) Hypothetical "ladder"
resulting from "untwisting"
of the double helix.
The base pairs act as rungs
of the ladder and are attached
to two deoxyribose phosphate
chains orientated in opposite
directions.

Fig. 14.2 The structure of DNA

molecules of adenine as of thymine, and as many molecules of guanine as
of cytosine. The hydrogen bond occurs most readily when the atoms
it links together are electronegative with respect to hydrogen, and the

molecules containing them are of such a configuration that the atoms concerned in the hydrogen bonding are no more than 0·34 mμ apart. All these conditions are fulfilled if hydrogen bonding occurs between the keto and amino groups of purines and pyrimidines participating in base-pairing, a similar bond being provided between imino-nitrogen in one ring and amino-nitrogen in the other. The probable positions of hydrogen bonds are indicated in Fig. 14.3; the suggestion is, therefore, that a 6-ketopurine will always bond with a 6-aminopyrimidine, and a 6-aminopurine with a 6-ketopyrimidine.

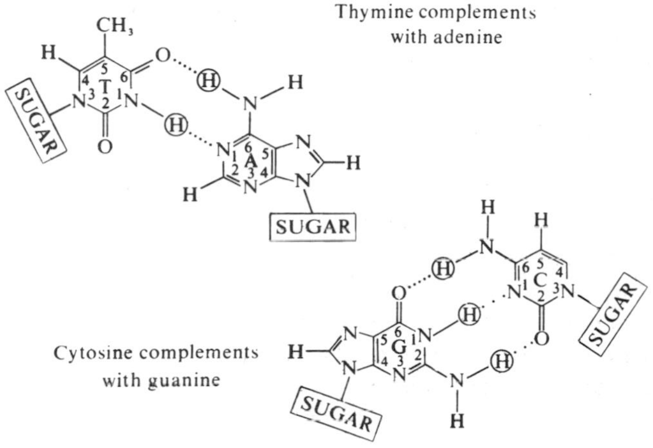

Fig. 14.3 Hydrogen bonding between base pairs of DNA

Hydrogen bonds can, more generally, be represented as resonance hybrids of two nearly equivalent structures:

$$A\!-\!\!-\!\!-\!H \cdots\cdots B \rightleftharpoons A \cdots\cdots H\!-\!\!-\!\!-\!B$$

where A and B are atoms of two electronegative elements (of which nitrogen and oxygen are the most important biochemically). The bond energy is low—about 4 kcal per mole for most biochemically important substances—but if it is repeated a very large number of times, as may well happen when macromolecules combine together, the total molecular stability contributed by hydrogen bonding may be very large (a typical covalent bond seldom exceeds a bond energy of 150 kcal per mole). Thus, in the case of DNA, two or three hydrogen bonds exist between bases paired on each 'step' of the staircase, and since its molecular weight

Polynucleotides and protein synthesis

is about 6×10^6, DNA typically has 8000–10,000 such 'steps' holding together a molecule which, if pulled out straight, could well be 3 μ long. The two hydrogen bonds holding adenine and thymine together can be regarded as resonance hybrids of the following partial structures involving atoms 1 and 6 of each of the bases in the base-pair (see also Fig. 14.3).

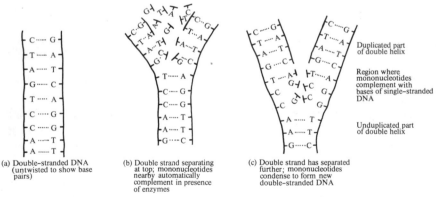

There are several slightly different crystalline forms of DNA, occurring at different humidities and when the acid is in the form of salts of different metals. The most interesting form has the base rings orientated perpendicularly to the axis of the helix in the way already described, and, for this form of DNA, the repeat distance in X-ray pictures is 0·336 mμ. The helix has a uniform diameter of 1·8 mμ and there are ten base pairs for one turn of the double helix (Fig. 14.2b). There is reason to believe that the structure of dissolved DNA does not differ materially from that of the crystalline substance.

Attention should be directed at this point to the special relationship existing between the two threads of a DNA double helix, for while the second thread is entirely different from the first, its structure is uniquely dependent on the base sequence in the first thread. If adenine occurs

(a) Double-stranded DNA (untwisted to show base pairs)

(b) Double strand separating at top; mononucleotides nearby automatically complement in presence of enzymes

(c) Double strand has separated further; mononucleotides condense to form new double-stranded DNA

Duplicated part of double helix

Region where mononucleotides complement with bases of single–stranded DNA

Unduplicated part of double helix

Fig. 14.4   Replication of a molecule of DNA.   (Shown here in 'ladder' form, the helix can be regarded as unwinding by rotating on its axis, the single-stranded DNA formed reforming a double helix by complementing with mononucleotides nearby)

in one spiral, the *complementary* base, thymine, will occur in the second; if guanine occurs in the first, then cytosine will be paired with it. This complementing means that, while one spiral can have any of the four bases in any order and occurring any number of times (giving 4^{8000} different possible DNA molecules for a structure 8000 nucleotides long), nevertheless the structure of the second spiral is predetermined by that of the first. Thus, if the two spirals were imagined as separating and each associating with mononucleotides (in the presence of appropriate enzymes and cofactors) so as to form a new partner thread, the two double helices so formed would be identical (Fig. 14.4). As will be seen later in the chapter, this self-templating mechanism partly explains the genetic importance of DNA.

14.3 Types of RNA; their structure and function

Chromosomal DNA not only acts as a carrier of information from generation to generation but, in addition, pre-determines the nature and quantity of the proteins (including enzymes) present within the cell during its individual lifetime. Thus, indirectly, DNA probably controls the composition of each of the components of the cell and hence determines all the activities of the cell from the moment it is formed until its death. DNA does not leave the nucleus, yet much cellular protein is synthesized outside the nucleus. Therefore, machinery must obviously exist for *transcribing* the message carried within the molecule of DNA, for carrying it out of the nucleus, and for *translating* it in such a way that proteins of a particular structure are eventually synthesized. The missing link between DNA and the proteins is RNA. It exists in three types, each of which functions in a different way in the total process of protein synthesis. They are called, respectively, *messenger RNA*, *ribosomal RNA*, and *transfer RNA*. Much less is known about the precise structures and functions of the different RNAs than about DNA and therefore the account which follows must be regarded as tentative; views now held on the subject will almost certainly need considerable revision in the next decade.

About 80% of cellular RNA is in the form of ribonucleoprotein particles known as *ribosomes*. These are scattered in the cytoplasm of the cell, many of them, at any time, being closely associated with the membranes of the endoplasmic reticulum. They are the seat of most of the protein synthesis occurring within the cell and, when protein is actively being made, they tend to aggregate in groups containing up to 20 individual

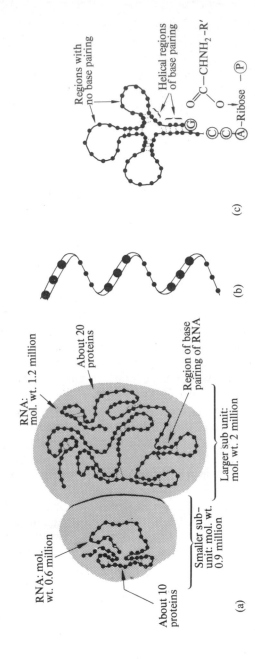

Fig. 14.5 Tentative structures of the three types of RNA

(a) Ribosomal RNA, mainly single-stranded but some base pairing. Closely associated with ribosomal proteins. The ribosome and its RNA is subdivided into sub-units as indicated. Very little is known of the detailed structure.

(b) Messenger RNA. Little is known of its secondary structure, but from its biological function, it is unlikely to have much base pairing.

(c) Transfer RNA. About 80 nucleotides, 'clover leaf' shape, with base pairing at the 'stalks'. No pairing in the 'leaflets'

Cellular biochemistry and physiology

ribosomes. Ribosomes are probably manufactured in the nucleolus, and another 15% of cellular RNA is normally present in this part of the nucleus.

The structure of *ribosomal RNA* is imperfectly understood. Some degree of base pairing occurs within a *single* thread of RNA, the thread being bent round upon itself, but the helical structure breaks down in places, so that the molecule may consist of a number of separate loops, perhaps held into one particular shape by the protein to which it is united (Fig. 14.5). As will be seen later, the ribosomal RNA provides machinery essential for protein synthesis; in this respect, it acts analogously to the pick-up head of a tape recorder—as a deciphering device—rather than itself conveying the 'message' which is to undergo 'translation' into the amino acid sequence of a specific protein.

The second kind of RNA present in the cell is called *messenger RNA*. It is far less abundant than ribosomal RNA, accounting for perhaps 3% of the total RNA in the cell. It is nevertheless of vital importance to an understanding of how nuclear DNA controls the activities and composition of the cell, for it is this material which carries within its structure a transcription of the information conveyed by chromosomal DNA. It is formed inside the nucleus (presumably as a result of base pairing with one of the strands of the DNA double helix) and moves out of the nucleus to the ribosomes. Its 'message' is then interpreted precisely (p. 344). at the ribonucleoprotein of the ribosome, its structure determining in what order amino acids are to be joined to produce a particular protein.

Messenger RNA is probably a single stranded molecule rather than a double helix. This conclusion follows in part from the nature of the biological role ascribed to messenger RNA—that it acts as a templating device for transfer RNAs (see below)—but it is supported by the fact that artificially-produced double-stranded molecules of messenger RNA are not readily destroyed by RNA-ase, whereas natural RNA is. Furthermore, the equality of purine and pyrimidine nucleotides, a characteristic of DNA, apparently does not occur for either messenger or ribosomal RNA. Messenger RNA has probably a rather short life; it is this property which accounts for the small amount of messenger RNA present in the cell at any one time and for the difficulty in obtaining experimental data about it. In many bacteria, the half-life of messenger RNA appears to be between 1 and 2 minutes, but this may not always be the case in cells of higher organisms. In mammalian erythrocytes, for example (cells which possess no DNA and so cannot code for new RNA) experiments have indicated that proteins are formed for many hours in systems that contain messenger RNA, ribosomal RNA, transfer RNAs and appropriate cofactors. The question of longevity of messenger RNA is of considerable importance in relation to the number of molecules of protein which are likely to be synthesized from a single molecule of the messenger. Finally, the size of the molecule of messenger RNA is still unsettled. Early work tended to give lower values for the molecular weight than are now being reported, presumably owing to contamination of preparations by RNA-ase. It may, in fact, have a molecular weight exceeding a million.

The molecules of the third and last form of ribonucleic acid differ from those of the other two kinds in several important respects. *Transfer*, *carrier*, or *soluble* RNA has molecules which are very much smaller than those of messenger and ribosomal RNA, the molecular weight being approximately 25,000. This corresponds to a sequence of about 80 nucleotides. The molecules terminate at one end with a nucleotide containing guanine (G) while at the other end is a sequence of three nucleotides containing, respectively, the bases cytosine, cytosine, and adenine (—C, C, A). Secondly, considerable parts of the molecule of a transfer RNA (there are, as will be seen later, many different molecular species of this sort in one cell) are probably in the form of a double helix. Since only one strand is present, this must mean that the thread is bent around upon itself, hairpin fashion.

The complete nucleotide sequence of several transfer RNAs has been determined. These include those for the amino acids alanine, tyrosine, phenylalanine, and valine. They range in length from 76 to 86 residues and each probably has a clover leaf arrangement with regions of base pairing (at the stem of each leaf and at the base of each 'leaflet') where the structure may be helical. The three 'leaflets' consist of unpaired nucleotides (Fig. 14.5).

The principal feature of the molecules of transfer RNAs, and one which is fundamental to their role in the cell, is that one end of the molecule can become *attached to an amino acid* in the presence of specific enzymes known as *aminoacyl synthases*. The nucleotide concerned in this attachment is the adenosine phosphate which terminates the sequence —C—C—A, the carboxyl group of the amino acid probably uniting with the 3' hydroxyl group of the ribose.

In practice, the amino acid must be activated before it unites with its specific transfer RNA; the details of this activation are discussed later. It is possible that the sequence of bases in the transfer RNA adjacent to the —C—C—A triplet may somehow allow a specific synthase to attach a particular amino acid to its own specific carrier. Since there are at

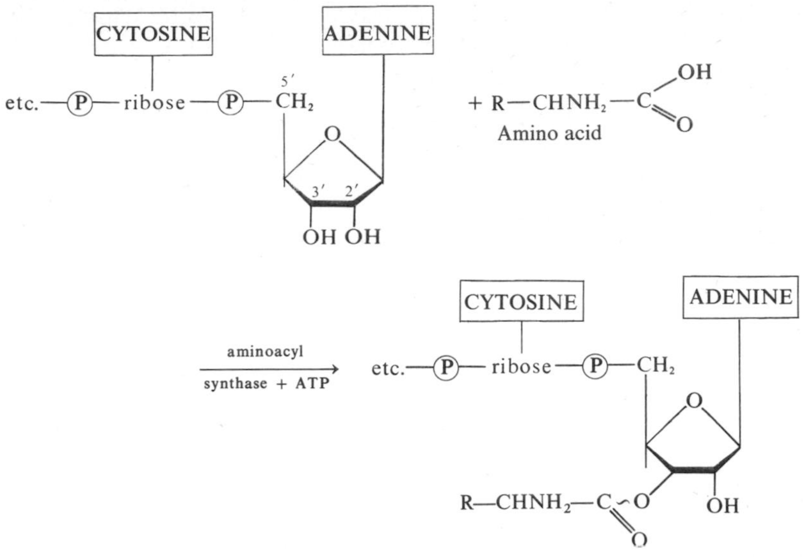

CYTOSINE
ADENINE
etc.—P—ribose—P—CH2
OH OH
+ R—CHNH2—C
OH
O
Amino acid
aminoacyl
synthase + ATP
CYTOSINE
ADENINE
etc.—P—ribose—P—CH2
R—CHNH2—C—O
OH
O
Aminoacyl transfer RNA complex

least twenty transfer RNAs, one (or more) for each of the twenty amino acids present in proteins there are at least twenty aminoacyl synthases.

At one position in the molecule of transfer RNA—probably at one of the bends where the bases are necessarily unpaired—a short sequence of bases exists which is different in each of the transfer RNAs. These bases are complementary to a sequence of bases in messenger RNA and are the means whereby a particular amino acid is inserted on to the end of a partly-formed polypeptide chain. An overall picture of how this is achieved is provided in section 14.5, but it is important to stress that each transfer RNA is unique in *two* quite separate respects. Each transfer RNA will attach itself to *one particular amino acid* and each aminoacyl-transfer RNA so formed will then *only complement with one particular base sequence* on messenger RNA (see Figs. 14.5c and 14.7).

14.4 The role of pyrophosphatases in the synthesis of nucleic acids and proteins

Before considering the synthesis of DNA, RNA, and of proteins it is worth drawing attention to an important general reaction mechanism.

Formally endergonic reactions are frequently re-routed by the activation of one (or more) of the reactants by conversion to a di- or triphosphate. One of the products is usually pyrophosphate, the hydrolysis of which is highly exergonic and therefore irreversible. Pyrophosphatases which catalyse the hydrolysis are common enzymes in cells and, by removing pyrophosphate, they facilitate many reaction sequences by disturbing the equilibrium. For example, if B and C are reactants which only lead to B—C by a highly endergonic direct route:

$$B + C \rightarrow B\text{---}C \qquad \text{(endergonic reaction)}$$

the reaction is often re-routed in the following manner, the overall reaction being exergonic:

$$B + ATP \xrightarrow[\text{AMP}]{} B \sim \textcircled{P} \sim \textcircled{P}$$

$$C + B \sim \textcircled{P} \sim \textcircled{P} \longrightarrow B\text{---}C$$

$$\textcircled{P} \sim \textcircled{P} \xrightarrow[-\Delta G° = 7\,\text{Kcal}]{\text{pyrophosphatase}} 2P_i$$

The synthesis of *nucleotides* is a good example of this type of reaction (p. 320). The pentose phosphate (ribose-5-P or deoxyribose-5-P) is first raised to a higher energy level by ATP, in a reaction which converts the sugar-5-P to 5′-phospho-(deoxy)-ribose-1′-pyrophosphate. The latter undergoes reaction with pre-existing purine or pyrimidine bases to form nucleotides (Fig. 14.6a) and inorganic pyrophosphate. The pyrophosphate is then hydrolysed to orthophosphate.

The synthesis of DNA and RNA is complicated, in the sense that a template device allows a highly improbable thing to occur—the assembly of bases in one particular order to form the polynucleotide chain (section 14.5). Energetically, however, the addition of nucleotides is similar to the mechanism of nucleotide synthesis described above. Mononucleotides of ribose or deoxyribose are first converted to *nucleoside-5′-triphosphates*, the phosphate groups being transferred from ATP (Fig. 14.6b). In the presence of a suitable DNA primer and cofactors (e.g., magnesium ions), the enzyme *DNA polymerase* catalyses the addition of nucleoside triphosphates (deoxyribose-type) to the end of a growing single DNA chain; attachment occurs at the 3′ hydroxyl group of the terminal unit of the partly-formed chain. The reaction would be energetically unfavourable if a second product, pyrophosphate, were not formed. The synthesis of RNA occurs in a similar way by the union of ribonucleoside triphosphates, the reaction being catalysed by RNA-polymerase (p. 343).

Fatty acids, it will be recalled, are activated by conversion to fatty acyl coenzyme A derivatives, the reaction involving the intermediate formation of acyl-AMP (p. 179 and Fig. 14.6c). It is noteworthy that, in the reaction which leads to the formation of this intermediate compound, pyrophosphate is also formed, for the overall reaction sequence is promoted by subsequent hydrolysis of the pyrophosphate. It is of considerable interest that the process whereby *amino acids* are activated prior to their incorporation into a growing peptide chain is catalysed by amino acyl synthase and begins with the synthesis of a compound in which the amino acid is united to AMP. ATP reacts with the amino acid, and pyrophosphate is a second product (Fig. 14.6d). The amino acyl-AMP is a high energy compound and reacts with the terminal adenine group of transfer RNA to form amino acyl-transfer RNA (compare fatty acyl coenzyme A). The amino acid so activated is subsequently added to the growing peptide by the formation of a new peptide bond. GTP appears

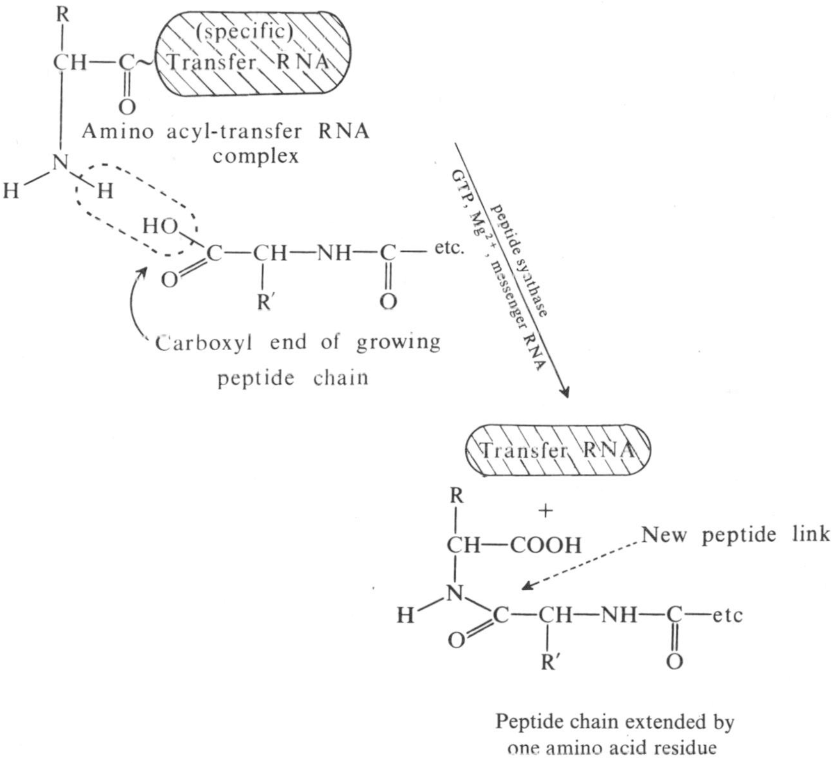

Amino acyl-transfer RNA complex

Carboxyl end of growing peptide chain

Peptide chain extended by one amino acid residue

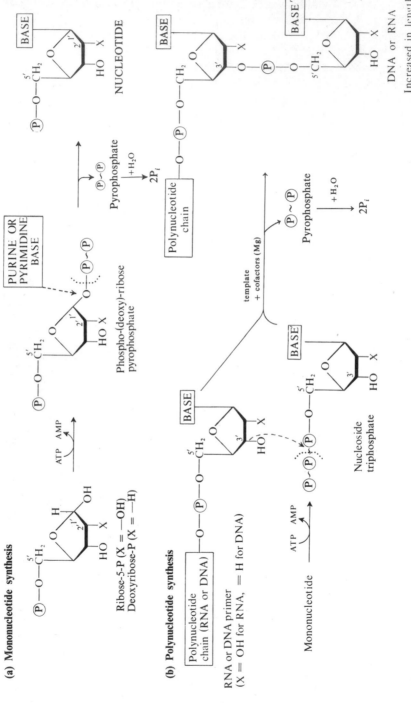

(a) Mononucleotide synthesis

Ribose-5-P (X = —OH)
Deoxyribose-P (X = —H)

PURINE OR PYRIMIDINE BASE

Phospho-(deoxy)-ribose pyrophosphate

Pyrophosphate

NUCLEOTIDE

(b) Polynucleotide synthesis

Polynucleotide chain (RNA or DNA)

RNA or DNA primer
(X = OH for RNA, = H for DNA)

Mononucleotide

Nucleoside triphosphate

template + cofactors (Mg)

Polynucleotide chain

Pyrophosphate

DNA or RNA

Increased in length by one nucleotide

(c) Fatty acyl coenzyme A synthesis

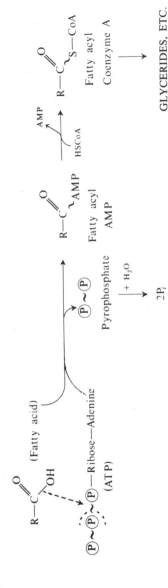

(d) Amino acid activation

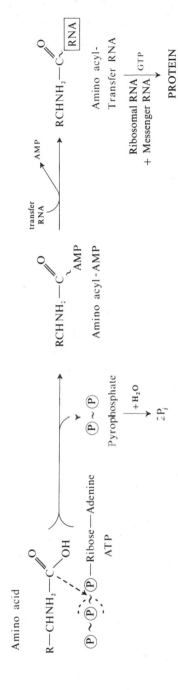

Fig. 14.6 Summary of some important reactions in which pyrophosphate is formed

to provide energy for this reaction (p. 337). Thus energy derived from one molecule of ATP and from one molecule of GTP is required for each peptide bond synthesized. Energy is necessary because the free energy change for synthesis of a peptide bond is about $+3$ kcal per mole.

14.5 Origin of molecules of DNA and RNA

The central biological role of DNA as a conveyor of information from parent to daughter cell implies not only that DNA is somehow synthesized in the nucleus, but that new molecules of DNA are identical to the old ones. This follows from the fact that longitudinal cleavage of chromosomes during the process of mitosis could not occur indefinitely if the rather limited number of threads of DNA constituting the chromosome were constantly halved, and from the fact that the daughter cells so produced retain the essential characteristics of the parent cell.

The complementarity of bases in a double helix offers an elegant explanation of how such self-duplication could occur, even though the precise way in which the two threads of the double helix unravel at the moment of duplication cannot be regarded as settled. For our present purposes, however, the unwinding hypothesis of Watson and Crick offers adequate explanation. According to this theory, the double helix rotates, the two threads unwinding at one end as it does so; as the two single helices separate, nucleoside triphosphates in the neighbourhood complement with the now unpaired bases to create two new double helices. This reaction is catalysed by DNA polymerase (section 14.4). Since this reaction cannot be readily illustrated without an animated diagram, the process is shown diagrammatically in the simpler 'ladder' form in which the helix of the molecule is imagined to have been 'untwisted' prior to the separation of the base pairs at one end of the ladder (Fig. 14.4). The hypothesis predicts that one thread of the original double-stranded molecule should be present in each of two daughter molecules. The suggestion has been substantiated by experiments in which heavy nitrogen (N^{15}) was used to label the two chains in the parent helix, for the daughter helices produced in the presence of unlabelled nucleoside triphosphates each had one half of the helix labelled and the other unlabelled.

Powerful support for the overall correctness of the 'complementary base' theory of replication has been obtained from *in vitro* experiments using the enzyme, DNA polymerase. If this enzyme is added to a mixture containing magnesium ions, the four deoxyribose nucleoside triphosphates and, significantly, a *trace of preformed DNA which acts as primer*, a considerable increase occurs in the total quantity of DNA present, the quantity of nucleoside triphosphates decreasing correspondingly. Even more significantly, analytical studies have shown that the base composition of the newly-formed DNA is very similar to that of the primer, and that this composition does not vary if the proportions of the four nucleoside triphosphates in the incubation mixture are varied within very wide limits. Although similarity of total base composition does not, in itself, presuppose an identical base *sequence* in primer and newly-formed DNA, nevertheless it is tempting to assume that this does, in fact, occur, at least for limited stretches of the total helix—especially since no synthesis occurs in the absence of DNA primer. RNA cannot replace DNA as a primer.

The mechanism by which RNA is synthesized is far less well understood. Most, if not all, the RNA within the cell is probably made on DNA templates; this supposition is especially necessary in relation to the synthesis of messenger RNA, for its base sequence codes directly for protein synthesis, and the nature and quantity of the proteins (and therefore enzymes) within the cell are known to be genetically determined.

The simplest possibility, which for a time was widely accepted, is that a process analogous to the replication of DNA occurs, except that the nucleotides of RNA are inserted into the two helices formed when DNA unwinds. In other words, the first step in the formation of RNA was considered to be the formation of a DNA-RNA hybrid, the double helix comprising one strand of each of these nucleic acids. The RNA strand was then imagined to leave the helix in some way and to pass to the ribosomes in the cytoplasm.

Although such a model is quite useful as a mental picture of the relation of large ribonucleotides to DNA, there are objections to it. Experimentally, it seems probable that the DNA which acts as a template is double-stranded, and remains so throughout the process of RNA formation, although the possibility cannot be excluded that it unwinds very transiently and locally during the process of transcription. Furthermore, the unwinding of a DNA helix in the way the simplest theory visualizes could lead to two entirely different, though complementary, messenger RNAs. In fact, it seems essential that the base sequence in *only one of the two* strands of the double helix of DNA should code for messenger RNA, and this, indeed, appears to be the case. For example, the two strands of DNA in a particular bacteriophage have been separated and, of these, only one is 'complementary' to the RNA which is formed when the bacteriophage infects its host.

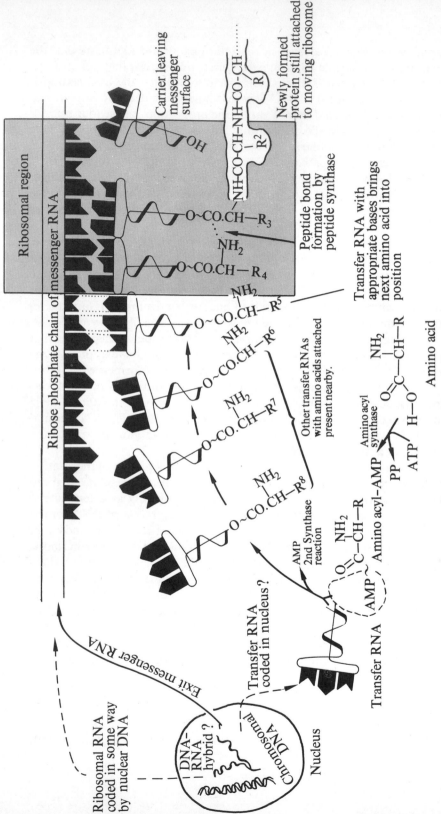

Fig. 14.7 The Adaptor Hypothesis: showing possible roles of messenger, transfer, and ribosomal RNA in protein synthesis

Another difficulty in relation to RNA synthesis is that several different enzyme systems have been studied *in vitro*, each capable of synthesizing RNA. Some of these require a primer of RNA and this primer may act either as a template or may become elongated in a way reminiscent of the manner in which starch and glycogen primers are lengthened in carbohydrate biosynthesis (p. 215). The most interesting system so far studied, however, is one in which DNA is an obligatory cofactor, acting as a primer for the synthesis of RNA. The enzyme involved is known as *RNA polymerase*, and its cofactor requirements are similar to those of the DNA polymerase described above. Thus *RNA is synthesized on a DNA primer* if the enzyme is added to a mixture containing the primer, magnesium ions, and the four ribonucleoside triphosphates.

Each of the major types of RNA present in the living cell almost certainly has its origin on some form of DNA template. In *E. coli*, for example, some 0·3% of the DNA acts as a code for ribosomal RNA and 0·02% for transfer RNA—the latter is enough to account for some fifty different types of transfer RNA. Some of the remaining DNA acts as a code for messenger RNA, the rest having an, as yet, undetermined function.

14.6 A model system for protein synthesis

Having seen in the previous sections something of what is known about the origin, structure, and function of polynucleotides, it is possible to provide a tentative scheme which incorporates many of the facts discussed already. Such a scheme is based on much experimental work, but many of the details are likely to need modification as further information becomes available.

The scheme is based on the *Adaptor Hypothesis* and is shown diagrammatically in Fig. 14.7. The DNA of the chromosome codes for a molecule of messenger RNA, possibly by way of some sort of DNA-RNA hybrid. This hybrid might involve a 'mixed' double helix, comprising one strand of DNA, and another of RNA, although, as we have seen, so simple an explanation seems unlikely.

The messenger RNA is single-stranded (p. 333) and, in order to transmit its message it moves out of the nucleus; this would appear to involve passing a lipoprotein barrier, for a nuclear membrane is present except at the time of cell division. It has been suggested that large pores apparently present in the membrane may allow the messenger RNA molecules to leave, thus avoiding the difficulty of a physical penetration or disruption of the lipoprotein membrane itself.

The messenger RNA molecules pass to the ribosomal granules present in the cytoplasm, possibly being directed there by folds or channels of endoplasmic reticulum. The ribosomes themselves, when actively engaged in protein synthesis, are aggregated on the endoplasmic membranes so producing 'rough' endoplasmic reticulum (Fig. 1.11, p. 30). The ribosomal RNA is very different in function from messenger RNA, for suitable preparative techniques can free ribosomal RNA from 'instructional' RNA, leaving a mass of ribonucleoprotein which, while devoid of any ability to synthesize protein on its *own* templates, nevertheless acts as the machinery for building proteins according to the *instructions of the messenger RNA* once the latter has reached it. It therefore acts, as was said earlier, in a way analogous to a tape-recorder head, the messenger RNA being the tape.

The ribosome then attaches itself to *one end* of the tape—it is known that there is a 'right' end and a 'wrong' end—the presence of the ribosome then allowing amino acyl derivatives of transfer RNAs to arrange themselves along the messenger RNA, fitment being decided by the base sequences on the coding region of the transfer RNA (p. 334) and by the bases of the messenger RNA in the neighbourhood of the ribosome. To change the analogy, the ribosome then acts rather like the tab of a zip-fastener, zipping into position the transfer RNA with the appropriate (complementary) base sequence (Fig. 14.7). This alignment automatically brings the amino acids attached to the carriers into position and determines the order of insertion of amino acids into the molecule of the newly-formed polypeptide chain. The peptide bond is formed from the *carboxyl group* of the growing peptide and the *free amino group* of the amino acid forming a part of the amino acyl-transfer RNA complex. A sulphydryl enzyme, *peptide synthase*, is necessary to catalyse the reaction. It is probable that several cofactors assist in the activation process leading to peptide bond formation, amongst them being GTP to act as energy source and magnesium ions, needed to activate the sulphydryl enzyme. There are at least twenty different synthases—one for each amino acid—but if, as is very probable, more than one transfer RNA exists capable of carrying one particular amino acid, there could well be one synthase for each *carrier* rather than one for each amino acid.

An interesting and perhaps very important question can now be asked. Since an amino acid is attached to its transfer RNA prior to being transported by the carrier to the polypeptide production line, what would happen if the amino acid were to undergo chemical change in the interim? Is the amino acid a passive passenger, or is it able to control its own destiny? The work of Chapeville in 1962 has indicated that at least one amino acid, cysteine, is, in these circumstances, quite unable to control its fate after attachment to its transfer RNA.   Using cell-free systems and a synthetic messenger RNA of such a composition that the ordinary carriers of alanine were unable to complement with it, it was shown that the transfer RNA for cysteine could introduce alanine into the peptide chain.   The transfer RNA for cysteine, with cysteine attached to it, was reduced by Ranay nickel to give a hybrid molecule in which alanine was attached to the cysteine carrier (Fig. 14.8).   This carrier acted normally in the sense that it transported the attached amino acid to the messenger RNA and then, by base pairing, brought the amino acid with which it was associated into the polypeptide sequence.   In consequence, instead of inserting cysteine into the chain, alanine was introduced.

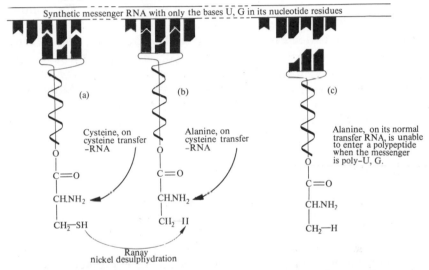

Fig. 14.8   Chapeville's proof (1962) that amino acids, after attachment to transfer RNAs, cannot determine their own destination.
(a) and (b) Cysteine-specific transfer RNA, with bases which complement with the U,G, present in the synthetic messenger, poly-U,G.   The bases of the nucleotide residues of poly-U,G occur in an order determined only by the laws of chance.   After Ranay nickel treatment, alanine is inserted into the polypeptide instead of cysteine.
(c) The bases of the usual transfer RNA for alanine are such that they cannot complement with the bases of the synthetic messenger RNA, so the carrier cannot insert alanine into polypeptides in these circumstances

When ribosomes are actively engaged in protein synthesis, they tend to be aggregated together in clusters consisting of from five to twenty or more ribosomes. Under normal conditions, ribosomes not aggregated in this way make little or no protein although, in the presence of an excess of messenger RNA, single ribosomes synthesize protein quite adequately. The explanation appears to be that more than one ribosome can occupy a place on the surface of a messenger RNA molecule; as one ribosome clears the first few bases of the 'tape', having meantime synthesized a peptide containing a few amino acid residues, another ribosome becomes attached to the end of the tape. In this way, several ribosomes can be attached to the messenger RNA at the same time, each having associated with it a polypeptide chain at varying stages of completion. Thus each ribosome probably acts independently from the others, each synthesizing a complete protein in some 10–100 seconds. The multiple occupation of the surface of the messenger can thus be regarded as a device leading to increased efficiency, rather than as a fundamental requirement for protein synthesis. The independence of the ribosomes constituting a polyribosome is indicated by the fact that the enzyme, RNAase, attacks messenger RNA, and, when it does so, the polyribosomes fall apart; i.e., polyribosomes are a chance collection of ribosomes which happen, for a while, to be moving along the same tape.

14.7 Coding ratio and the genetic code

This subject has been extensively investigated in recent years. The literature upon it is enormous, and a detailed discussion is beyond the scope of this book. Nevertheless, a brief account is necessary to show how the 4-character language of the bases of RNA is translated into the 20-character language needed for the synthesis of proteins.

Since there are only four bases but twenty amino acids, a unique coding for one particular amino acid must involve a sequence of at least three nucleotide bases—four bases, taken one at a time could only code for 4^1 amino acids and a sequence involving two of them could only account for 4^2 different amino acids. A triplet comprising any one of the four bases, occurring in any of the three positions, could, on the other hand, code for up to 64 different amino acids. A *coding ratio* of three bases for one amino acid is now regarded as a near-certainty; purely in terms of economy of effort, a ratio greater than three would seem unlikely, a supposition supported by experimental work on mutant bacteria and investigations into the protein-RNA composition of certain viruses (p. 351). All these studies lead to the conclusion that the coding ratio is indeed three, as portrayed in Figs. 14.7 and 14.8.

If a distinct triplet of bases codes for each amino acid, only twenty of the 64 possible triplets need code for the amino acids and the problem then arises as to what function, if any, is served by the other 44 possible triplets. The answer appears to be that the code is degenerate—that is to say, more than one triplet of bases is able to code for a single amino acid. In all probability, for example, two triplets, or *codons*, code for phenylalanine—UUU and UUC. Similarly, it is possible that four or more codons exist for each of the amino acids alanine, cysteine, arginine, valine, and leucine. By 1968, claims had appeared that functions had been ascribed to all the 64 possible triplets. Apart from the possible presence of commencing and terminating instructions, there is unlikely to be any 'punctuation'; almost certainly there are not, for example, any 'commas' separating two adjacent codons.

It is as yet impossible to assess very adequately the biological significance of the existence of multiple codons for one amino acid. A current hypothesis proposes that certain codons occur rather rarely and, were they to occur in the coding mechanisms for proteins possessing a central role in metabolism, their availability could act as a control mechanism of great importance. Whether or not this is so, degeneracy may have evolutionary advantages by conferring a certain safeguard against mutation. This would particularly apply if nearly-related codons coded for the same (or even similar) amino acids, so that the consequence of an 'error' following mutation (which in some cases replaces one base by another) would be minimized.

Another significant feature of the code is that it appears to be *universal*, or nearly so. That is to say, the same codon, or codons, translate for the same amino acids in most of the organisms so far investigated. One type of investigation which has been particularly useful in relation to studies on the universality or otherwise of the code has involved preparing cell-free systems containing messenger RNA and ribosomes from one organism and adding a preparation containing transfer RNAs and synthases derived from a quite different type of organism. In one remarkable experiment, rabbit haemoglobin was synthesized by adding to rabbit ribosomes (associated with messenger RNA coding for haemoglobin) a preparation from *E. coli* which contained amino acids, transfer RNAs, and the appropriate synthases. Clearly, in this case, the complementary 'anti-codons' of the bacterial transfer RNAs fitted the rabbit's messenger RNA as satisfactorily as transfer RNAs present in normal rabbit erythrocytes.

Finally, a study of the genetic code has provided some useful information about the probable nature of the mutation caused by various chemical and physical agents. Nitrous acid, for example, typically removes an amino group and puts a hydroxyl group in its place. The nucleotide bases are no exception; cytosine, for example, is deaminated to uracil. Consequently, if CUU codes for one amino acid, after deamination the original triplet will have become UUU, which may code for a different amino acid. In fact, CUU codes for leucine and UUU for phenylalanine, and it is of interest that nitrous acid has been known to produce a mutant virus in which phenylalanine was incorporated instead of leucine into its characteristic protein. Other chemical reagents are known which similarly transform one nucleotide base into another with the result that 'unusual' amino acid sequences occur in the protein for which the transformed RNA is the messenger. Other mutations occur as a result of the alteration of a 'natural' nucleotide base to an unnatural one, rather than to a different 'natural' base. Physical means, such as X-rays and ultraviolet light, may insert an additional base into a DNA molecule, or may delete one which ought to be there. In either case, far-reaching consequences are to be expected, for, in the absence of 'commas' in the message, a whole sequence of wrong amino acids may be inserted. Thus if an extra base, X, is inserted into the sequence CCA.UGA. . . . so as to form a messenger CXC.AUG.A. . . . , not only is the first triplet incorrect, but so are all subsequent ones.

14.8   The structure of bacteria and viruses

Bacteria and viruses are often regarded as being primitive compared with animal and plant cells. They certainly represent a different level of organization and, although it is not intended to give a detailed description of them here, some mention of them is necessary since many current theories of the genetic code are based on studies of these life-forms.

The bacterial cell is much smaller than the typical animal or plant cell. *Escherichia coli*, for example, is rod shaped, 2 μ long and 1 μ wide. The cell lacks any clearly-defined nucleus or nuclear membrane but nevertheless contains 2 to 4 chromosomes which consist of DNA and protein. There are numerous ribosomes but no mitochondria. This is not surprising, since a cell of *E. coli* is about the size of a typical mitochondrion.

The cell has a well defined plasma membrane which is about 10 mμ thick and it is thought that the electron transport particles which subserve the functions of mitochondria in higher organisms are attached to it. Outside the plasma membrane bacterial cells are typically surrounded by a rigid wall. In *E. coli* this protective wall consists of lipid, protein, and polysaccharide.

E. coli is extremely versatile biochemically. It can grow and thrive in glucose solution provided inorganic salts are present. It is able to synthesize all the amino acids present in proteins and also the appropriate cofactors (many of which must be provided in animals as vitamins). If lactose replaces glucose as the main nutrient, *E. coli* synthesizes a specific permease which carries lactose into the cell and a lactase, β-galactosidase, which hydrolyses the lactose to glucose and galactose. This synthesis of what appears to be completely new proteins has aroused considerable interest. It is thought that *E. coli* normally has the genetic potentiality to synthesize these proteins but that synthesis only actually occurs in the presence of lactose. This *adaptive enzyme synthesis* (p. 106) may result from the exercise of some sort of control on the formation of the appropriate messenger RNA, or it may be caused by regulation occurring at the ribosomal level.

Viruses are much simpler in structure than even the bacterial cell. In size, most of them lie within the range 10–200 mμ. They comprise a core of nucleic acid, which in some is DNA and in others RNA, surrounded by a shell of protein. The relative amounts of nucleic acid and protein vary, but typically only 20–40% of the mass is contributed by the nucleic acid. The protein shell may, in simple cases, consist of only one molecular species but, in more complex viruses, such as the T-2 bacteriophage, there may be up to fifteen different types of protein surrounding the core of nucleic acid.

Viruses only replicate within the cells of their host, taking over the machinery of the host cell to supply energy and all the cofactors necessary for their replication. They can be considered to substitute a messenger coding for virus protein for that normally supplied to the ribosomes and, in addition, to utilize nucleotides and other substances present within the cell for the synthesis of more molecules of virus nucleic acid. At the moment of infection, much of the protein coat is often discarded by the virus, the nucleic acid alone being infective. Thus the bases of the nucleic acid of the virus must provide all the information needed for replication— i.e., the protein shell does not consist of enzymes essential for the process of replication.

Geometrically, viruses are of three types. Some of them show cubic symmetry, others helical symmetry, whilst those in a third group exhibit complex symmetry. Examples of each are illustrated in Fig. 14.9. Those with cubic symmetry are often based on the icosahedron, the protein sub-units or *capsomeres* being arranged in a geometrical pattern on each of its faces. Those with a helical structure have a central core of nucleic acid, around which capsomeres of protein are again regularly arranged. The structure of one virus of this type—the tobacco mosaic virus—has

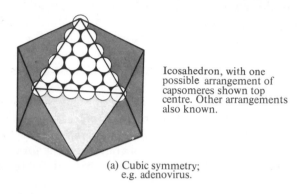

Icosahedron, with one possible arrangement of capsomeres shown top centre. Other arrangements also known.

(a) Cubic symmetry; e.g. adenovirus.

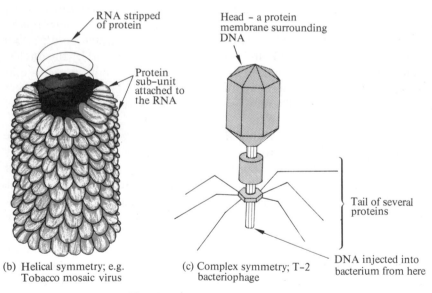

RNA stripped of protein

Protein sub-unit attached to the RNA

Head – a protein membrane surrounding DNA

Tail of several proteins

DNA injected into bacterium from here

(b) Helical symmetry; e.g. Tobacco mosaic virus

(c) Complex symmetry; T-2 bacteriophage

Fig. 14.9 Structure of viruses

been extensively studied and has proved to be a helix of RNA surrounded by a helix of elongated capsomeres.

The T-2 bacteriophage, which exhibits complex symmetry, infects cells of *E. coli*. Despite its greater complexity, the proteins associated with its DNA probably play no part in the process of infection, for less than 3 % of the protein enters the cell. In all probability the protein serves a protective role and also stabilizes the DNA in its biologically active form. In addition, however, its structure strongly suggests that the proteins in the narrow region below the head and above the base plate act as a trigger mechanism which assists the DNA to gain access to the cell.

Virus nucleic acid may be either DNA or RNA. *Animal viruses* built upon DNA reproduce in the nuclei of their host cells and those containing RNA replicate in the cytoplasm—it should be observed that virus RNA differs from ordinary cellular RNA discussed earlier, for there is little evidence that macromolecular RNA within the cell is normally able to replicate itself. Sometimes the DNA is single-stranded, sometimes it is a double helix and in some others (e.g., the polyoma virus) it is a double helix, bent round on itself to form a ring.

All *plant viruses* so far investigated have a core of RNA. This is single-stranded and, as was seen above, may either occupy the centre of an icosahedron (or similar structure) or may form a regular helix which is coated with capsomeres of protein. A plant virus, the tobacco necrosis satellite virus, has the distinction of being one of the smallest viruses so far studied. It is of considerable interest, in relation to the coding ratio of the genetic code, that the RNA of this virus contains about 1200 nucleotide bases and that the single protein which forms the virus coat contains some 370 amino acid residues. *Bacterial viruses*, or bacteriophages, are extremely diverse in both their shape and in the nature of their nucleic acids. Many contain double-stranded DNA but others are known which contain single-stranded RNA. One bacteriophage which infects *E. coli* even possesses a one-stranded circular DNA. Clearly, whilst viruses are often ideal for studying many aspects of the biochemistry of the nucleic acids, they are probably, in certain respects, also very specialized structures.

15

Movement of dissolved substances into and out of cells

In living cells the intracellular composition usually differs with respect to both inorganic and organic substances from that of the immediate extracellular environment. This is true not only of simple unicellular organisms, but applies also to the mammalian cell where the extracellular fluid (such as blood plasma) is itself a solution of highly controlled composition. The individual cell is only able to carry out its numerous activities if the properties of its internal environment are carefully maintained. A number of 'unwanted' substances must be excreted, and controlled amounts of water, inorganic ions, and often organic metabolites must be taken into the cell. The major function of the plasmalemma is to fulfil this task, but it does this in conjunction with processes occurring in the underlying cytoplasm.

15.1 Some general considerations

In simple solutions, thermodynamic considerations require that movement of both solute and solvent particles will occur *down* a concentration gradient until an even distribution is reached throughout the system. That is to say, molecules in solution tend to mix rather than to unmix. Even when a restriction is placed on the free movement of solute particles by the presence of a semipermeable membrane, the behaviour of the system can be predicted in terms of relatively simple physical laws (a semipermeable membrane is one which, ideally, allows solvent molecules to pass through it in both directions, but acts as a barrier to solute molecules). A *living membrane* is semipermeable, in the sense that it impedes the free diffusion of ions and simple solutes, but it is also far more than this. For, unlike most (imperfectly) semipermeable non-living membranes it is *selectively permeable*, allowing some substances to pass through it much faster than others, and, unlike all non-living membranes, the ease with which certain substances can pass through the plasmalemma alters in accordance with the metabolic requirements of the cell. In other words, the membrane participates as a living entity, its integrity being preserved at the expense of such metabolic factors as ATP, and its permeability-semipermeability characteristics become modified according to the needs of the cell.

The movement of ions and molecules through the cell membrane is sometimes *passive* but often involves an *active* process. Passive processes include simple diffusion and require no output of energy on the part of the cell. Active processes, on the other hand, are formally endergonic (p. 132) and involve the intervention of metabolic energy and often of carriers, enzymes, and cofactors. They usually operate in such a way that transport occurs against a concentration gradient and sometimes they assist the transport across the membrane of a substance to which the membrane would otherwise be nearly impermeable. A third general process can be distinguished which is known as *facilitated transfer*. This process is passive in the sense that it does not require a supply of energy, though energy may be needed indirectly to maintain the membrane integrity indispensible for the continuance of the process. It only occurs, however, because of the presence in the cell, or at its surface, of a protein, which reversibly combines with the substance. Haemoglobin in red blood cells is a particularly well known example of a protein which enables the facilitated transfer of oxygen into the erythrocyte.

The following sections illustrate important examples of the processes mentioned above and also briefly describe cellular aspects of intestinal absorption of food and the carriage of oxygen and carbon dioxide by the blood.

15.2 Osmosis and the uptake of water by cells

Cell membranes are normally freely permeable to water, the passage of water either into or out of cells taking place by the passive process of osmosis. If cells are placed in distilled water, the water enters the cells more quickly than it leaves them, with the result that they begin to swell. For example, when the egg of the sea-urchin is suspended in normal sea water, it maintains a constant size, but if transferred to sea water diluted with distilled water, the volume of the cells increases. The curvilinear relationship between the concentration of the sea water and the cell volume is depicted in Fig. 15.1. When the cells are placed in 50% sea

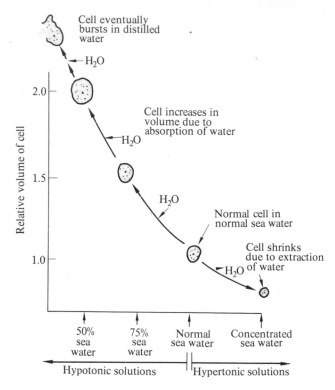

Fig. 15.1 The osmotic behaviour of the single-celled sea-urchin egg

water, their volume almost doubles; if the eggs are placed in distilled water, so much water diffuses into them that the cells burst, or *lyse*.

The movement of water by osmosis is a typical diffusion process from a region of high concentration *with respect to water* (distilled water, or a solution dilute with respect to a solute) to a region of low concentration with respect to water (a solution more concentrated with respect to a solute). Thus, if N is the number of moles of water present in an aqueous solution, and n is the number of moles of a dissolved solute present, the fraction of solvent molecules present is $N/(N + n)$ and that of solute molecules is $n/(N + n)$. Each of these fractions is termed a *mole fraction*[#]—of solvent and solute respectively—and their sum is necessarily 1·0. In pure water, $n = 0$, and the mole fraction of water is unity; in all other cases, the concentration of solvent, expressed as a mole fraction, is less than unity, and the solution is more dilute with respect to the solvent. If, then, a system exists in which solute is prevented from diffusing by the presence of a semipermeable membrane, water will flow towards the side of the membrane of lower *solvent* concentration.

For this reason, if a cell is placed in a *hypotonic* medium—i.e., one with a lower osmotic pressure than its cytoplasm—water will enter the cell. When the external solution has the same osmotic pressure as the cell sap the cell is said to be in an *isotonic* medium, and no net movement of water will occur into or out of the cell. When the cell is placed in a *hypertonic* solution, the mole fraction of water is less in the external medium than it is in cell sap; water will therefore leave the cell and the cell will shrink.

In practice, cytoplasm is a complex mixture of molecules and ions, rather than a binary mixture, and each particle of each solute contributes to the total osmotic pressure operating within the cell, for osmotic pressure is a colligative property[#] of a solution. (6×10^{23} particles of solute, in 22·4 litres of solution, at 0°C, exert ideally an osmotic pressure of one atmosphere.) Biological solutions are usually far from ideal and it is often impracticable either to measure their osmotic pressure or to calculate it. Instead, the osmotic pressure of solutions is frequently estimated indirectly by the measurement of the depression of the freezing point.

15.3 Osmotic forces in plant cells

The osmotic behaviour of plant cells is comparable to that of animal cells such as sea-urchin eggs or red blood cells, except that plant cells possess a cell wall composed largely of cellulose fibrils. This cellulose wall, which surrounds the selectively permeable plasmalemma, is freely permeable to both water and solutes. Since it possesses physical strength, it offers a physical resistance which prevents indefinite swelling of the cytoplasm under hypotonic conditions. When a plant cell is placed in distilled water, the cytoplasm increases in volume till it presses against the cell wall. The cell wall is slightly elastic and starts to bulge outwards, at the same time exerting a back pressure which opposes increasingly the further entry of water. Eventually the wall pressure is equal and opposite to the osmotic pressure, and net flow of water then ceases for water then enters and leaves the cell at the same rate. When this happens, the cell is said to be *turgid* and, in this state, has a considerable physical strength.

If a plant cell is placed in a hypertonic solution water diffuses out of the cell and the cytoplasm decreases in volume, eventually contracting away from the cell wall. The *plasmolysed* cell has much less mechanical strength than has the turgid cell, and is said to be *flaccid*.

15.4 Osmoregulation in animals

The tissues of marine invertebrates tend to have an osmotic pressure similar to that of sea water, so loss or gain of water by osmosis presents no great problem to these animals. On the other hand fresh water, having a very low salt content, has also a very low osmotic pressure and fresh water animals tend therefore to absorb water from their environment. They usually have mechanisms for reducing water uptake or for eliminating excess of it. For example, the eggs of fresh water animals tend to be almost impermeable to water, and in this respect they differ from sea-urchin eggs. Fresh water protozoa (e.g., amoeba) are faced with the task of eliminating water constantly and they do this by accumulating water in vacuoles—a process which requires expenditure of energy. This is probably provided by ATP, for the vacuoles in amoeba are often surrounded by mitochondria. The water is eliminated by the discharge of the vacuolar contents to the outside of the animal.

In the vertebrate animal, most of the body cells have the same osmotic pressure as the extracellular fluids, but this pressure is usually very different from that of the environment. Water uptake or loss is largely avoided by the presence of an impermeable skin, often supplemented by scales or other protective features. Fresh water fish, however, constantly absorb excessive water through their gills, for these must be thinly-walled to enable gaseous exchange to occur. The body cells and fluids are hypertonic compared with fresh water, and the excess water absorbed by osmosis is eliminated by the kidneys which excrete a very dilute urine.

The situation is quite different in marine teleost (or bony) fish, for these animals are hypotonic with respect to their environment and there is a constant tendency for water to be lost by osmosis. Marine teleosts survive by drinking sea water and excreting the excess salt. They can only produce urine which is isotonic with their body fluids, the excess salt being secreted by the gills. This is achieved by special cells which pump salts into the surrounding water, and since this involves moving ions *against* a concentration gradient, a considerable amount of energy is required.

Marine elasmobranchs (cartilaginous fish, e.g., sharks) do not have quite the same problem as the teleosts. Their body fluids would be hypotonic compared with sea water were it not for the fact that they contain high concentrations of urea (1–2 moles per litre), and this raises the osmotic pressure to a value very close to that of sea water.

Vertebrates adapted to a terrestrial environment, the birds and mammals, are able to excrete hypertonic urine, a process which is necessary for their survival during periods of drought since water is often scarce on land. Excessive loss of water in dilute urine would of necessity restrict their activities.

As pointed out in chapter 12 (p. 278), the form in which animals excrete nitrogen is often closely linked to their need to control the osmotic concentration of their body fluids.

15.5 Intracellular inorganic cations and the sodium pump

The concentrations of the various inorganic ions present in the cell are in most cases quite different from their concentrations in the extracellular fluid. In particular, the cytoplasm usually contains more potassium ions and less sodium ions than the bathing fluid. This is illustrated, for example, by the eggs of the sea-urchin, and by cells of the alga *Valonia*, both of which live in sea water (Table 15.1).

The sodium concentration in these cells is maintained at a lower level than in sea water, sodium ions being pumped out of the cell through the outer membrane. This activity is, in part, related to the fact that cells contain numerous soluble organic compounds, all of which contribute to the total osmotic pressure; the cells would absorb excessive amounts

Table 15.1 Concentration of principal ions in sea water, in sea-urchin eggs, and in *Valonia* (mM/kg)

	Sea water	Sea-urchin eggs	*Valonia*
K^+	10	210	500
Na^+	470	52	90
Ca^{2+}	11	4	2
Mg^{2+}	53	11	—
Cl^-	540	80	507

of water by osmosis if the inorganic ion content of cells was not correspondingly lower than in sea water, and, in practice, it is usually sodium ions which are selectively eliminated.

Similarly, most mammalian cells contain considerably less sodium but more potassium than the surrounding extracellular fluid (e.g., plasma). This is illustrated for erythrocytes and muscle cells in Table 15.2. Chapter

Table 15.2 Concentration of principal ions in mammalian plasma, erythrocytes, and muscle cells (mM/kg)

	Plasma	Erythrocytes	Muscle cells
K^+	4	150	100
Na^+	140	6	25
Ca^{2+}	5	3	3
Mg^{2+}	2	5	22
Cl^-	100	80	16
HCO_3^-	16	4·3	3

16 describes how the transmission of impulses along nerves, and the stimulation of muscles to contract both depend on a control of the intra- and the extracellular levels of these inorganic ions.

Experiments employing radioactive isotopes of the principal ions present in biological material suggest that water and hydrated potassium, chloride, and bicarbonate ions are able to pass relatively easily across the plasma membrane of cells. Larger ions such as sulphate, phosphate, calcium, and magnesium pass through the membrane less readily. Sodium ions slowly enter active cells, but the intracellular sodium concentration is maintained at a low level by a mechanism which pumps them out again.

The extrusion of sodium ions and the accumulation of potassium ions by cells probably occur by active transport. The inside of the plasmalemma of many cells is negatively charged with respect to the outside; this is a consequence of the unequal distribution of potassium ions on the two sides of the cell membrane (p. 383). The ejection of sodium ions must occur therefore not only in a direction which opposes the concentration gradient, but one which opposes an electrical gradient as well. Very little is known about how sodium ions are eliminated from cells in general, although many studies have been carried out upon erythrocytes, nerve cells, and muscle cells—all of which are in different ways very specialized. Nevertheless, there can be little doubt about the *existence* of a transport mechanism known as the *sodium pump*, even though suggestions as to its mode of operation are still speculative.

The sodium pump is known to be linked to the metabolism of the cell; treatment with metabolic inhibitors such as cyanide, dinitrophenol, or fluoride stops the outward transport of sodium ions. A consequence of a failure of the sodium pump is that continued passive diffusion inwards causes the internal concentration of sodium ions to increase. Moreover, when red blood cells are cooled so as to diminish the rate of metabolism, not only does the intracellular sodium ion concentration rise, but internal potassium ions also leak out. If such cooled cells are then very carefully warmed to blood temperature in a medium containing glucose, the original ionic gradients are rapidly restored.

Red cell membranes have ATP-ase activity and there is often a striking correlation between the level of this activity and the rate of transport of sodium ions. This has led to the suggestion that some or all of the ATP-ase of membranes could form part of the carrier system. Presumably, although ATP is hydrolysed to ADP and inorganic phosphate *in vitro*, the functional living system involves some such sequence of events as (a) the priming of a 'carrier' molecule by phosphorylation, which then (b) moves an ion across a membrane, and (c) undergoes dephosphorylation as the ion is released from the carrier. Some current ideas about the sodium pump are summarized in Fig. 15.2.

A specific poison, ouabain (formula, Fig. 1.6), inhibits the sodium pump. This poison is of interest in that it is a steroid glycoside; the sugar,

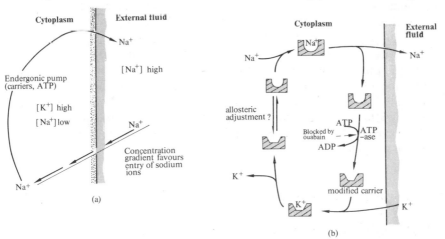

Fig. 15.2 (a) The sodium pump and (b) a possible mechanism for it

Movement of dissolved substances into and out of cells

rhamnose, is somewhat unusual in that it is a methyl pentose. Ouabain is closely related to digitalis and, like this alkaloid, has a powerful action on the heart. When, however, the action of ouabain is on the sodium pump, it inhibits the transport of sodium ions out of blood cells and the influx of potassium ions. At the same time it decreases the amount of inorganic phosphate released from ATP in a red cell system provided with optimal amounts of sodium and potassium ions.

It has been suggested that ATP activates a carrier which transfers potassium ions across the plasmalemma from the outside. Maybe some change (such as dephosphorylation) leads to a decrease in carrier affinity for the ions being carried, with the result that potassium ions are released inside the cell, the now unoccupied carrier associating with sodium ions, which are then carried to the outside. Other models are possible—the precise point in the cycle where activation occurs is still not settled—but the preferential association with potassium ions outside (where the medium is richer in sodium ions) and with sodium ions inside (where potassium ions predominate) suggests the possibility of an allosteric modification of a carrier protein (p. 99). Certainly, a direct link seems to exist between the movement of potassium and sodium ions, for the sodium pump does not extrude sodium ions if potassium ions are absent from the external medium. In red blood cells, some three sodium ions are removed for every two potassium ions carried into the cell; the nature of the carrier is unknown except that it is probably protein or lipoprotein which operates within the lipoprotein structure of the plasmalemma.

Magnesium, like potassium, is generally present in mammalian cells in higher concentration than in the extracellular fluid (plasma), while calcium is present within the cell at a lower concentration than it is externally. The concentration of *ionic* calcium and magnesium is only known approximately; up to 60% of both metals may be bound in non-ionic form to proteins in plasma. Even more may be complexed within the cells. Magnesium is, of course, an extremely important intracellular cation, being a cofactor for many enzymes, including phosphatases and kinases (p. 197). Both calcium and magnesium ions play an important role in the process of muscular contraction (p. 402).

15.6   Donnan equilibrium

Erythrocytes and muscle cells contain lower concentrations of chloride and bicarbonate ions than extracellular fluid such as plasma, yet these anions are able to pass relatively easily across the plasmalemma of these cells in both directions. The sodium pump is indirectly one of the reasons for this difference, for when positively-charged sodium ions pass out of the cell, the electrostatic attraction between them and the oppositely charged chloride and bicarbonate ions has the effect of allowing anions to leave the cell faster than they enter.

However, a second important effect—the Donnan equilibrium—results in the intracellular concentrations of these anions being lower than their concentrations outside the cell. Cells contain several types of substances which possess large charged molecules, principal amongst them being proteins and organic phosphate esters. These materials have molecules which are far too large to be able to pass through the outer membrane— they are, in effect, immobile, multivalent anions at physiological pH values. Now, in *any* solution, the total number of negatively charged particles must equal the total number of positively charged particles; in other words, the number of equivalents of anions on one side of the membrane must equal the number of equivalents of cations on that side of the membrane.

The intracellular cations—potassium, together with smaller amounts of magnesium, calcium, and sodium—balance both the negative charges of the large non-diffusible anions and those of the mobile anions such as chloride and bicarbonate. The mobile anions therefore become distributed in such a way that their intracellular concentrations are less than their concentrations outside the cell. For any phase within an organism the following relationship exists:

$$([K^+] + [Mg^{2+}] + [Na^+] + [Ca^{2+}] + [\text{organic cations such as choline}])$$
$$= ([\text{protein}^{-}] + [\text{organic phosphates}^-] + [Cl^-] + [HCO_3{}^-]$$
$$+ [\text{organic acid anions such as citrate}])$$

The concentrations here are expressed as equivalents. Clearly, the concentrations of chloride and bicarbonate ions in the cell will be less than in the external medium, where the complex anions are present in smaller concentrations. The passage of diffusible anions across the plasmalemma takes place *passively* to fulfil the requirements of Donnan equilibrium.

15.7 Absorptive cells of the small intestine

All living cells in the vertebrate have the ability to absorb selected organic nutrients such as glucose and amino acids. In most cases these are absorbed from the extracellular fluid. The cells which line the intestine, however, are highly specialized for the absorption of food materials from the lumen of the alimentary canal. The processes of absorption by these absorptive cells have been extensively investigated and are discussed here because they provide an excellent illustration of a number of general features of cellular absorption. In addition, these absorptive cells show how elegantly cellular structure can be related to function (Fig. 15.3).

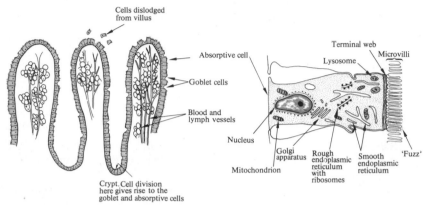

Fig. 15.3   Structure of intestinal villi and of an absorptive cell

The wall of the intestine is convoluted to form villi, which project into the lumen of the intestine. These are separated by crypts, the epithelial cells at the base of which are constantly undergoing mitosis. All the epithelial cells of the villus arise in this way, the newly-formed cells gradually moving up and over its surface. This is a continuous process, for cells at the top of the villus are constantly being replaced, the movement of chyme down the intestine probably helping to dislodge older ones. Dislodged cells are subsequently broken down in the intestine, the enzymes they contain being released into the lumen; in this way the cells contribute a valuable but secondary source of enzymes (the main source of digestive enzymes being the pancreas). The cell division which occurs at the base of the crypt gives rise mainly to absorptive cells, although a second type of cell, the *goblet cell*, is also present. These cells acquired their name from the appearance they presented to the early microscopists, for within them are vacuoles which are filled with mucus. This mucus is eventually secreted into the intestinal lumen and may protect the general epithelial cells from the action of digestive enzymes.

Intestinal absorptive cells are columnar, the nucleus being positioned near the base of each cell. The free outer surface of the cell, bordering the intestinal lumen, has its area increased about twenty-fold by membrane invagination to give numerous slender finger-like projections called *microvilli* (Fig. 15.3).   Each microvillus is about 1 μ in length and 0·1 μ across. Described as the 'brush border' by light microscopists, the microvilli have been shown by the electron microscope to be bounded by a lipoprotein membrane (p. 21).   Such a membrane is, of course, typical of the plasmalemma of all types of cells, but in the present instance the protein layers are somewhat thicker than they are in parts of the plasmalemma elsewhere in the absorptive cell.   A probable explanation is that numerous enzymes and carrier proteins, which together are responsible for the transport of materials across the cell membrane, are concentrated in the region of the microvilli.   Such an accumulation is to be expected, for the absorptive cells are able to absorb all the major products of digestion— amino acids, monosaccharides, disaccharides, free fatty acids, and monoglycerides—and, as will be seen in section 15.8, the absorption of many of these materials involves processes far more complex than simple physical diffusion.

The microvilli have a rather indefinite or 'fuzzy' appearance, owing to the presence, in the 'micro-crypts' between them, of a mucopolysaccharide. This material probably decreases bacterial attack on the cells and may, in addition, assist in some way in the initial binding of substances about to undergo absorption.   Each microvillus contains a number of fine central filaments which run along its length and which possibly help to strengthen it.   At the base of the microvilli is a region containing further fibrous elements, and known as the *terminal web*.   This web separates the apex of the cell, which bears the microvilli, from the remainder of the cell.   It has been suggested that one function of the web may be to help regulate the passage of materials towards the base of the cell after their absorption.

It is possible to separate the microvilli, together with the web and a little of the underlying cytoplasm, from the rest of the cell and thus to study the absorptive apparatus and the enzyme systems associated with it.   It contains most of the intestinal disaccharidases (e.g., lactase and maltase), as well as an alkaline phosphatase and the proteolytic enzyme, aminopeptidase.   These enzymes are probably involved in the uptake of materials from the intestinal lumen (section 15.8).   An enzyme is also present which, *in vitro*, hydrolyses ATP to ADP, and, for this reason, is called ATP-ase.   This enzyme might well participate, *in vivo*, in the absorption of one or more of the products of digestion, for absorption is frequently a formally endergonic process which is coupled to the exergonic breakdown of ATP. The cells contain numerous mitochondria which are capable of providing the ATP which somehow enables the transfer and carrier proteins to function.

The membranes of the endoplasmic reticulum of absorptive cells have less ribosomes associated with them than those of the enzyme-secreting cells of the stomach and pancreas. This is to be expected, since absorptive cells, unlike the cells just mentioned, do not have to synthesize large quantities of protein for purposes other than their own metabolism. On the other hand, intestinal absorptive cells do possess considerable amounts of *smooth* reticulum, and this probably reflects the fact that fat absorption is followed by the synthesis of triglycerides and phospholipids. Some steps in these syntheses occur in the neighbourhood of endoplasmic membranes.

15.8 Absorption of sugars and amino acids by cells of the intestinal epithelium

Much dietary carbohydrate is absorbed as disaccharides, for starch and glycogen are broken down to maltose and isomaltose (p. 77) while sucrose and lactose only undergo partial hydrolysis to monosaccharide. Nevertheless, there is an appreciable absorption of carbohydrate in monosaccharide form. Disaccharases located in the microvilli of absorptive cells act as specific binding sites for particular disaccharides, which are thus able to enter the cell in the form of enzyme-substrate complexes. The hydrolase action of the carrier enzyme follows entry of the disaccharide into the cell, the substrate being split into its monosaccharide units before the enzyme surface is vacated. Thus *maltose* and *isomaltose* are absorbed bound to maltase and isomaltase and, after transfer to the inside of the cell, glucose units are released. Similarly, *lactose* is carried into the cells complexed to lactase, a β-galactosidase, glucose, and galactose being released within the cell. *Sucrose* enters attached to invertase and is only released after hydrolysis to glucose and fructose.

The absorption of *glucose* by intestinal absorptive cells has been the subject of much research. An important feature of the process is that D-glucose is transferred from the lumen of the intestine, through the absorptive cells and eventually into the blood, and this passage occurs *against a glucose concentration gradient*. The entry of glucose into epithelial cells is, therefore, an endergonic process; it is analogous to using machinery to drive water uphill. Thus glucose must be *actively* transported into the cell. The energy of ATP directly or indirectly provides the motive power for this active transport. For this reason it is found that inhibitors of cell metabolism such as fluoride, cyanide, fluorocitrate, and dinitrophenol, all of which, in different ways, inhibit the synthesis of ATP, are able to reduce the rate of absorption of glucose at concentrations well below those at which they damage the cell.

A specific carrier is responsible for the transport of glucose across the cell membrane. Its precise nature is unknown, although it is probably a protein with which the glucose forms a complex. The same carrier also transports D-galactose into the cell, for an excess of galactose on the side of the cell membrane nearest the intestinal lumen reduces the rate of glucose transport. It thus appears that the two sugars occupy the same sites on the same carrier, and compete with one another for these sites. The situation is clearly similar to that whereby two substrates compete with one another for an enzyme surface and give rise to the phenomenon of competitive inhibition (p. 104).

The mechanism of absorption of glucose by intestinal cells is probably similar to that for the uptake of glucose from blood by muscle and liver cells and also to that which leads to the reabsorption of glucose from the lumen of kidney tubules. Some evidence for this is provided by the observation that *phlorizin*, which specifically inhibits the intestinal absorption of glucose or galactose, also interferes with glucose reabsorption in the kidney and so causes excretion of glucose in the urine.

A small proportion of the carbohydrate taken up by the absorptive cells is metabolized by the cells themselves, but nearly all of it is transferred to the intestinal blood, which in turn carries it, by way of the portal venous system, to the liver. This organ then converts most monosaccharides to glycogen by means of the pathways described in chapter 10.

Dietary protein is almost completely hydrolysed to L-amino acids by the proteolytic enzymes (p. 109). These amino acids are *actively* absorbed, for uptake can occur against the concentration gradient. Metabolic inhibitors are again effective in blocking the transport mechanism, so the ultimate energy source is almost certainly ATP. It is of interest in this respect that when the (un-natural) D-amino acids are fed to an animal they enter the absorptive cells at only one-fifth the rate of the normal dietary L-forms.

There appear to be four distinct carriers involved in the absorption of the amino acids. The first and least specific carrier is able to transport all the neutral amino acids. A second is responsible for facilitating the absorption of three basic amino acids—lysine, arginine, and ornithine.

A third carrier transports the two *imino* acids and the last carries the amino acids with branched aliphatic side-chains. This specificity suggests that amino acid carriers, like those which transport carbohydrates, have active centres which form more or less specific attachments with their substrates. Clearly, they resemble enzymes which have greater or lesser specificity for their substrates.

After absorption, amino acids are carried by the blood stream to the liver. The process of absorption, both of amino acids and of sugars, is extremely efficient and occurs throughout the length of the small intestine. Only very small amounts of either type of compound remain unabsorbed by the time the residue of the food leaves the small intestine.

15.9 The digestion and absorption of glycerides

Neutral triglycerides are almost insoluble in water and for this reason their digestion and absorption follow a pattern somewhat different from those of proteins or carbohydrates. Digestion takes place mainly in the duodenum, the *lipase* which catalyses the hydrolysis being secreted by cells of the pancreas. Since the enzyme is water-soluble, effective hydrolysis is only possible if the surface of contact between aqueous and lipid phases is very large. Thus fats and oils must first be dispersed in the form of exceedingly small droplets and several regulatory mechanisms exist which ensure that *emulsification* occurs. Bile assists in the emulsification process, for it contains several powerful emulsifying agents— lecithin, glycocholate, taurocholate, and cholesterol (chap. 1). Together these substances disperse the fat into droplets of about 500 mμ diameter. Particles of this size are large enough to scatter light and therefore the emulsion is milky in appearance. The droplets are stabilized by the charges on the anions of the bile acids and by the polar groups of lecithin, the hydrophobic parts of the emulsifying agents being directed towards the inside of the droplets. As in the cell membrane, cholesterol may facilitate a closer packing of the phosphoglyceride and glyceride molecules (p. 20).

Pancreatic lipase plays a major role in processing the triglyceride to a point where absorption is possible by intestinal cells. It selectively removes the fatty acids from the two α-positions of each glyceride molecule, producing a mixture containing a little residual triglyceride, but which is mainly composed of fatty acids and β-monoglycerides.

The milkiness of the duodenal contents, apparent when fats are first emulsified, gradually clears as lipase hydrolyses the triglyceride, for the fatty acids and β-monoglycerides break away from the droplets to form new and much smaller aggregates known as micelles (Fig. 15.4). One of the original droplets of the emulsion can give rise to some 10^6 micelles, each with a diameter of about 5 mμ. Such *micelles* are too small to scatter light and the colloidal solution containing them is consequently almost clear. These micelles are, however, still stabilized by the presence of bile acids and lecithin. Their total surface area is enormous compared to that of the emulsion droplets from which they were formed and consequently many of them can simultaneously come into contact with the microvilli of the absorptive cells.

Free fatty acids and monoglycerides are readily absorbed but some di- and triglycerides also enter. In all cases, the entry appears to be passive in nature, for the rate of entry is almost independent of temperature. In contrast, active transport, like other kinds of chemical reactions, increases as the temperature rises within the limits imposed by a biological environment. The fatty acids and glycerides penetrate through the lipoprotein surface of the microvillus, perhaps by diffusion, or perhaps by slipping between the lipid molecules of the membrane by a process known as 'two-dimensional diffusion'.

Once within the absorptive cells, the fatty acids and monoglycerides diffuse through the terminal web to the abundant smooth endoplasmic reticulum below it. This region of the cell is rich in the enzyme systems responsible for the synthesis of triglycerides and phospholipids, and electron micrographs of active absorptive cells suggest oil droplets amongst the reticulum. An enzyme located there then converts free fatty acids to fatty acyl coenzyme A derivatives which, in the presence of a second enzyme known as *monoglyceride transacylase*, react with β-monoglycerides to form triglycerides.

$$RCOOH + ATP + CoASH \rightarrow R.CO \sim SCoA + AMP + PP$$

$$2R.CO \sim SCoA + \beta\text{-monoglyceride} \xrightarrow[\text{transacylase}]{\text{monoglyceride}} \text{triglyceride} + 2CoA.SH$$

Other reactions leading to the synthesis of phospholipids and triglycerides (sections 8.7 and 8.8) also take place in the intestinal cells.

The ribosomes attached to the rough endoplasmic reticulum also participate indirectly in the formation of the oil droplets which appear within the cell, for they synthesize specific proteins which form an envelope around the droplets. These proteins, known as β-globulins, are characterized by the fact that they possess a more than usual proportion of

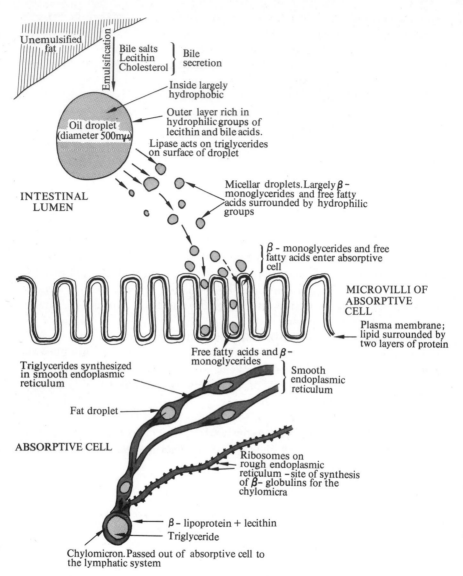

Fig. 15.4 Outline of the mechanism of fat digestion and absorption

amino acid residues with hydrocarbon side-chains. The side chains form weak apolar bonds with molecules of lipid in the droplets to produce what are, in effect, lipoproteins. The droplets, complete with their lipoprotein envelope, are discharged from the endoplasmic reticulum into the cell sap (Fig. 15.4).

The droplets, which have a diameter of 75–1000 mμ are transferred to the base of the absorptive cell, from where they are discharged into the lymph duct. In addition, an appreciable quantity of short-chain free fatty acids passes into the blood. Lymph is turbid in appearance after the animal has digested a fatty meal because of the presence of these fat globules. They are known as *chylomicra*; each chylomicron usually contains some 86% triglyceride, 9% phospholipid, 3% cholesterol, and 2% lipoprotein. The presence of surface active materials allows the water-insoluble lipids they contain to be transported to the liver in aqueous body fluids.

15.10 Michaelis kinetics applied to the transport of compounds into cells

In a number of instances, including the absorption of glucose and of amino acids by intestinal cells, and by proximal tubular cells of the kidney, it is found that the rate of transport of a compound into the cells, v, is related to the concentration S of the substance in the extracellular fluid, so that as the concentration is increased the rate of uptake of the compound by the cells rises. Above a certain concentration, however, a maximum rate of absorption, V_{max}, is observed. This is explicable if it is assumed that carriers combine with the substance being absorbed, rather as an enzyme combines with its substrate. The maximum rate of absorption is reached when the substrate concentration is high enough to keep all the carriers of the transport mechanism fully employed. In many cases it is found that

$$\left(\frac{V_{max}}{v} - 1\right)[S] = \text{a constant, } K_t$$

The similarity of this expression to the Michaelis equation (p. 99) suggests a close parallel can be drawn between the active absorption of a compound by the cell and enzyme catalysis. In the case of absorption, K_t is numerically equal to the concentration of the substance being absorbed when the rate is half the maximum rate. This should be compared with the value of K_m in the Michaelis equation of enzyme kinetics (equation (4.6), p. 102).

In many cases, the kinetics of absorption are best expressed by applying an equation similar to that of Lineweaver and Burk (equation (4.7), p. 102). Thus, if the reciprocal of the initial rate of absorption ($1/v$) is plotted against $1/[S]$ for several concentrations of the substance being absorbed, K_t and V_{max} can readily be estimated (Fig. 15.5).

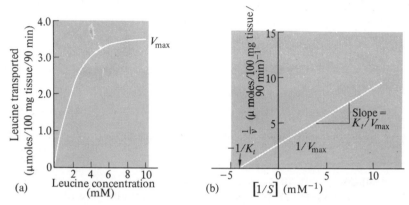

Fig. 15.5 Michaelis kinetics applied to leucine absorption by rat small intestine. (Graph (b) is a Lineweaver and Burk plot using the equation

$$\frac{1}{v} = \frac{K_t}{V_{max}} \cdot \frac{1}{S} + \frac{1}{V_{max}}$$

$$V_{max} = \frac{1}{2 \cdot 9 \times 10^5} = 3 \cdot 4 \times 10^{-6}\text{M}$$

$$K_t = \frac{1}{4 \cdot 6 \times 10^2} = 22 \times 10^{-3}\text{M}$$

(From Larsen, Ross, and Tapley, *Biochem. Biophys. Acta.*, **88**, 570 (1964))

15.11 Blood and interstitial fluid

The spaces between tissue cells are typically filled with fluid known as *interstitial* or *tissue fluid*. Interstitial fluid bathes the surface of the tissue cells and is very similar in ionic composition to blood plasma except that while plasma is very rich in dissolved protein (about 70 g per litre of plasma), tissue fluid contains very little protein.

In vertebrates, the blood vascular system reaches the fine structure of the tissues by very small blood vessels, the capillaries. These are lined by very thin flattened epithelial cells forming a membrane about 0.1μ thick. Nutrient materials, such as glucose and amino acids, diffuse readily through the capillary membranes and reach individual cells by diffusion through the tissue fluid. The movement takes place down a concentration gradient created by the absorption of the nutrients by the tissue cells. Elimination of unwanted materials by a cell takes place by their active, passive, or facilitated transport out of the cell into the tissue fluid where they diffuse, again down a concentration gradient, to the blood capillaries.

Tissue fluid undergoes continuous formation and circulation, processes largely maintained by *ultrafiltration* of the plasma in the capillaries. The blood pressure at the arterial end of the capillary is great enough to force ions and relatively small molecules such as water, glucose, and amino acids, through the capillary wall into the tissue fluid. The process is known as ultrafiltration because large molecules, such as proteins in the plasma, filter through to only a very small extent.

The ultrafiltration is caused by the force resulting from the pressure of the blood in the capillary but a second process tends to oppose the passage of the fluid out of the capillary. The opposing force is that part of the osmotic pressure of the blood plasma which is related to those of its constituents which are too large to pass through the capillary wall. These are colloidal particles, the principal ones being proteins. They produce what is known as the *colloid osmotic pressure*, and, in mammals, this has a typical value of about *30 mm Hg*.

The blood pressure in the capillaries decreases from about *40 mm Hg* at the arteriolar end to only *10 mm Hg* at the point where the blood reaches the small veins which carry the blood away. Initially, as the blood enters the capillary, the net force tending to cause the ultrafiltration is about *10 mm Hg*, since the pressure leading to ultrafiltration is the blood pressure less the colloid osmotic pressure. Further along the capillary, the colloid osmotic pressure of the plasma exceeds the blood pressure so tissue fluid is drawn *into* the capillary by osmosis. The combined effects of the ultrafiltration of plasma fluid to form tissue fluid and the subsequent osmotic reabsorption of the fluid into the capillary, tend not only to promote the movement of compounds to the cells but to facilitate the removal of waste products.

15.12　The transport of oxygen into animal cells

Active cells consume considerable quantities of oxygen. In most cases the dissolved gas simply diffuses into the cell by passive means for the consumption of oxygen in the cell creates a concentration gradient, the extracellular fluid having the higher oxygen level. The great majority of animals need an ancillary transport system which, in effect, connects the extracellular fluid to the organ where oxygen is absorbed and, in vertebrates, there is a highly-developed blood vascular system which transfers oxygen from gills or lungs to the tissue capillaries. The oxygen diffuses out of the blood capillaries in the tissues, again down a concentration gradient, and by doing so maintains at a high level the oxygen concentration of the fluid immediately bathing the cells. Vertebrate blood is able to absorb large quantities of oxygen in the lungs or gills because of the presence of haemoglobin in the red cells. The passage of oxygen into the erythrocytes occurs by a process of *facilitated transfer* which is dependent on the haemoglobin within the cells. Mammalian erythrocytes are remarkable in lacking nuclei; they are flattened discs

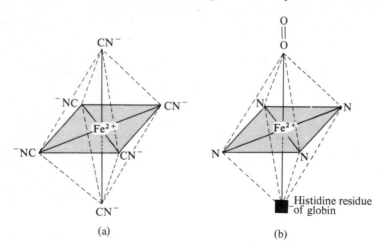

(a)   (b)

Fig. 15.6　Comparison of the structures of (a) the ferrocyanide ion, $[Fe(CN)_6]^{4-}$, and (b) oxyhaemoglobin. In (b) the nitrogen atoms shown are those of protohaem

having a diameter of about 8 μ in man and occupying some 40% of the blood volume.

The haemoglobin molecule of higher vertebrates consists of four peptide chains each associated with one protohaem group (p. 56). Each protohaem group contains a ferrous ion at its centre held by resonating covalent and dative covalent bonds to the four nitrogen atoms of the five-membered pyrrole rings. The ferrous ions are the immediate oxygen carriers and they are thought to link to the molecular oxygen in the manner shown in Fig. 15.6. It should be observed that the iron remains permanently in the ferrous form. This contrasts with the iron of the cytochromes which is reversibly converted from the ferrous to the ferric state (p. 143). Since there are four haem groups in each haemoglobin molecule, one mole of haemoglobin, when fully oxygenated, carries four moles of oxygen. In fact, the oxygenation takes place in four stages and therefore fully oxygenated haemoglobin is best represented as $Hb(O_2)_4$, although the equilibrium between unoxygenated and oxygenated haemoglobin is often shown more simply as:

$$Hb + O_2 \rightleftharpoons HbO_2$$

Hb		HbO_2
Unoxygenated		Oxygenated haemoglobin,
haemoglobin		or Oxyhaemoglobin

The initial step in the transfer of oxygen across the erythrocyte plasmalemma is passive diffusion. This would soon cease in the absence of haemoglobin as the oxygen concentration gradient evened out. However, the presence of unoxygenated haemoglobin facilitates a far greater uptake of oxygen. Note, however, that the process is still passive. The importance of haemoglobin in mammalian blood is demonstrated by the fact that haemoglobin carries about 98% of the oxygen; only 2% is in simple solution.

An essential feature of the oxygenation of haemoglobin is its reversibility. Oxygen is readily taken up by the red cell haemoglobin in the lungs or gills of animals, and when the blood is pumped to active tissues, the oxyhaemoglobin dissociates to yield oxygen, which diffuses passively out of the red cells into the extracellular fluid. The dissociation of oxyhaemoglobin is not only geared to the requirements of the tissues through which the blood flows, but, in addition, it is linked to the transport of carbon dioxide (section 15.14).

The most important factor determining the extent to which haemoglobin is in the oxygenated form is the oxygen concentration to which it is exposed. This is shown in Fig. 15.7 where the *solid line* indicates that at a partial pressure of oxygen (P_{O_2}) greater than 80 mm Hg, the haemoglobin is very nearly completely saturated with oxygen. The curve is S-shaped; at a partial pressure of oxygen of 40 mm Hg, the haemoglobin is about 75% saturated (i.e., 75% is in the oxygenated form), while at a partial pressure of 20 mm Hg, it is about 20% saturated.

In the lung capillaries of mammals, the blood becomes almost completely saturated with oxygen because the partial pressure of oxygen in lung alveolar gas is about 100 mm Hg and oxygen readily diffuses through

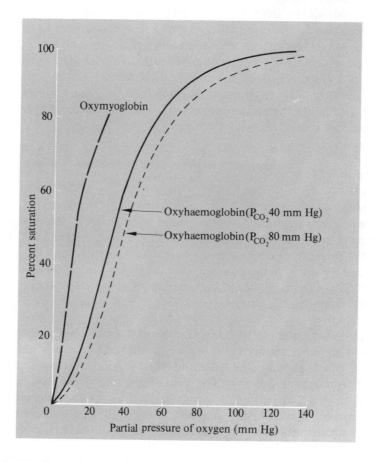

Fig. 15.7 The oxygen dissociation curve of human haemoglobin (at a partial pressure of CO_2 of 40 and 80 mm Hg) and the dissociation curve for oxymyoglobin

Cellular biochemistry and physiology

the thin alveolar membrane into the blood via the thin capillary walls (Fig. 15.8).

In active tissues such as contracting muscle the partial pressure of oxygen is low (30 mm Hg or less) largely owing to oxygen utilization by mitochondrial cytochrome oxidase. Oxygen diffuses into the tissue cells from the surrounding extracellular fluid where the oxygen concentration is higher; simultaneously, as the extracellular fluid loses oxygen to the tissue cells, oxygen diffuses out of the red cells in the blood capillaries to

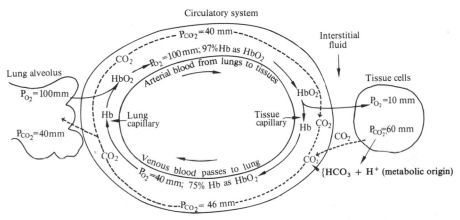

Fig. 15.8 Diagram to show that the movement in lung and tissue capillaries of both oxygen and carbon dioxide is determined by partial pressure (*P*) gradients

replace it. In effect, a concentration gradient exists between the erythyrocytes and the tissue cells so that oxygen diffuses to where it is being consumed. The diffusion of oxygen from the blood constantly lowers the oxygen concentration there, and this causes the dissociation of oxyhaemoglobin. Venous blood returns to the lungs with oxygen at a partial pressure of about 40 mm Hg (75% saturation). In the lungs it is re-oxygenated and so is ready to be recycled (Fig. 15.8).

Myoglobin is another haem pigment (p. 57) which, in many animals, is of physiological importance because of the ease of its (reversible) association with oxygen.

$$Mb + O_2 \rightleftharpoons MbO_2$$

It is present in the skeletal muscles of higher vertebrates, its single protohaem group reacting with oxygen in the same way as each of the four protohaem groups of haemoglobin.

The affinity of myoglobin for oxygen is greater than that of haemoglobin. Therefore, in a system containing oxyhaemoglobin and myoglobin, the former readily dissociates, the oxygen passing to the myoglobin, so oxygenating it. This is shown by the position of the myoglobin line in respect to that for haemoglobin in Fig. 15.7. Clearly, the greater affinity for oxygen possessed by myoglobin greatly enhances the extraction of oxygen from blood into the muscle cells—especially in active muscular tissue, where the oxymyoglobin has in part dissociated to supply oxygen to the muscle mitochondria.

15.13 Intracellular pH, and the passage of hydrogen ions and carbon dioxide out of tissue cells

Many of the processes occurring in the cell, especially those catalysed by enzymes are only effective under conditions where the hydrogen ion concentration is maintained within a certain physiological range. Typical intracellular pH values are 6·8–7·0 in animals. Yet active cells are constantly producing acids. The catabolism of amino acids containing sulphur yields sulphuric acid, the breakdown of nucleic acids produces phosphoric acid. In addition, actively contracting muscle produces lactic acid. Clearly there must be an effective buffering system in the cell, and also a mechanism for excreting excess hydrogen ions to the extracellular fluid.

Within the cell there is a number of important buffering systems, the principal ones being proteins, phosphate (both inorganic orthophosphate and phosphate esters), bicarbonate, and organic acids such as lactate. These act as buffers by combining with hydrogen ions:

$$\text{Protein}^{n-} + \text{H}^+ = \text{H-Protein}^{(n-1)-}$$
$$\text{HPO}_4{}^{2-} + \text{H}^+ = \text{H}_2\text{PO}_4{}^-$$
$$\text{HCO}_3{}^- + \text{H}^+ = \text{H}_2\text{CO}_3 \rightarrow \text{H}_2\text{O} + \text{CO}_2 \text{ (dissolved)}$$
$$\text{Lactate}^- + \text{H}^+ = \text{Lactic acid}$$

In general, a buffer is most effective at resisting changes in hydrogen ion concentration at pH values close to the pK_a of its acid. Within cells, where the pH is about 6·8, phosphate probably plays an important role in stabilizing pH since the pK_a of the system $\text{HPO}_4{}^{2-} + \text{H}^+ = \text{H}_2\text{PO}_4{}^-$ is 6·8 (p. 38).

The proteins of cells and blood act as buffers mainly because of the part played by acidic and basic amino acid residues such as those of glutamic acid and lysine. These can reversibly accept or release hydrogen ions.

The bicarbonate buffering system in the cell is of particular importance because hydrogen ions combine with bicarbonate to produce carbonic acid, which largely breaks down to form water and dissolved carbon dioxide. The latter, together with carbon dioxide produced as a result of the numerous decarboxylation reactions (e.g., those of the TCA cycle), diffuses out of the cell down a concentration gradient for extracellular fluid typically contains less dissolved carbon dioxide than the intracellular fluid. There is also a hydrogen ion concentration gradient, the pH of the extracellular fluid being greater than inside the cell. This enables hydrogen ions to diffuse out passively. In the next section the processes which maintain the pH and carbon dioxide gradients will be discussed.

There is one further way in which the cell eliminates hydrogen ions. Undissociated organic acids tend to pass far more readily across the cell plasmalemma than the corresponding anions, perhaps on account of their greater solubility in the lipid part of the membrane. When lactate, for example, buffers hydrogen ions in the cell, the protons are transferred to extracellular fluid as undissociated lactic acid. The lactic acid dissociates again on the outside of the cell.

15.14 How the erythrocyte participates in the transport of carbon dioxide and in buffering the blood

The blood vascular system of vertebrates not only carries oxygen and various nutrients to the capillaries which supply the tissue cells but also carries away waste products. Of these, carbon dioxide is excreted by the lungs or gills. The transport of carbon dioxide by the blood is, however, inextricably linked to the control of the hydrogen ion concentration of the body fluids and therefore these two functions are best considered together.

The pH of blood plasma is normally about 7·4, the hydrogen ion concentration being less than the intracellular level where the pH is about 6·8. As indicated in the previous section, this enables hydrogen ions to diffuse from cells to the extracellular fluid. When hydrogen ions diffuse into the blood they are buffered by the plasma proteins and by inorganic phosphate:

$$\text{Protein}^{n-} + H^+ = \text{Protein}^{(n-1)-}$$
$$HPO_4^{2-} + H^+ = H_2PO_4^-$$

In addition, the bicarbonate present in the plasma provides a remarkable form of buffering to resist an excessive fall in the pH. Hydrogen ions entering the blood combine with bicarbonate ions to form carbonic acid. However, the latter is unstable and almost all of it breaks down to form dissolved carbon dioxide and water:

$$H^+ + HCO_3^- \rightleftharpoons H_2CO_3 \rightleftharpoons H_2O + CO_2 \text{ (dissolved)}$$

This has a particular physiological significance. Now, applying the Henderson-Hasselbalch equation (p. 38):

$$pH = pK_a + \log \frac{[A^-]}{[HA]}$$

The pK_a value of the HCO_3^-/H_2CO_3 system is 6·1. But carbonic acid is almost all converted quickly to dissolved carbon dioxide. For this reason, the equation can be re-written in the form

$$pH = 6·1 + \log \frac{[HCO_3^-]}{[\text{dissolved } CO_2]}$$

In the normal mammal, the rate of breathing (or more precisely the rate at which carbon dioxide is eliminated by the lungs) is regulated to maintain the plasma molar ratio of $[HCO_3^-]/[\text{dissolved } CO_2]$ at approximately 20. So the equation above becomes:

$$pH = 6·1 + \log 20$$

and since $\log 20 = 1·30$,

$$pH = 6·1 + 1·3 = 7·4 \quad \text{(i.e., the normal physiological value)}$$

Bicarbonate is the only base whose conjugate acid can be very rapidly eliminated, a fact which partly explains the unique importance of bicarbonate as a buffer in plasma. But there is a second reason why bicarbonate occupies a special position in the buffering of plasma and other body fluids, namely, that it is readily formed from carbon dioxide, a constant product of cell metabolism. Dissolved carbon dioxide appears to pass freely across the plasmalemma of most types of cells but this is especially so in the case of erythrocytes. These cells, unlike the surrounding plasma, contain *carbonic anhydrase*, an enzyme which contains zinc and which catalyses the hydration of carbon dioxide. Carbon dioxide, dissolved in the plasma, diffuses into the erythrocytes as the blood flows through the tissue capillaries and is hydrated to form carbonic acid.

Being a weak acid, this immediately dissociates forming bicarbonate ions and hydrogen ions.

$$H_2O + \underset{\text{(dissolved)}}{CO_2} \xrightleftharpoons{\text{carbonic anhydrase}} H_2CO_3 \xrightleftharpoons{\text{spontaneous}} H^+ + HCO_3^-$$

Carbonic anhydrase is important in catalysing the hydration reaction which is, in its absence, relatively slow.

The reaction above has a number of important implications. Carbon dioxide, although a very soluble gas, could not be carried in sufficient quantities by the blood in simple solution. By partly converting it to bicarbonate ions the erythrocyte effectively lowers the plasma carbon dioxide concentration so maintaining a concentration gradient between the blood and the intracellular fluid of the tissues. Moreover, the essential buffering power of the plasma is regenerated, for as discussed later, the bicarbonate ions diffuse in large measure out of the erythrocyte into the plasma.

The hydrogen ions resulting from the dissociation of carbonic acid in the erythrocyte are partly buffered by reaction with phosphate, but most of them react with the imidazole groups of histidine residues within the molecules of haemoglobin:

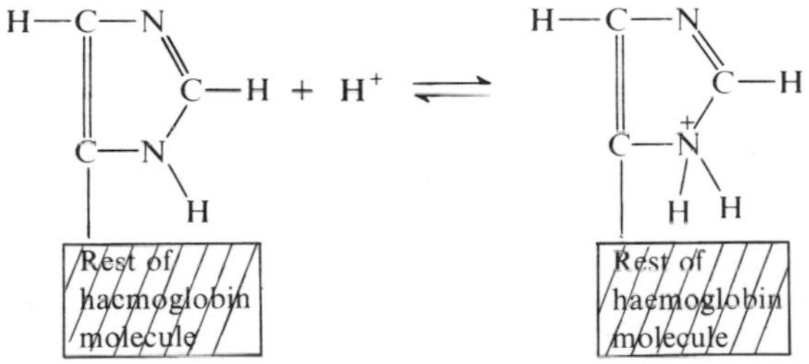

Within the erythrocyte, some of the carbon dioxide unites with haemoglobin to form carbamino-haemoglobin. When this complex is formed, hydrogen ions are again liberated, and are buffered by the erythrocyte proteins and phosphate. If haemoglobin is represented as Hb—NH$_2$, the reaction with molecular carbon dioxide can be summarized in the form:

$$\underset{}{Hb-NH_2 + CO_2 \rightleftharpoons \underset{\text{Carbamino-haemoglobin}}{Hb-NH.COO^- } + H^+}$$

The addition of hydrogen ions to oxyhaemoglobin greatly reduces its affinity for oxygen. In other words, as bicarbonate and carbamino-compounds are formed in the blood in those capillaries where there is a high carbon dioxide concentration, there is a simultaneous production of hydrogen ions which induce oxyhaemoglobin to release its oxygen:

$$HbO_2 + H^+ \rightarrow H\text{---}Hb + O_2$$

This is a partial explanation of what is known as the *Bohr Effect*, namely that raised levels of carbon dioxide enhance the dissociation of oxyhaemoglobin (see Fig. 15.7). It can be regarded as a physiologically-useful bonus accompanying the main transport reactions, for it ensures that oxygen is most available in actively respiring tissues where it is, of course, most needed.

We turn now to the fate of the bicarbonate ions formed from carbon dioxide within the erythrocyte. An equilibrium exists between the bicarbonate within the corpuscles and the quantitatively much larger amount (but not higher *concentration*) in the plasma. Consequently, as bicarbonate ions are synthesized within the corpuscles, many of them diffuse out. Since the cell membrane is relatively impermeable to cations such as potassium, the exit of negatively charged bicarbonate ions must be countered by the ingress of chloride ions in order that electrical neutrality should be preserved (Fig. 15.9). In fact the total quantity of the small diffusible anions in the erythrocyte increases since as the proteins buffer hydrogen ions the extent of their net negative charge decreases

$$\text{Protein}^{n-} + H^+ = \text{Protein}^{(n-1)-}$$

Consequently, the total amount of chloride and bicarbonate ions increases so as to balance the intracellular cations so fulfilling the requirements of the Donnan equilibrium (section 15.6). Thus as the blood passes through the tissue capillaries, carbon dioxide diffuses into the red cells, and is accompanied by chloride ions, a process known as the *chloride shift*.

In the capillaries of the lungs the processes described above and illustrated in Fig. 15.9 are reversed. Oxygen reacts with haemoglobin to form oxyhaemoglobin. The latter has a lower affinity for hydrogen ions than has haemoglobin—i.e., it is a stronger acid; it also forms carbamino-derivatives far less readily than haemoglobin does. Carbon dioxide is formed by the reversal of the carbonic anhydrase reaction and leaves the erythrocytes, diffusing through the plasma to reach the alveolar gas space. Bicarbonate ions enter the erythrocytes from the plasma, to replace the bicarbonate ions which have been converted to carbon dioxide, and chloride ions compensate for this movement by moving from the erythrocytes into the plasma.

Figure 15.8 summarizes the overall process of carbon dioxide transfer from the tissue cells to the lung alveoli. Within the tissue cells the partial pressure of carbon dioxide may be of the order of 60 mm Hg. It therefore tends to diffuse to the capillary blood where the partial pressure is only about 40 mm Hg. Carbon dioxide thus enters the blood so raising the partial pressure to some 46 mm Hg. This is accompanied by the changes

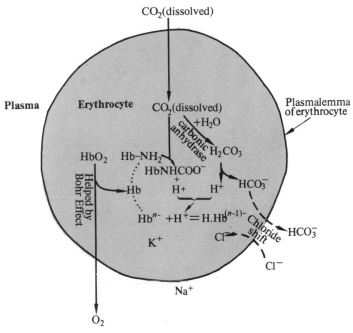

Fig. 15.9 Outline of some of the major events occurring in the red blood cell in a tissue capillary. (Each of the processes shown is reversed in the capillaries of the lungs)

in the plasma and erythrocytes indicated in Fig. 15.9. When the venous blood reaches the lung capillaries carbon dioxide diffuses down a concentration gradient into the alveolar gas where the carbon dioxide has a partial pressure of only 40 mm Hg.

Movement of dissolved substances into and out of cells

16

Excitable cells of nerve, muscle, and retina

A remarkable attribute of cells is their capacity, in various ways and to different extents, to respond to changes in their environment. The speed and effectiveness of response is most marked in animal cells and especially the cells of nerve and muscle. The lipoprotein membranes of all cells probably carry an electrical charge, but the electrical characteristics of the membranes of cells of nerve and muscle are developed to an extraordinary degree. It is upon these properties that the physiological functions of the cells ultimately depend. This chapter therefore begins with a discussion of the origin of the electrical potentials associated with cell membranes and describes how these potentials are altered when an impulse passes along a nerve or muscle. Finally, two related problems are considered; these are the mechanism by which proteins of muscle induce muscular contraction and the mechanism of the remarkable photochemical response shown by the rod cells of the eye.

16.1   The resting potential of cells

It has long been known that nerves and muscles respond to applied electrical stimuli and sensitive electronic equipment now available enables the electrical properties of cells to be investigated. Fine capillary electrodes (diameter $0.5\,\mu$) can be inserted into single cells to measure the potential between the inside of the cell and a reference electrode situated in a suitable buffered solution surrounding the cell. The potential so measured, the *membrane potential*, is the potential across the plasmalemma. This technique has been applied not only to nerve and muscle cells but also to cells of other types (e.g., large algal cells). In almost every case investigated it has been observed that the inside of a cell is negatively charged when the cell is in its normal physiological environment. This charge is known as the resting potential and, for nerve or muscle cells, the charge alters in a remarkable manner when the cell is activated (section 16.3).

In most kinds of cells, the intracellular level of potassium ions is greater than the normal extracellular concentration (p. 358). It is this gradient—associated with the fact that sodium is actively eliminated from cells—which leads to the existence of a resting potential. Cell membranes tend to be relatively permeable to potassium ions, and since the intracellular potassium ion concentration (K_{int}^+) is greater than the extracellular level (K_{ext}^+) there is a strong tendency for potassium to leak out of cells. When potassium ions diffuse out they tend to leave the intracellular environment negatively charged since indiffusible anions, such as protein anions, are left in excess. In the resting state at equilibrium, the negative intracellular charge tends to attract positive charges into the cell and since the sodium pump keeps down the sodium ion concentration within the cell, potassium ions are exposed to two opposing influences—a tendency to diffuse out and an attraction towards the internal negative charge (Fig. 16.1). The result is a steady state system where the interior of the cell has a particular resting, or *potassium diffusion*, potential.

The *magnitude* of the resting potential E depends mainly on the concentrations of K_{int}^+ and K_{ext}^+. In many of the cases so far investigated, including nerve and muscle cells, there is good agreement between the measured potential and the value calculated from what is known as the *Nernst equation:*

$$E = 1000 \frac{RT}{F} \ln \frac{[K_{ext}^+]}{[K_{int}^+]}$$

$$= 1000 \frac{RT}{F} \times 2 \cdot 303 \log_{10} \frac{[K_{ext}^+]}{[K_{int}^+]} \tag{16.1}$$

where E is the potential across the membrane, expressed in mV, R is the gas constant (8·3 joules per degree per mole), T is the absolute temperature, and F is the Faraday in coulombs per g equivalent. At 18°C (291K), the equation becomes

$$E = \frac{1000 \times 8 \cdot 3 \times 291}{96{,}500} \times 2 \cdot 303 \log \frac{[K_{ext}^+]}{[K_{int}^+]}$$

$$= 58 \log_{10} \frac{[K_{ext}^+]}{[K_{int}^+]} = -58 \log_{10} \frac{[K_{int}^+]}{[K_{ext}^+]} \tag{16.2}$$

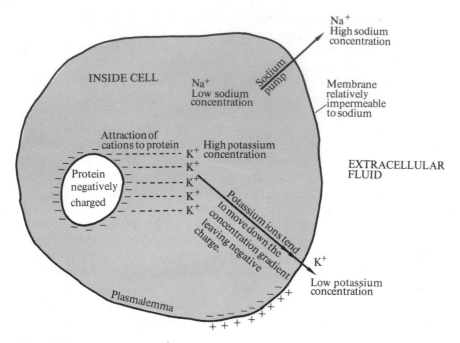

Fig. 16.1 The major factors producing the resting potential of nerve and muscle cells

If then, at 18°C, the intracellular potassium ion concentration of frog muscle is 60 mEq/l. and the extracellular level is 3 mEq/l., the potential across the membrane, according to the Nernst equation, is

$$E = -58 \log_{10} \tfrac{60}{3} = -58 \times 1{\cdot}301 = -75 \text{ mV}$$

The negative sign associated with the 75 mV implies that the inside of the membrane is to this extent negative with respect to the outside, but the most convincing explanation is obtained by reasoning from first principles, in the following way. The Nernst equation relates to a *diffusion* potential, and so requires a tendency of ions to move down a concentration gradient. Since, in this case, the internal concentration of potassium is higher than the external, positively-charged ions tend to move outwards, leaving the inside of the membrane negative with respect to its outside.

The concentration of potassium ions is readily measured by flame photometry, and in many cases a good correlation has been found between calculated values of the resting potential and the measured value. For example, the resting potentials for frog sartorius muscle and for rat skeletal muscle have been calculated from analytical results to be $-99{\cdot}2$ and $-90{\cdot}1$ mV; the corresponding values measured with electrodes are $-101{\cdot}1$ and -91 mV. The squid, a marine mollusc, has certain nerve cells which are particularly large. For this reason it is both relatively easy to put electrodes into one of these cells and to squeeze the cytoplasm out for chemical analysis. The measured resting potential has been found to be -77 mV while the value calculated by inserting the analytical values into the Nernst equation is -76 mV.

It might be predicted from the Nernst equation that, if the external potassium level $[K_{ext}^+]$ is varied, the resting potential should change correspondingly. This is indeed what happens, for when muscle is incubated in a suitable oxygenated buffer containing varying amounts of potassium, a straight line relationship is obtained when E is plotted against $\log_{10}([K_{ext}^+]/[K_{int}^+])$. At 18°C the slope is about -58 as predicted by equation (16.2) (Fig. 16.2). Since the activity of nerves and

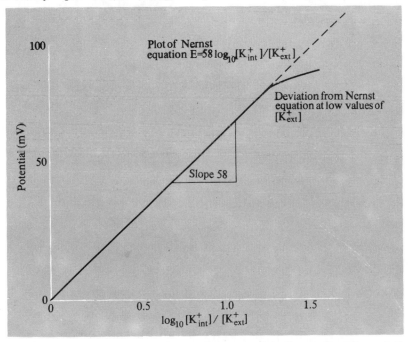

Fig. 16.2 Relation between the resting potential of muscle cells and the ratio $[K_{int}^+]/[K_{ext}^+]$ when $[K_{ext}^+]$ is varied. Temperature 18°C

Excitable cells of nerve, muscle, and retina

muscles depends on the existence of a particular resting potential, it is crucial to the life of the animal that the extra- and intracellular levels of potassium ions should be carefully controlled—i.e., the composition of interstitial fluid is as vital to animal life as the composition of the cells themselves.

16.2 The general structure of a nerve cell

There is considerable variation in the structure of nerve cells (neurones) in different species of animals and even within a single animal, but Fig. 16.3 depicts a fairly typical peripheral neurone of a vertebrate. The

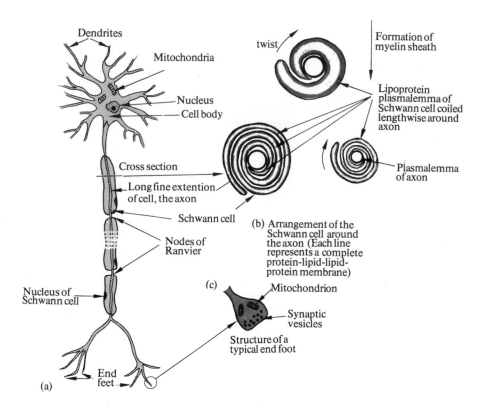

Fig. 16.3 Structure of a nerve cell

essential function of a neurone is to transmit signals from one end of the cell to the other. The signals are usually *received* by fine processes known as dendrites; these are near to the part of the cell containing the nucleus. Also present in this, the main part of the cell, are mitochondria and bodies known as Nissl granules; the latter are rich in RNA. The typical neurone has a fine extension known as an *axon*. This, like the rest of the cell, is surrounded by a typical lipoprotein plasmalemma. Axons vary greatly in length, some being less than a millimetre in length while others extend several metres. The axon normally divides into fine branches which terminate in structures known as *end feet*. Electron microscopy has shown that these contain mitochondria and numerous small synaptic vesicles, which are vacuoles of about 30 mμ diameter. The vesicles probably contain *chemical transmitters* which relay the signal to neighbouring cells (section 16.4).

The axon of a neurone is a very delicate structure often protected and strengthened in vertebrates by being surrounded by cells known as *Schwann cells*. Some nerves have these cells developed in a remarkable manner, the cells being wound around the axon several times. The result is that the axon is surrounded by *numerous layers* of the plasmalemma of the Schwann cell (Fig. 16.3). The layers of lipoprotein membrane are known collectively as the *myelin sheath* of the neurone. The sheath provides electrical insulation and *myelinated* nerves usually conduct signals faster than non-myelinated ones. However, neurones are not entirely covered with the sheath, for there are small gaps between the Schwann cells, known as *nodes of Ranvier*. Here the axon comes into direct contact with the normal extracellular fluid of the animal. Not all nerves have elaborate myelin sheaths though electron microscopy suggests that more rudimentary insulating sheaths may be more widespread than light microscopists had hitherto supposed.

The account which follows describes how a signal is carried along a nerve cell, the apparently more simple case of the non-myelinated nerve being considered first.

16.3   The action potential of a nerve cell

The passage of impulses along non-myelinated nerves has often been compared with the transmission of signals along the wires of a telegraph system. The analogy must not be pressed too far, however, for nerve fibres behave not so much as 'wires' but as 'batteries'. By the mechanism outlined below, a flow of cations *across* the membrane creates a current which simultaneously alters the potential difference across an adjacent section of membrane and by so doing 'switches on' the next battery, the process being repeated along the whole length of the axon.

In the resting state, the nerve cell is negatively charged inside (by about 75–90 mV) compared with its immediate environment (section 16.1). The passage of an impulse along the length of a non-myelinated axon is a rapid wave of depolarization moving at some 1–20 metres/second. As the impulse passes any particular position along the axon, the inside of the plasmalemma at that point rapidly changes from being negatively charged and becomes *positive* (about 40 mV compared with the outside of the membrane). Recovery takes place in 2–3 μ-second, the negative charge inside the cell being restored (Fig. 16.4a). This chain of events triggers off the same process a fraction further along the axon and so the impulse (often called a spike owing to its appearance on a cathode-ray tube) travels progressively down the cell.

Much of the present knowledge about the passage of the nerve impulse is based on work done by A. L. Hodgkin and A. F. Huxley on the large neurones of the squid. The following hypothesis appears to explain the electrical changes accompanying the movement of the nerve impulse in these giant axons, and, with small modifications, it is probably applicable to the action potential of other nerves and to muscle fibres as well (section 16.6).

The formation of the action potential at a given point along the axon is preceded by a change in the permeability of the plasmalemma to sodium ions. It will be recalled (p. 358) that the membrane is normally of relatively low permeability to sodium ions and, in any case, these are actively extruded by the sodium pump. The sudden change in the membrane potential (from negative to positive) is caused by a rapid influx of sodium ions through the membrane. Just as the resting potential of the axon is effectively a *potassium diffusion potential*, so the action potential is largely a *sodium diffusion potential*. The magnitude of the positive intracellular potential at the peak of the spike is related to the concentration of sodium ions outside the cell $[Na_{ext}^+]$ and to their concentration inside $[Na_{int}^+]$. Quantitatively, the relationship is again expressed by the Nernst equation, this time applied to sodium ion concentrations:

$$E = \frac{1000 \times 8 \cdot 3 \times 290}{96,500} \ln \frac{[Na_{ext}^+]}{[Na_{int}^+]} = 58 \log_{10} \frac{[Na_{ext}^+]}{[Na_{int}^+]} \quad (16.3)$$

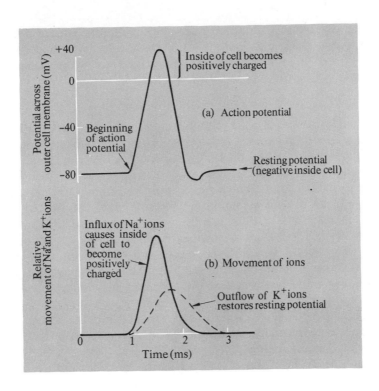

(a) Action potential

(b) Movement of ions

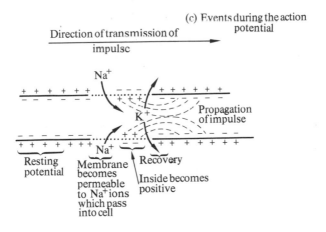

(c) Events during the action potential

Fig. 16.4   Action potential of the nerve cell

Excitable cells of nerve, muscle, and retina

In the case of the squid neurone, the sodium ion concentration of the axoplasm is some 50 mEq/l. while the extracellular fluid contains 500 mEq/l. The action potential may therefore be predicted from equation (16.3):

$$E = 58 \log_{10} \tfrac{500}{50} = 58 \log_{10} 10 = +58 \, \text{mV}$$

It should be noted that the relative magnitude of the intra- and extra-cellular concentrations of sodium ions is the opposite to that for potassium ions (equation (16.2)), so sodium ions, by entering the cells, tend to make the inside of the membrane *positively* charged (by about 58 mV) with respect to the outside. The change in potential from a resting value of about $-80 \, \text{mV}$ to $+58 \, \text{mV}$, or about 140 mV, represents the *theoretical* maximum magnitude of the action potential. In practice, this maximum is never reached, since potassium ions begin to diffuse out of the axon shortly after the sodium ions begin to diffuse in (Fig. 16.4b). This partly offsets the positive potential resulting from the inward movement of the sodium ions, with the result that the potential in practice only reaches about $+40 \, \text{mV}$.

Equation (16.3) predicts that, if the external sodium ion concentration is reduced experimentally, the height of the spike (i.e., the magnitude of the action potential) should decrease. This, in fact, is exactly what happens. The correctness of the prediction provides further striking evidence that the physiological function of nervous tissue depends ultimately upon the nature and the concentrations of ions on each side of a membrane, the membrane not being one with unchanging properties, but one which is capable of varying in permeability.

The potential inside the cell membrane quickly returns to the normal negative resting level. By a mechanism which is not yet fully understood, the axon membrane loses its temporary permeability to sodium ions, and the resting potential is restored by the efflux (outward movement) of potassium ions which was mentioned before. There is often a slight 'overshoot' during the recovery and the membrane momentarily adopts a slightly more negative charge than the true resting potential (Fig. 16.4a).

The rapid influx of sodium ions which occurs when the membrane becomes more permeable to sodium ions during the development of the action potential, does not lead, short-term, to a significant increase in the intracellular concentration of sodium ions. In fact, when a single impulse is transmitted along an axon, only about 3×10^{-12} gram ions of sodium enter every square centimetre of cell surface. A similar quantity of potassium is lost as the resting potential is restored (Fig.

16.4b).   Since a gram ion of sodium is 23 g, only about 69×10^{-12} g sodium ions pass through one square centimetre of membrane, and, in consequence, an axon can carry hundreds of impulses before any appreciable change in ionic composition is apparent.   On the other hand, the process can obviously not continue indefinitely without metabolic intervention to restore the *status quo*, for otherwise the ionic gradients would eventually be annulled.   Indeed, if the sodium pump is prevented from working, such a run-down is exactly what is observed—after some thousands of impulses have been transmitted, the axon no longer transmits action potentials.   Similarly, if the axons are bathed in a medium which contains lithium instead of sodium, the ionic influx functions normally but the sodium pump cannot pump lithium ions out again, and eventually the axon ceases to function.

The propagation of the nerve impulse in the *myelinated* nerve is similar to that described above but is slightly more complicated.   The electrical changes are similar to those of non-myelinated nerves except that they occur only at the nodes of Ranvier.   The impulse jumps, or saltates, from node to node and the speed of conduction along myelinated axons is therefore greater than it is along those which are non-myelinated—the time taken for the impulse to travel between nodes is negligible and typical of electrical conduction along a metal wire.   Indeed, if the axon is severed in the inter-nodal region, it is possible to insert a metal conductor between the severed ends without impeding the transmission of the impulse.   Thus the myelinated part of the axon appears to act as an insulated electric cable.   Since vertebrate nerves are often myelinated, such nerves conduct signals more quickly than typical invertebrate nerves.

16.4   How a nerve cell carries a message

In effect, neurones carry coded messages, the types of information transmitted being enormously varied.   For example, neurones from some sense organs carry messages about light intensity or colour, from others about pitch and loudness of sound.   Yet others carry information about temperature, injury, tastes, or smells.   Nevertheless, the action potential which is propagated along the appropriate neurones is very similar in all these cases for it is only the *termination* of the stimulus in some particular region of the brain which distinguishes the different types of information.

It might perhaps be imagined that the *intensity* of the stimulus which produces a signal carried by a neurone might determine the magnitude of the action potential. This, however, is not the case, for the action potential carried by each and every nerve cell is virtually the same in shape and magnitude as that carried by every other. However, the number of signals conveyed per second (i.e., their frequency) is highly variable. In general, the greater the intensity of a stimulus, the more *frequently* are action potentials transmitted along the appropriate neurone.

When the stimulus which induces a neurone to propagate a signal is very weak, no action potential is produced and it is described as being *sub-threshold*. If the intensity of the stimulus is gradually increased, a point is reached, the *threshold intensity*, at which the neurone begins to transmit its impulses. Thereafter, the greater the stimulus the greater the rate of impulse transmission. Either a neurone carries a signal or it does not; this is sometimes called the '*all-or-nothing*' law.

16.5 Transmission of the nervous impulse from one nerve cell to another

Once the action potential has travelled along the length of the axon, it must stimulate some other cell, or cells, if it is to produce any subsequent response. The essential function of nerve cells is to transmit information between sensory cells and the cells of organs which produce the appropriate response to the stimulus. The latter are known as *effector cells*, and, of these, gland cells and muscle cells are two principal types. A simplified outline of the general arrangement of sense cells, neurones, and effector cells is indicated in Fig. 16.5.

Theoretically, the simplest way an impulse can travel from a sense cell to an effector is via two neurones, an *afferent* neurone which carries the message to the central nervous system and brain and an *efferent* neurone which conveys it to the muscle or gland. Transmission of this kind is known as a *simple reflex*, although it is rare in higher animals. In practice at least one, and often large numbers, of intermediate *relay neurones* participate in the transmission within the central nervous system. The intervention of relay neurones enables information received from numerous sensory cells to be co-ordinated so as to produce a simultaneous response from large numbers of effector cells. The central nervous system is developed to greatly differing degrees in different Phyla of the animal kingdom, according to the complexity of the relay neurone system.

When an impulse passes from one neurone to the next, or from a neurone to an effector organ, the action potential does not simply 'jump'

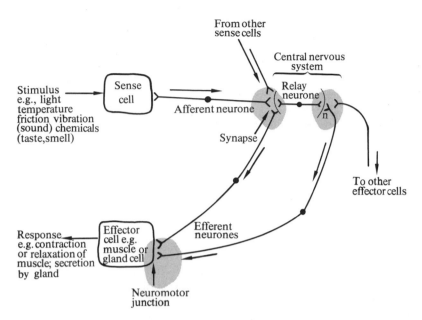

From other
sense cells

Central nervous
system

Stimulus
e.g., light
temperature
friction vibration
(sound) chemicals
(taste,smell)

Sense
cell

Relay
neurone

Afferent neurone

n

Synapse

To other
effector cells

Response
e.g. contraction
or relaxation of
muscle; secretion
by gland

Effector
cell e.g.
muscle or
gland cell

Efferent
neurones

Neuromotor
junction

Fig. 16.5 The reflex arc. (Arrows indicate the direction of movement of
the impulse. The number of relay neurones (n) involved is
extremely variable)

from one cell to the next. The link between connecting neurones is
known as a *synapse*. Although the end feet and dendrites of the cells
concerned are in very close proximity at a synapse, there is nevertheless
a small gap between them, the plasmalemma of each cell being distinct
and separate. The signal is usually carried across the narrow gap by
the secretion by the end-feet terminating the axon of a *chemical transmitter or mediator*. The transmitter then diffuses across the synapse,
inducing an action potential in the receiving neurone. Similarly, a
chemical transmitter carries the signal across the narrow gap separating
the final efferent neurone and the muscle or other effector which responds
to the stimulus. This gap is called the *neuro-motor* junction, or, in the
usual case of a muscle being the effector, a neuro-muscular junction.

The two principal synaptic and neuro-motor transmitters are *nor-
adrenalin* and *acetylcholine*. These are stored in the synaptic vesicles or
neurones (Fig. 16.3). Vertebrates have two types of neurone; cholinergic
nerve cells initiate a new signal by liberating acetylcholine at synapses
and adrenergic nerve cells secrete nor-adrenalin. Acetylcholine is

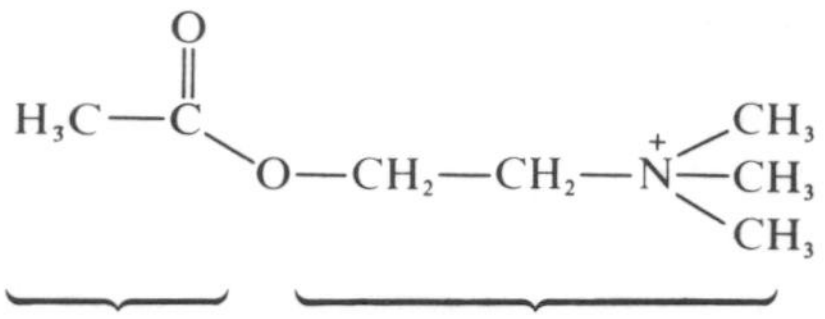

$$H_3C-C\overset{O}{\underset{O-CH_2-CH_2-\overset{+}{N}\overset{CH_3}{\underset{CH_3}{\diagdown CH_3}}}{\Big\|}}$$

HO—⟨benzene ring with OH⟩—CH—CH$_2$—NH$_2$ (with OH on CH)

Nor-adrenalin

Acetyl.................Choline

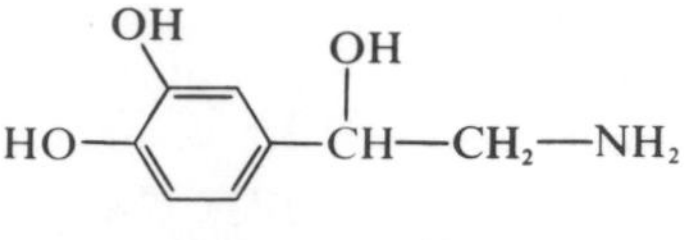

liberated at most neuro-motor junctions. Brief mention has been made earlier of several typical *hormones* (e.g., insulin, p. 232; cortisol, p. 15; oxytocin, p. 49). These 'chemical messengers' are secreted into the blood by the cells of endocrine glands and the blood carries them to other cells, the activities of which alter in response to their presence. The neuro-motor transmitters are, in a sense, *local hormones*, for they enable one cell to induce a response in a second cell situated nearby.

The existence of *two* different chemical transmitters has an important consequence, for the actions of acetylcholine and nor-adrenalin are often antagonistic. For example, certain cholinergic nerves, cause a decrease in the size of the pupil of the eye, while other nerves which are adrenergic, induce dilatation of the pupil. Similarly, the vertebrate heart rate is decreased by acetylcholine but is increased by nor-adrenalin. *Adrenalin* is closely related to nor-adrenalin and is a hormone secreted by the adrenal gland (p. 322). It is secreted when the animal is in a state of fear or excitement and many of the effects it evokes are similar to those produced by adrenergic nerves.

16.6   How a nerve stimulates a muscle

Most vertebrate *skeletal muscles*—the muscles which move the limbs, backbone, and jaws—are controlled largely by neurones for which acetylcholine is the chemical transmitter. The end-feet of the neurones are closely attached to the muscle at the neuromuscular junctions. Small indentations facilitate contact between neurone and muscle at the *motor end-plate* (Fig. 16.6), but a minute gap nevertheless remains and it is across this gap that the chemical mediator diffuses.

The vesicles of the end-foot are about 30–40 mμ in diameter and contain acetylcholine in an inactive form; conjugation with protein may be the cause of this lack of activity. The vesicles are thought to be synthesized in the cell body of the neurone and reach the end-foot by passage along the axon. When impulses reach the end-foot they induce the secretion of acetylcholine from the nerve cell into the neuromuscular gap, from where it diffuses to the membrane of the muscle. Extremely small quantities of acetylcholine (about 10^{-16} moles) are sufficient to cause a *partial* depolarization of the membrane of the muscle in the region of the end plate (Fig. 16.7). The result is that, in this region, the charge on the membrane falls with respect to that of the surrounding membrane. When sufficient acetylcholine has reached the end-plate to produce an end-plate potential of about half the resting potential, the end plate acts as an 'electron sink', current flowing to it from the neighbouring muscle membrane. This flow of current renders the neighbouring membrane permeable to sodium ions, which pass in to produce a typical action potential (section 16.3). This explains why the arrival of small discrete amounts of acetylcholine produce an 'all-or-nothing' muscular response by muscle cells, for once the 'electron sink' reversal effect has commenced, the action spreads over the muscle cell and induces the process of contraction (section 16.8).

Associated with the vesicles of the neurone is an enzyme, *choline acetyltransferase* which catalyses the *synthesis* of acetylcholine from acetyl coenzyme A and choline (Fig. 16.8). The passage of acetylcholine from the neurone can be inhibited by a specific protein, *botulinus toxin*, which is one of the most poisonous substances known. It is of interest that the mode of action of *curare*, used by South American Indians as an arrow poison, also depends on interference with neuromuscular transmission,

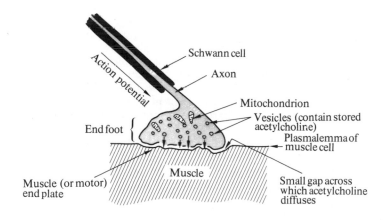

Fig. 16.6 Structure of neuromuscular junction

Excitable cells of nerve, muscle, and retina

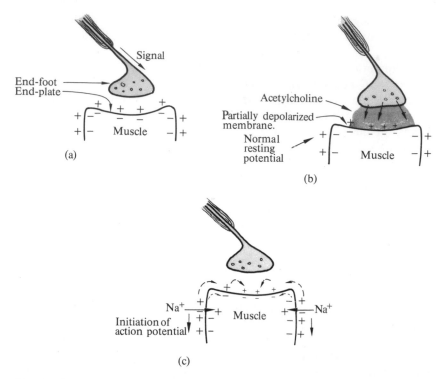

Fig. 16.7 How partial depolarization of the muscle and plate initiates an action potential in muscle membrane. (a) Signal reaches end-foot. Potential difference across muscle membrane the same in the end-plate region as elsewhere. (b) Acetylcholine released from end-foot diffuses to end-plate. Here, it lowers end-plate potential. Partial depolarization indicated by the smaller positive and negative signs. (c) General membrane potential disturbed when end-plate potential has fallen sufficiently. Charge readjustment in an attempt to equalize alters permeability to sodium ions. Entry of Na^+ triggers off a typical action potential as the inside of the membrane becomes positively charged (Fig. 16.4)

for curare prevents acetylcholine from depolarizing the muscle and so causes fatal paralysis. Drugs which interfere with the release of acetylcholine or which, in various ways, block (or simulate) its mode of action, are of great practical importance in medical science.

For a local chemical transmitter of signals to be effective it must be removed as soon as it has done its allotted task—if it were not, the effect would be similar in its consequence to unreleased depression of a Morse key in a telegraph system. Acetylcholine is hydrolysed to choline and acetic acid at the muscle motor end-plate by an enzyme, *acetylcholinesterase* (Fig. 16.8). Some potent poisons inhibit the action of this enzyme so that acetylcholine accumulates causing a continued muscular contraction. Di-isopropyl fluorophosphonate (DFP) inhibits cholinesterase and also certain other hydrolytic enzymes including trypsin and chymotrypsin. Numerous similar compounds, used commercially as

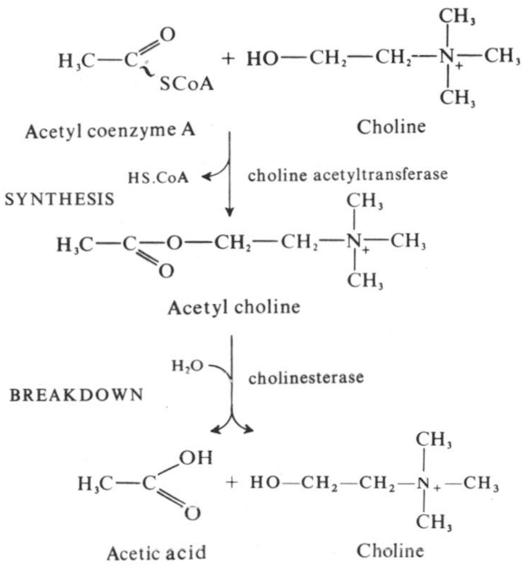

Fig. 16.8 Synthesis and breakdown of acetyl choline

insecticides, function by inhibiting cholinesterase. They attach themselves to the enzyme surface in a rather similar way to the enzyme's natural substrate, but, after hydrolysis, the section of the molecule containing the phosphorus atom remains attached to the enzyme, phosphorylating part of its active site. This phosphorylation prevents the access of natural substrate, with the result that the organism dies as a result of the accumulation of its own acetylcholine. Nerve gases (e.g., tabun and sarin) designed for use in chemical warfare, act in a similar manner.

16.7 Structure of muscle

Muscle provides an excellent example of the way structure and function of cells are inter-dependent. Basically, there are two types of muscle.

$$\begin{array}{c} \mathrm{CH_3} \\ \mathrm{CH_3} \end{array}\!\!>\!\mathrm{CH-O} \qquad \mathrm{O}$$
$$\underset{\mathrm{P}}{\big\backslash}$$
$$\begin{array}{c} \mathrm{CH_3} \\ \mathrm{CH_3} \end{array}\!\!>\!\mathrm{CH-O} \qquad \mathrm{F}$$

DFP

$$\begin{array}{c} \mathrm{CH_3} \\ \mathrm{CH_3} \end{array}\!\!>\!\mathrm{CH-N} \overset{\mathrm{H}}{} \qquad \mathrm{O}$$
$$\underset{\mathrm{P}}{\big\backslash}$$
$$\begin{array}{c} \mathrm{CH_3} \\ \mathrm{CH_3} \end{array}\!\!>\!\mathrm{CH-N} \underset{\mathrm{H}}{} \qquad \mathrm{F}$$

Mipafox
(an insecticide)

$$\begin{array}{c} \mathrm{CH_3} \\ \mathrm{CH_3} \end{array}\!\!>\!\mathrm{CH-O} \qquad \mathrm{O}$$
$$\underset{\mathrm{P}}{\big\backslash}$$
$$\mathrm{CH_3} \qquad \mathrm{F}$$

Sarin
(a nerve gas)

Striated muscle has characteristic bands when viewed microscopically, and muscles of this type are responsible for the movement of the skeleton of vertebrates; in invertebrates, striations are often apparent in the muscles which operate very rapidly, such as those moving the wings of insects. *Unstriated* or smooth muscle, which occurs in almost all Phyla of the animal kingdom, will not be discussed since it has been studied less intensely than striated muscle. As the name suggests, it lacks the characteristic appearance of striated muscle. In vertebrates, unstriated muscle participates in activities which are not under direct conscious control, e.g., the movement of muscles in the gastrointestinal tract and in the walls of the blood vessels.

Striated muscle consists of bundles of individual *fibres*, about 50 μ thick, and varying in length from a millimetre to ten or more centimetres, and running more or less in the longitudinal direction of the whole muscle. The cellular structure of the fibre is ill-defined, for each fibre is composed of many cells joined end-to-end, with no clear boundary between them. The fibre therefore looks like a very long multinucleate cell, the membrane surrounding the fibre (the sarcolemma) being, in effect, the shared plasmalemma of several cells. This is indented at regular intervals along its length by small cavities which are the external openings of a transverse system of tubular channels present within the fibre. This system is involved in the control of muscular contraction (section 16.8). The fusion of muscle cells produces a functional syncytium which enables a synchronous response to be achieved when the muscle contracts.

The muscle fibre has a very elaborate sub-structure, the most conspicuous feature being an extensive system of fibrils (*myofibrils*). These are bounded by numerous mitochondria and the membranes of a specialized 'smooth' endoplasmic reticulum known as the *sarcoplasmic reticulum*. The myofibrils are about 1 μ thick and each is subdivided into *myofilaments* which are 5–10 mμ in diameter. The structure of the myofilaments, discernible in electron micrographs, reflects the arrangement and orientation of the molecules involved in the process of muscular contraction (Fig. 16.9).

Under the light microscope, the myofibrils are seen to have characteristic bands which correspond in adjacent fibrils. The two principal bands are the larger, dark A-bands, which consist principally of *myosin* and the narrower, light I-bands, which contain *actin* (Figs. 16.9, 16.11). The initials of these bands stand for anisotropic and isotropic regions, the names indicating whether the regions are doubly refracting or not.

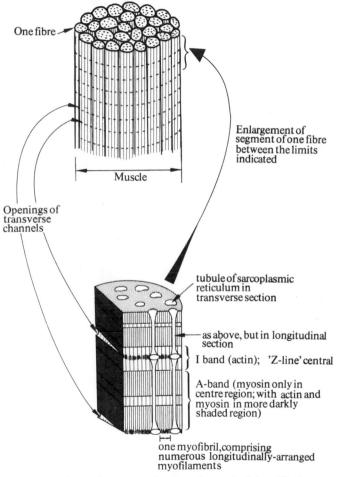

One fibre

Muscle

Enlargement of
segment of one fibre
between the limits
indicated

Openings of
transverse
channels

tubule of sarcoplasmic
reticulum in
transverse section

as above, but in longitudinal
section

I band (actin); 'Z-line' central

A-band (myosin only in
centre region; with actin and
myosin in more darkly
shaded region)

one myofibril, comprising
numerous longitudinally-arranged
myofilaments

Fig. 16.9 Structure of striated muscle (simplified)

The smallest functional unit of the myofibril can be regarded as lying between two Z-bands, each lying in the centre of an I-band. About 15% of the total protein in muscle is *actin* and it is the functional protein in the I-band. Only the I-band shortens when the muscle contracts, the A-band maintaining a constant length. The *myosin* amounts to 40–45% of the protein of muscle and is the main protein in the A-band when the muscle is relaxed. However, when the muscle contracts, more and more actin combines with it in the A-band. The actin actually slides *into* the myosin A-band as it unites with the myosin, a fact which explains the reduction in the size of the I-band during muscular contraction.

The work of H. E. Huxley has revealed much of the finer structure of the actin-plus-myosin A-band and of the actin I-band. Myosin has a molecular weight of 500,000 and some 60% of its structure is in the form of an α-helix. The molecules appear to have globular heads with actin-binding activity, the rest of the molecule being more or less linear (Fig. 16.10). These molecules may be arranged in the filaments of the A-band

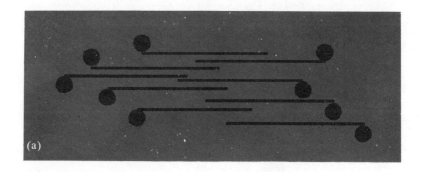

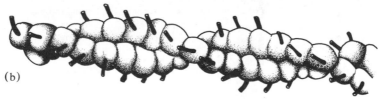

Fig. 16.10   Structure of myosin and actin (after H. E. Huxley).   (a) *Myosin*. Molecules aggregate with chains parallel and projecting heads. Chains in the form of an α-helix.   (b) *Actin*.   Two chains in a double helix; protruding, regularly-arranged side-groups may react with heads of myosin

Cellular biochemistry and physiology   **400**

in such a way that the linear tails are almost parallel, with the actin-binding heads projecting towards filaments of actin. The filaments of actin in the I-band (many of which project into the A-band) comprise two strands of protein, twisted around one another to form a double helix and associated with bound ADP. Active centres regularly arranged on this double helix are presumed to engage with the globular heads of the myosin and so, by some sort of ratchet device, to draw the actin filaments towards the centre of the myosin band (Fig. 16.11). The actin-myosin bridges are about 40 mμ apart and it has been suggested that the contraction may be a sliding process involving the successive forging and breaking of these links, thereby drawing the actin filaments further into the myosin band and so shortening both the I-band and the myofibril.

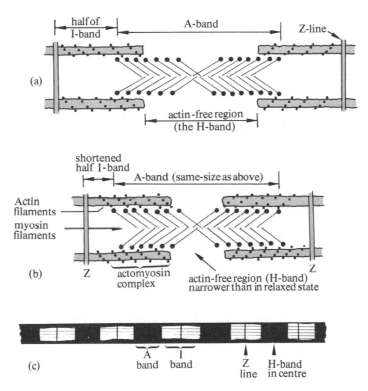

Fig. 16.11 'Ratchet' mechanism of muscle contraction
        (a) Arrangement of filaments when muscle is relaxed
        (b) Arrangement of filaments when muscle is contracted
        (c) Arrangement of bands along the myofibril

16.8 How muscular contraction is initiated

Before discussing the chemistry of muscular activity, attention must be given to the mechanism which 'triggers off' the process of contraction. It will be recalled that acetylcholine, released from the end-feet of the nerve axon, disturbs the electrical potential of the motor end-plates of the muscle fibres and that, in consequence, a wave of action potential spreads over the whole muscle. This excitation is coupled with muscular contraction. The depolarization spreads to the interior of each fibre by way of the membranes lining the transverse channel system which has already been mentioned. This system is separated by only thin membranes from the longitudinal channels of the sarcoplasmic reticulum, the membranes of which, upon depolarization, release calcium ions into the myofibrils. These calcium ions appear to initiate the contraction for they probably trigger off the process which leads to the formation of actin-myosin bridges. Relaxation of muscle begins as soon as nervous stimulation ceases. The calcium ions are re-absorbed by the membranes of the sarcoplasmic reticulum in an endergonic process which requires the presence of ATP.

It will be apparent that several sorts of ions play a part in muscular activity. Sodium and potassium ions are responsible for the resting and action potentials. The release of calcium ions when the membranes of the sarcoplasmic reticulum are depolarized causes the actual process of contraction. In addition, the concentration of both calcium and magnesium ions regulates the excitability of the membrane of the motor end-plate—if too few calcium ions are present, the membrane 'fires' the muscle too readily, and repetitive contractions take place. Magnesium ions, while they tend, like calcium, to stabilize the motor end-plate, oppose the action of calcium ions after their release from the sarcoplasmic reticulum.

16.9 The chemistry of muscular contraction

Muscular contraction is an *endergonic process* driven by energy derived from ATP. In vertebrate skeletal muscle, the ADP so formed is re-phosphorylated to form ATP, the enzyme creatine kinase catalysing the transfer of phosphate from the energy store, creatine phosphate (p. 134). In some invertebrates, (e.g., molluscs and arthropods) arginine phosphate replaces creatine phosphate, but the process is otherwise similar.

It has been shown *in vitro* that actin and myosin combine together to form a complex known as *actomyosin*. This is thought to be the basis of the formation of the bridges between I- and A-filaments during the process of contraction (Fig. 16.11). ATP is changed to ADP and inorganic phosphate during contraction, the lysis being catalysed by the myosin molecule. Myosin is therefore an enzyme and the active centre of the ATP-ase is situated in the globular part of the molecule.

Various hypothetical models of muscular contraction have been proposed and the following is attractive, if only for its simplicity. ATP first phosphorylates myosin (M) giving ADP and a myosin-phosphate (M $\sim$ (P)).

$$\text{ATP} + \text{M} \longrightarrow \text{M} \sim \text{(P)} + \text{ADP}$$

The latter then reacts with actin to produce actomyosin (AM) and inorganic orthophosphate.

$$\text{A} + \text{M} \sim \text{(P)} \longrightarrow \text{A} - \text{M} + \text{P}_i$$

It has been suggested that the formation of links between the actin and the myosin occur, not at right-angles, but obliquely across between the two proteins. Since the mechanism is believed to be by a telescopic movement, or 'ratchet' device, it is also necessary to postulate that the actin-to-myosin links are of a very transient nature; having pulled the actin into the myosin A-band, the linkages must break and new, more oblique unions must be formed ready for the next sliding movement.

Any hypothesis designed to explain the mechanism of muscular contraction needs to include a consideration of: (1) how the process is initiated by calcium ions, (2) how cross links are formed between the actin and myosin, (3) the way in which the energy of ATP is harnessed to promote the sliding movement, and (4) how cross links, once formed, are cleaved ready for a subsequent sliding movement. For a hypothesis to be useful it should, in addition, both stimulate and be capable of experimental investigation. A very ingenious (but as yet *unproven*) scheme was suggested by R. E. Davies (1963) and warrants a brief description.

It is suggested that, in the *resting form*, the globular heads of the myosin filaments consist of two parts, one containing the ATP-ase, and the other comprising an extended polypeptide side-chain. The two parts tend to repel one another for each carries a negative charge, the charge on the side-chain resulting from the presence of one molecule of *bound ATP*

(Fig. 16.12a). The actin is known to contain *bound ADP*, and, at physiological pH, it bears a negative charge. When the sarcoplasmic reticulum releases calcium ions to initiate the process of muscular contraction, an electrostatic link is formed between these two negatively charged groups and the positively charged calcium ions, thus forming a bridge between the actin and myosin (Fig. 16.12b). This diminishes the electrostatic repulsion between the two parts of the myosin molecule, so enabling the side-chain to contract to form the more stable α-helix structure. As the side-chain shrinks it draws the actin filament along, for the actin and myosin entities are still attracted to one another by the electrostatic calcium bridge (Fig. 16.12c).

So far the hypothesis explains the role of calcium ions, and suggests that the formation of hydrogen bonds as the myosin side-chain assumes the α-helical form provides the immediate source of energy for the sliding movement. But the contraction of the myosin side-chain, according to the theory, brings the actin-bound ATP into contact with the ATP-ase with the result that it is hydrolysed to ADP (still bound to the myosin) and inorganic phosphate (Fig. 16.12d). This has the effect of breaking the link between the actin and the myosin for, it is suggested, the terminal phosphate of the ATP is the one principally involved in the electrostatic bridge.

Finally, the myosin is restored to its resting extended form by the re-phosphorylation of bound ADP. This occurs as a result of a transfer of high energy phosphate from *free ATP*, in solution in the cytoplasm, which is thereby converted to ADP. The myosin side chain thus regains its negative charge, and is forced back to the random form (Fig. 16.12a), the forces of mutual repulsion between the two parts of the myosin molecule being greater than the forces associated with hydrogen bonding in the α-helix.

During the contraction of a muscle this cycle of events may be imagined to occur repeatedly and simultaneously at numerous points along the myofibrils. When the sarcoplasmic reticulum reabsorbs the free calcium ions, muscular relaxation takes place for calcium is necessary for the link between the actin and myosin.

It is probable that this hypothesis will have to be modified many times before the mechanism of muscular contraction is fully understood but it is a challenging exercise in molecular biology and offers a possible explanation of the way physiological and biochemical events could be linked to the alteration of the three-dimensional structure of protein molecules.

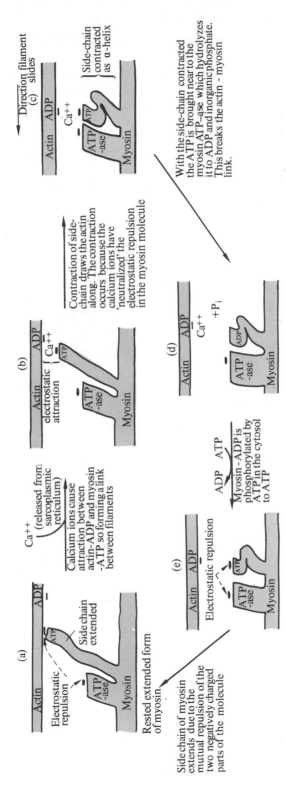

Fig. 16.12 A possible mechanism by which muscle contracts, based on a suggestion of R. E. Davies (1963)

16.10 Structure and function of the rod cell of the eye

Both the *transmission* of the signal and the *response* of muscle depend in a special way on the excitability of cells. In this last section attention is drawn to a type of sensory cell which *induces* nerve impulses as a result of its peculiar excitability to light.

Of the many types of sensory cells to be found in higher animals, the rod cells of the retina, the light-sensitive layer of the vertebrate eye, have been most thoroughly investigated. Millions of them are orientated side by side, forming part of the retina of the eyeball. They have a very characteristic structure (Fig. 16.13). A so-called *outer segment* contains some 11,000 layers of lipoprotein, stacked on top of one another in a manner resembling a pile of coins. Electron microscopy

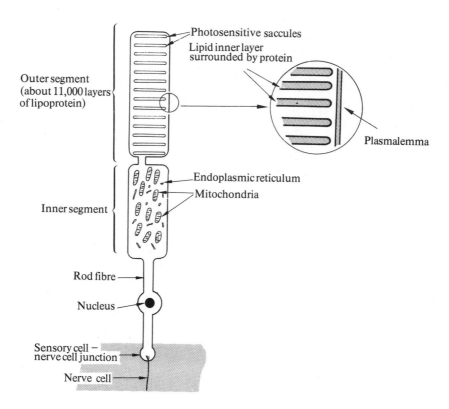

Fig. 16.13 Structure of a typical rod cell of a vertebrate retina

indicates that each is a flat saccule comprising an outer layer of protein about 3 mμ thick surrounding a lipid layer or layers some 7 mμ thick. It will be noted that this protein-lipid-lipid-protein double layer, being 13 mμ thick, is, somewhat thicker than the typical lipoprotein membrane, perhaps because the saccules also contain the light-sensitive pigment, *rhodopsin*.

The inner segment of a rod-cell contains numerous long, slender mitochondria. Very probably, these provide the energy which drives the as yet little-understood train of events which causes an action potential to be produced in the afferent nerve cell when light strikes the pigment in the outer segment of the rod cell. It will be seen from the figure that the junction of the sensory cell and the nerve cell is situated within the base of the rod cell.

Rhodopsin is a magneta-coloured protein, consisting of an apoprotein, opsin, conjugated with a prosthetic group called *11-cis-retinene*. The latter can be regarded as the initial reactant in the visual process. It is closely related to carotenoids (formula of β-carotene Fig. 11.10, p. 257), many of which are involved in other photochemical reactions—carotene itself is a pigment present in the chloroplasts of leaves. 11-*cis*-retinene is formed from β-carotene as follows:

$$\beta\text{-carotene} \rightarrow \text{vitamin A} \rightarrow \text{all-}cis\text{-retinene} \rightarrow 11\text{-}cis\text{-retinene}$$

β-Carotene is a major dietary source of vitamin A, for it is quite common in plant material, including the tap root of carrots. The intestinal cells of vertebrates are able to split the β-carotene molecule hydrolytically to give vitamin A, which is an alcohol. In the presence of NAD$^+$ as electron acceptor, it is oxidized by *alcohol dehydrogenase* to its aldehyde, known as '*all-trans*'-retinene, since all the double bonds are in the *trans* configuration (Fig. 16.14). The prosthetic group of the visual pigment is similar to this compound, except that the double bond between carbon atoms 11 and 12 has the groups around it arranged in the *cis* configuration. This *cis*-isomer of retinene is formed from the all-*trans* compound by an isomerase which occurs in the retina.

When visible light (wavelength 400–750 mμ) falls on rhodopsin, it becomes bleached. This happens because the energy of the light causes a photo-isomerization of the 11-*cis* compound to give all-*trans* retinene. This isomerization dramatically alters the shape of the molecule, so that, while the former compound fits neatly onto the opsin, the *trans* compound does not (Fig. 16.15a). Consequently, the conjugate protein which has

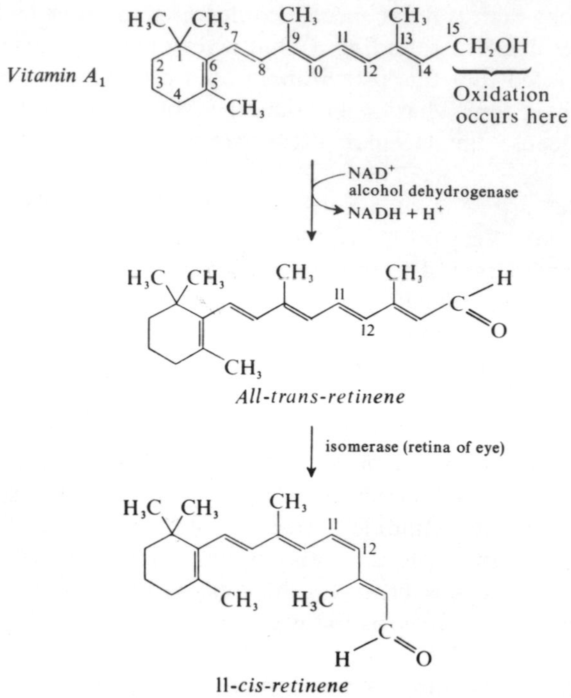

Fig. 16.14 Structural relationship between the retinenes and vitamin A

all-*trans* retinene attached to it, and which is known as lumi-rhodopsin, is very unstable, and the prosthetic group rapidly hydrolyses away from it.

Rhodopsin is regenerated from opsin and all-*trans*-retinene by the two reactions shown in Fig. 16.15b. In the first of these the isomerase already mentioned converts all-*trans*-retinene to the 11-*cis* compound. The latter is then conjugated to an opsin molecule by formation of a Schiff base between the aldehyde group of the prosthetic group and an amino group of the protein:

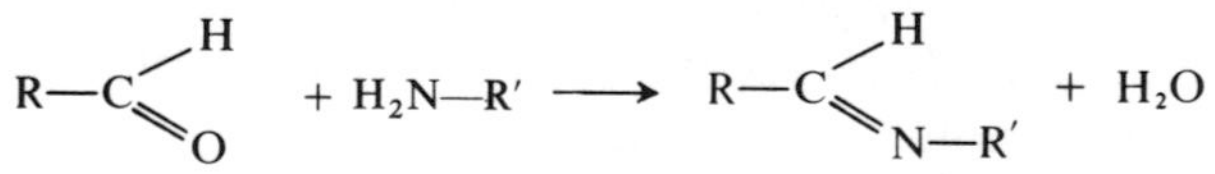

11-*cis*-Retinene + Opsin   Schiff base
                                         (rhodopsin)

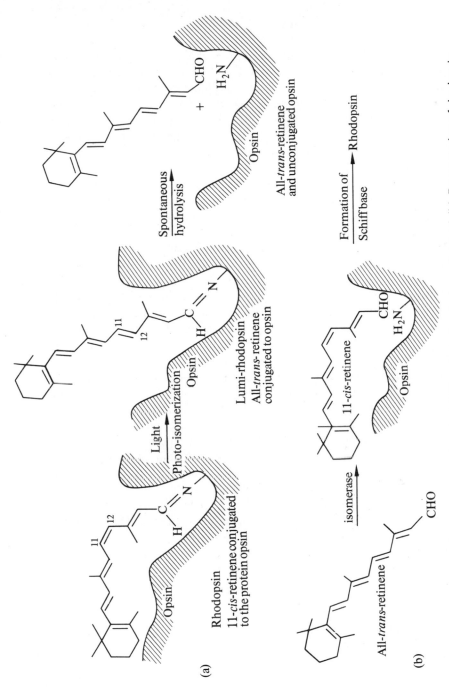

Spontaneous hydrolysis

All-*trans*-retinene and unconjugated opsin

Opsin

Formation of Schiff base → Rhodopsin

Lumi-rhodopsin
All-*trans*-retinene conjugated to opsin

Light
Photo-isomerization

11-*cis*-retinene

Opsin

isomerase

Rhodopsin
11-*cis*-retinene conjugated to the protein opsin

Opsin

All-*trans*-retinene

(a)

(b)

Fig. 16.15 Reactions of the visual cycle (a) The photochemical reaction and (b) Regeneration of rhodopsin

16.11   Concluding remarks

The biological properties of complex organic molecules are intimately linked to molecular geometry as well as to chemical reactivity, as will be apparent from examples to be found in almost every chapter of this book. For example, the active centres of enzymes and carriers are usually highly specific for the substrates whose fate they determine.   If the molecular topography of these proteins is altered by denaturation, by chemical reagents, or by changes of pH, their ability to function is diminished or even totally abolished.   The specificity of enzymes for substrates or cofactors such as NAD or NADP is a reflection of the importance of spatial factors.   Similarly, coenzyme A cannot be replaced by other sulphydryl compounds equally capable *in vitro* of forming thioester linkages.   Again, the detailed structure of the various lipoprotein membranes of the cell and of the base composition of the nucleic acids are examples of molecular topography of the most fundamental biological importance.

More recently, a related (and in some ways even more exciting) aspect of the structure of complex biologically-active molecules has engaged the attention of molecular biologists.   Interest has been focussed upon the possibility that many complex molecules may actually *alter their shape* in a cyclic or reversible manner as they fulfil their physiological role.   It is suggested, in fact, that alteration in shape is essential to the mechanism whereby their cellular function is fulfilled.   A superficially-simple example is that of rhodopsin, where the photochemical reaction causes a remarkable change in the structure of the *prosthetic* group.   But it would be surprising if so drastic an alteration in the shape of a prosthetic group did not impose new forces on the apoprotein, and any alteration in the conformation of this protein could well affect the structure of the lipoprotein membrane where the rhodopsin is located.

Similar arguments may be applied to the mechanisms of other important physiological processes.   It is known, for example, that a significant change occurs in the shape of haemoglobin when one of the four oxygen molecules becomes attached during its oxygenation.   The ATP-ase activity of myosin and the mechanism of the sodium pump provide two other instances where it is likely that a connection exists between cellular function and change in molecular topography.   These proposed alterations in shape are, of course, analogous to the allosteric effects (p. 99) postulated by the biochemists to explain how the activity of certain enzymes is geared to, and controlled by, the cellular requirements for the products of a reaction sequence. Perhaps, in a more general sense, each and every enzyme reaction is a dynamic event with respect to the shape of the enzyme's surface, as well as with respect to the substrate.

It is with such ideas as these that we should like to leave our readers—some to concentrate on various biological disciplines, others to pursue biochemistry and physiology beyond the confines of the cell to the level of the whole organism. It is certain that the tertiary structures of many more biological macro-molecules will be elucidated in the near future. Such studies must inevitably provide future generations of biologists with an ever clearer insight into how a cell functions and, therefore, into the nature of life itself.

Appendix 1 Systematic classification of enzymes

Based on the 1964 recommendations of the International Union of Biochemistry. The following list is only a small selection of those enzymes considered in this book. For a complete list see Florkin and Stotz, *Comprehensive Biochemistry*, Vol. 13 (2nd ed.), Elsevier Publishing Co. (1965). For most general purposes, the trivial name may be used. See also chapter 4, p. 92.

Number	Systematic name	Trivial name	Reaction	Notes
1	**OXIDOREDUCTASES**			
1.1.1.1	Alcohol:NAD oxidoreductase	Alcohol dehydrogenase	Alcohol + NAD$^+$ = Aldehyde + NADH + H$^+$	Cofactor:Zn
1.1.1.27	L-Lactate:NAD oxidoreductase	Lactate dehydrogenase	Lactate + NAD$^+$ = Pyruvate + NADH + H$^+$	Cofactor:Zn
1.1.1.35	L-3-Hydroxyacyl-CoA:NAD oxidoreductase	3-Hydroxyacyl-CoA dehydrogenase	L-3-Hydroxyacyl-CoA + NAD$^+$ = 3-keto acyl CoA + NADH + H$^+$	Involved in fatty acid catabolism
1.1.1.37	L-Malate:NAD oxidoreductase	Malate dehydrogenase	L-Malate + NAD$^+$ = oxaloacetate + NADH + H$^+$	Acts in TCA cycle
1.1.1.41	$threo$-D$_s$-Isocitrate:NAD(P) oxidoreductase (decarboxylating)	Isocitrate dehydrogenase	Isocitrate + NAD$^+$ = α-keto-glutarate + CO$_2$ + NADH + H$^+$	Cofactor Mn^{2+} Acts in TCA cycle
1.1.1.49	D-Glucose-6-phosphate:NADP oxidoreductase	Glucose-6-phosphate dehydrogenase	Glucose-6-P + NADP = glucono-δ-lactone-6-P + NADPH + H$^+$	Acts in pentose-P pathway
1.2.4.1	Pyruvate:lipoate oxidoreductase	Pyruvate dehydrogenase	Pyruvate + oxidized lipoate = Acetyl hydrolipoate + CO$_2$	Requires thiamine pyrophosphate
1.3.99.1	Succinate:(Acceptor) oxidoreductase	Succinate dehydrogenase	Succinate + FAD = Fumarate + FADH$_2$	Contains non-haem iron Acts in TCA cycle
1.4.1.3	L-Glutamate:NAD oxidoreductase (deaminating)	Glutamate dehydrogenase	Glutamate + NAD$^+$ + H$_2$O = α-keto-glutarate + NH$_3$ + NADH + H$^+$	Co factor may be NAD or NADP
1.9.3.1	Ferrocytochrome **c**:Oxygen oxidoreductase	Cytochrome oxidase	4 Cytochrome **c**(Fe^{2+}) + O$_2$ = 4 Cyt. **c**(Fe^{3+}) + H$_2$O	Contains Cu Terminal oxidase of electron transport chain
2	**TRANSFERASES**			
2.1.2.1	L-Serine:tetrahydrofolate 5,10-hydroxymethyl transferase	Serine hydroxymethyl transferase	L-Serine + Coenzyme F = glycine + 'Active formate'	Contains pyridoxal-P and Mn^{2+}
2.2.1.1	D-Sedoheptulose-7-phosphate:D-glyceraldehyde-3-phosphate glycoaldehydetransferase	Transketolase	Sedoheptulose-7-P + Glyceraldehyde-3-P = Ribose-5-P + Xylulose-5-P	Contains thiamine pyrophosphate and Mg^{2+}

TRANSFERASES (*continued*)

2.2.1.2	D-Sedoheptulose-7-phosphate:D-glyceraldehyde-3-phosphate dihydroxyacetonetransferase	Transaldolase	Sedoheptulose-7-P + Glyceraldehyde-3-P = Erythrose-4-P + Fructose-6-P	This and transketolase are involved in pentose-P-cycle
2.3.1.6	Acetyl-CoA:Choline O-acetyltransferase	Choline acetyltransferase	Acetyl CoA + Choline = CoA . SH + acetylcholine	Occurs in nervous tissue
2.4.1.11	UDPglucose:glycogen α-4-glucosyltransferase	UDPglucose-glycogen glucosyltransferase	UDPglucose + $(glycogen)_n$ = UDP + $(glycogen)_{n+1}$	Involved in glycogen synthesis
2.6.1.1	L-aspartate:2-oxoglutarate aminotransferase	Aspartate aminotransferase (Aspartate transaminase)	Aspartate + α-ketoglutarate = oxaloacetate + glutamate	Contains pyridoxal-P
2.6.1.3	L-cysteine:2-oxoglutarate aminotransferase	Cysteine aminotransferase (Cysteine transaminase)	Cysteine + α-ketoglutarate = mercaptopyruvate + glutamate	Contains pyridoxal-P
2.7.1.1	ATP:D-hexose 6-phosphotransferase	Hexokinase	ATP + glucose = glucose-6-P + ADP	Contains Mg^{2+} Takes part in glycolysis
2.7.1.40	ATP:pyruvate phosphotransferase	Pyruvate kinase	ATP + pyruvate = ADP + Phosphoenolpyruvate	Contains Mg^{2+} Takes part in glycolysis
2.7.3.2	ATP:creatine phosphotransferase	Creatine kinase	Creatine + ATP = ADP + creatine-P	Contains Mg^{2+} Occurs in vertebrate muscle
3 HYDROLASES				
3.1.1.3	Glycerol-ester hydrolase	Lipase	Triglyceride + H_2O = diglyceride + fatty acid	
3.1.1.7	Acetylcholine acetyl-hydrolase	Acetylcholinesterase	Acetylcholine + H_2O = choline + acetate	
3.1.3.1	Orthophosphate monoester phosphohydrolase	Alkaline phosphatase	$R—\textcircled{P} + H_2O = R—OH +$ orthophosphate	pH optimum, about 9 wide specificity
3.1.3.2	Orthophosphate monoester phosphohydrolase	Acid phosphatase	$R—\textcircled{P} + H_2O = R—OH +$ orthophosphate	pH optimum, about 5 wide specificity
3.1.4.5	Deoxyribonucleate oligonucleotidohydrolase	Deoxyribonuclease (DNAase)	$DNA + (n-1) H_2O =$ n oligodeoxyribonucleotides	Hydrolyses DNA

415

Appendix 1

Number	Systematic name	Trivial name	Reaction	Notes
HYDROLASES (*continued*)				
3.2.1.1	α-1,4-Glucan 4-glucanohydrolase	α-Amylase	Hydrolyses starch to maltose	Contains Ca^{2+}
3.2.1.20	α-D Glucoside glucohydrolase	α-Glucosidase (maltase)	Maltose + H_2O = 2 glucose or generally: α-glucoside + H_2O = glucose + alcohol	Another enzyme (3.2.1.21) acts on β-glucosides
3.4.2.1	Peptidyl L-amino acid hydrolase	Carboxypeptidase A	Peptidyl-L-amino acid + H_2O = a peptide + L-amino acid	Contains Zn^{2+}
3.5.1.5	Urea amidohydrolase	Urease	Urea + H_2O = CO_2 + $2NH_3$	Not involved in the *synthesis* of urea
3.5.3.1	L-Arginine amidinohydrolase	Arginase	L-Arginine + H_2O = urea + ornithine	One of ornithine cycle enzymes for urea synthesis
3.6.1.3	ATP phosphohydrolase	ATP-ase	ATP + H_2O = ADP + P_i	Many enzymes; energy probably coupled to endergonic processes
4 LYASES				
4.1.1.1	2-Oxoacid carboxy-lyase	Pyruvate decarboxylase	Pyruvate = acetaldehyde + CO_2	Contains thiamine pyrophosphate and Mn^{2+}
4.1.1.32	GTP:oxaloacetate carboxy-lyase	Phosphopyruvate carboxylase	GTP + oxaloacetate = GDP + phospho-enol pyruvate + CO_2	Important in gluconeogenesis
4.1.2.7	Ketose-1-phosphate aldehyde-lyase	Ketose-1-phosphate aldolase	Ketose-1-P = dihydroxyacetone-P + an aldehyde	Acts on fructose-1,6-di-P in glycolysis called 'aldolase'
4.2.1.2	L-Malate hydro-lyase	Fumarate hydratase	L-Malate = fumarate + H_2O	Takes part in TCA cycle
5 ISOMERASES				
5.1.3.1	D-Ribulose-5-phosphate 3-epimerase	Ribulosephosphate 3-epimerase	Ribulose-5-P = Xylulose-5-P	Takes part in pentose-P cycle
5.1.3.2	UDPglucose 4-epimerase	UDPglucose epimerase	UDPglucose = UDPgalactose	Contains NAD
6 LIGASES				
6.2.1.2	Acid:CoA ligase (AMP)	Acyl-CoA synthetase (Fatty acid thiokinase)	ATP + an acid + CoA = AMP + Pyrophosphate + acyl-CoA	Involved in many preliminary activations of fatty acids
6.4.1.1	Pyruvate:carbon-dioxide ligase (ADP)	Pyruvate carboxylase	ATP + pyruvate + CO_2 + H_2O = ADP + P_i + oxaloacetate	Contains biotin

Appendix 2 References

The following books provide the necessary background reading for the topics indicated by the symbol #.

Brown, G. I. (1964) *Introduction to Physical Chemistry*, Longmans, Green and Co., London.
  Chapter 5. Gram-atom.   Chapter 25. Colligative properties, mole fraction.   Chapter 26. Osmotic pressure.   Chapter 28. Equilibrium constant.   Chapter 30. Activation energy.   Chapter 32. Catalysts. Chapters 36, 37. pH and acid-base equilibria.

Buttle, J. W., Daniels, D. J. and Beckett, P. J. (1966) *Chemistry: a Unified Approach.*   Butterworths, London.
  Chapter 5. Colligative properties, mole fraction.   Chapter 6. Thermochemistry, velocity constant.   Chapter 7. Electrochemistry, pH, electrode potentials.

Ellis, G. P. (1966) *Modern Textbook of Organic Chemistry.*   Butterworths, London.
  Chapter 3. Addition reactions.   Chapter 9. Aldehydes and ketones. Chapter 10. Stereochemistry.   Chapter 16. Carbohydrates.

Hassall, K. A. and Dobinson, C. H. (1966) *Essential Chemistry, Vol. 2.* Iliffe Books Ltd., London.   (For a very elementary approach.)
  Chapter 17–25. Electrochemical series, Redox reactions.   Chapter 28. Soap and detergents.

Holmyard, E. J. and Palmer, W. G. (1958) *A Higher School Inorganic Chemistry*, J. M. Dent and Sons Ltd., London.
  Chapter 7. Osmotic pressure.   Chapter 8. Chemical equilibria, equilibrium constant.

MacKenzie, C. A. (1962) *Unified Organic Chemistry*, Harpers, New York.
  Chapter 12. Stereochemistry.   Chapter 23. Carbohydrates.

Roberts, J. D. and Caserio, M. C. (1964) *Basic Principles of Organic Chemistry*, W. A. Benjamin, New York.

417

Chapter 14. Schiff's base. Chapter 15. Addition reactions, aldol addition. Chapter 17. Isomerism. Chapter 18. Carbohydrates.

Wallwork, S. C. (1960) *Physical Chemistry for Students of Pharmacy and Biology*, 2nd Edition, Longmans, Green and Co., London.
Chapter 8. Buffer solutions, weak acids. Chapter 9. Electrode potentials and redox reactions. Chapter 11. Colloids.

Williams, V. R. and Williams, H. B. (1967) *Basic Physical Chemistry for the Life Sciences*, W. H. Freeman and Co., San Francisco and London.
Chapter 2. Thermodynamics. Chapter 3. Mole fraction and colligative properties. Chapter 4. Acid-base equilibria and pH. Chapter 6. Kinetics. Chapter 7. Colloids.

The following books are recommended for wider, though not necessarily more advanced, reading.

Baldwin, E. (1964) *An Introduction to Comparative Biochemistry*, 4th Edn., Cambridge University Press. (Ionic composition of blood and osmotic regulation in relation to evolutionary progress; nitrogen excretion in different organisms; variation in respiratory pigments from species to species, and many similar aspects of comparative biology.)
Cohen, G. N. (1968) *The Regulation of Cell Metabolism*, Hermann, Paris; Holt, Rinehart and Winston, NY. (Some chapters deal with the methods which have been used to study biochemical pathways and others with biological control mechanisms.)
Katz, B. (1966) *Nerve, Muscle, and Synapse*, McGraw Hill, London. (The topics mentioned in the title are discussed in an interesting and thought-provoking manner.)
Kendrew, J. C. (1966) *The Thread of Life*, G. Bell & Sons, Ltd., London. (An introduction to molecular biology, the text based on a BBC Television Lecture series.)
Needham, J., Editor (1970) *The Chemistry of Life*, Cambridge University Press. (Eight lectures by specialists, describing developments in their fields of biochemistry and physiology.)
Watson, J. D. (1965) *Molecular Biology of the Gene*, W. A. Benjamin, Inc., NY. (An account of experimental investigations which have led to modern ideas about the genetic control of protein synthesis.)
In addition, *Scientific American*, a popular Journal appearing monthly, carries many articles on recent developments in biochemistry, physiology and other scientific subjects. The coloured illustrations are often particularly useful.

INDEX

A-band, in myofibrils, 399–401
Absorption,
 of amino acids, by intestine, 363–366, 369
 of fatty acids and glycerides, 363, 366–369
 of sugars, by intestine, 232, 363–365, 369
 of water, 353–357
Absorptive cells of small intestine, 361–364
 see also Intestinal absorptive cells
Acetaldehyde, 195, 207–209
Acetoacetic acid, 228, 229, 299, 309, 310
Acetyl choline,
 breakdown, 397 (Fig. 16.8)
 effect of drugs on, 396
 formula, 394
 secretion at neromuscular junction, 395
  (Fig. 16.7)
 synthesis, 395 (Fig. 16.8)
Acetyl choline esterase see Cholinesterase
Acetyl coenzyme A,
 a high energy compound, 135, 161, 163
 activity of its methyl hydrogen atoms, 163
 central metabolic position of, 182, 183
 comparison with succinyl coenzyme A,
  169
 conversion to carotene and cholesterol,
  255–257
 conversion to ketone bodies, 228–229
 conversion to oxaloacetate, 225–229
 formation following glycolysis, 194–195
 formation from amino acids, 271, 291,
  292, 299
 formation from palmitic acid, 175
 formation in fatty acid oxidation spiral,
  180
 formation outside mitochondrion, 183,
  184 (Fig. 8.3)
 in nitrogen fixation, 264

Acetyl coenzyme A,
 intra and extra-mitochondrial, 183, 184
 production, by oxidation of pyruvate, 205
 role in fatty acid biosynthesis, 185, 188
 why it is not glucogenic in animals, 229
N-Acetyl glucosamine, 79
Acetyl phosphate, 265
Acid,
 buffering of, in cells, 376, 377
 denaturing of protein by, 58
 effect on ionic form of amino acids, 37
 hydrolysis of peptide bonds by, 41
 production by cells, 376
Acidic and basic amino acids, 44, 45
Aconitate hydratase, 164
cis-Aconitic acid, 164
Actin, 399, 401 (Fig. 16.10)
Action potential see Potential
Activation energy (E_A), 94 (Fig. 4.3), 117
Activators of enzymes, 106
'Active carbon dioxide', 185–188 (Fig. 8.4),
  223, 225
Active centres,
 cysteine and serine in, 44
 function of, 97 (Fig. 4.4)
 of actin, 401
 of carriers and enzymes, 366, 410
'Active formaldehyde',
 in conversion of serine to glycine, 304
 meaning of, 293
 structure, 295
'Active formate',
 in histidine synthesis, 295
 in porphyrin synthesis, 295
 in purine synthesis, 295
 in tryptophan synthesis, 297
 meaning of, 293

419

'Active formate',
  structure, 295
'Active formimino groups', 293
'Active methyl groups', 293, 294
'Active oxygen', 260, 261
Active transport, 353, 365
  see also Transport
Actomyosin, 32, 403–405
Acyl carrier protein (ACP), 186, 188
Acyl chlorides, 86
Acyl coenzyme A, 86
Acyl coenzyme A derivatives,
  and lipoic acid, 168
  in fatty acid oxidation spiral, 178, 180
  in lipid absorption, 367
  ketoacids as, 166, 167
  oxidative decarboxylation of its ketoacid
    derivatives, 166, 167
Adaptive enzyme synthesis (induction), 106,
    349
Adaptor hypothesis, 342–346 (Fig. 14.7)
Adenine, 81, 83
  as a base in polynucleotides, 324–326
  in coenzyme A and in NAD, 84–86
  pairing with thymine in DNA, 329
Adenosine diphosphate see ADP
ADP,
  association with actin, 401 (Fig. 16.2)
  conversion to ATP in coupled reactions,
    118, 119, 123, 131
  formation by sodium pump, 359, 360
  relation to AMP and ATP, 83
Adenosine monophosphate see AMP
Adenosine triphosphate see ATP
S-Adenosyl methionine, 293, 294, 315, 323
Adenovirus, 350
Adipate, 149
Adipose tissue, 13, 175
Adrenal gland, 393
Adrenalin,
  as a chemical transmitter or mediator, 393
  effect on blood glucose level, 232
  formation, from phenylalanine, 292, 322,
    323 (Fig. 13.15)
Aerobic respiration,
  energy release by complete oxidation of
    glucose in, 209, 210
  fate of pyruvate during, 193–195, 205, 206
L-Alanine, 35, 36 (Fig. 4.2), 308, 311, 345
β-Alanine, 315, 321
Albumen, 33
Alcohol, 207–209
Alcohol dehydrogenase,
  action on vitamin A, 407 (Fig. 16.14)

Alcohol dehydrogenase,
  function of, 208, 209
  mechanism of reaction, 108 (Fig. 4.9)
Aldehyde mutarotase, 65
Aldehydic properties of sugars,
  D-glucose, 65
  ketoses, 62
Aldolase,
  equilibrium concentrations of reactants
    and products, 200, 201
  in gluconeogenesis, 221
  in pentose phosphate pathway, 252
  in Stage 2 of glycolysis, 199, 201
  reaction of, 107, 108 (Fig. 4.8)
Aldol condensation, 135, 163, 199, 200
Aldoses,
  examples of, 66–69 (Fig. 3.2)
  general formula, 61
  reducing properties, 62
Alimentary canal,
  absorption of food materials by, 362 et seq
    see also Intestinal absorptive cells,
      Absorptive cells
Aliphatic amino acids, 42, 43, 366
Alkaline phosphatase, 363
Alkanoic acids (fatty acids), 12
Allantoin, 277, 278
Allosteric effects, 99 (Fig. 4.5), 106, 410
'All or nothing' responses, 392, 395
Alveolus, 374, 375
D-Amino acids,
  in bacterial cell walls, 36
  intestinal absorption of, 365
  structure, 35, 36
L-Amino acids,
  absorption of, by intestinal cells, 361, 363,
    365, 366
  activation, for protein synthesis, 335, 339
  carriers of, in cell absorption, 365, 366
  catabolism, to produce sulphate, 376
  conversion to other compounds, 292
  decarboxylation mechanism, 296–298
  determination in peptides and proteins,
    48–52
  essential; meaning of, 291
  formation from other compounds, 292
  formation of ammonia from, 291
  general metabolism of, 270–272, 298, 299
    (Fig. 13.1)
  incorporation into proteins, 334, 335, 343
  in plant proteins, 273
  interconversion of, by transamination,
    268
  intestinal absorption of, 365, 366, 369

L-Aminoacids,
linkage of their metabolism with those of fats and sugars, 291
linkage, to form peptides, 32, 40, 41 (Fig. 2.3)
metabolism of carbon skeletons, 291 *et seq*
number of, 32
origin and fate of carbon skeletons, 299
stereoisomers, 35, 36
structure, 35–39, 41–46
transamination, mechanism of, 296, 297
Amino sugars, 72, 73
see also chitin, hyaluronic acid
Aminoacyl synthases, 334, 335
Aminoacyl-transfer RNA complexes, 335, 339, 345
p-Aminobenzoic acid, 294
Amino groups,
formation from ammonia by root nodule bacteria, 263
synthesis of, from ammonia, 267–270
transfer, by transamination, 269
δ-Amino laevulinic acid, 287, 289
Aminopeptidase, 110, 363
Aminotransferases (*see also* transamination), 268
Ammonia,
connexion with ornithine cycle, 279–282
excretion, by some aquatic animals, 277, 278
formation from amino acids, 291, 292, 305
formation of organic nitrogenous compounds from, 267–282
toxicity of, 270
Ammoniotelic animals, 278
Amoeba, 22, 24, 356
AMP (adenosine monophosphate),
formation, during amino acid priming, 339
formation, during fatty acid priming, 179
formula, 80 (Fig. 3.9), 83
relation to RNA, NAD, ATP, 83
Amylase, 77, 106, 109
Amylopectin, 77, 78 (Figs. 3.8, 10.6)
see also Starch
Amylose, 75–77 (Fig. 3.7)
see also Starch
Anabolism, meaning, 2
Anaerobic conditions,
development in exercised muscle, 195
fate of pyruvate under, 206–209
fermentation during, 193, 195
Anoxia, 206, 207

Anthranilic acid, 311
Antidiuretic hormone *see* Vasopressin
Apoenzyme, 92
Apoprotein of rhodopsin *see* Opsin
Arginase, 280, 281
L-Arginine,
absorption of, 365
a member of α-ketoglutarate family of amino acids, 301
formation of creatine from, 315, 316
formula of side chain of, 42 (Fig. 2.4)
in ornithine cycle, 279–281, 301
test for, 47
Arginine phosphate, 402
Argininosuccinic acid, 280, 281
Aromatic amino acids, 45, 46, 309
Aromatic bonds, 45, 143
Arrow poison *see* Curare
Artefacts of electron microscopy, 4
L-Asparagine,
biosynthesis, 307
formation from aspartic acid, 274
role as a nitrogen store, 306, 307
side group of, 42 (Fig. 2.4)
Asparaginase, 306, 307
L-Aspartic acid,
biosynthesis, 307
formation of β-alanine from, 315
formation of fumarate from, 275
formation of ionic bridges, 45 (Fig. 2.5)
in coenzyme A synthesis, 318
in glycogen synthesis, 223
in purine synthesis, 282–285
in pyrimidine synthesis, 287, 288
role in nitrogen excretion, 276, 280, 281
role in transamination reactions, 269
side group of, 42 (Fig. 2.4)
Astbury, W. T., 326
ATP (adenosine triphosphate),
analogy to energy trapped by waterwheel, 114
as source of atoms for histidine, 298
biosynthesis of AMP, ADP and ATP, 285
central role in cell energetics, 132 (Fig. 5.5)
formation, 5, 24, 28, 136–138, 161, 202
formation by coupled reactions, 118, 123
formation by glycolysis, 193–196, 202–206
formation by photosynthetic phosphorylation, 29, 236, 239, 240
formation by oxidative phosphorylation, 28, 161, 162

ATP (adenosine triphosphate),
  formation by substrate phosphorylation, 162
  formation during oxidation of fats, 181–184
  formula, 83, 133
  functions, 24, 83, 403–405
  hydrolysis of, 170
  in activation of amino and fatty acids, 337
  in glutamine and asparagine synthesis, 274, 275
  in purine synthesis, 283, 284
  in sodium pump, 359, 360
  in thiokinase reaction, 170
  in urea formation, 279–282
  in uric acid formation, 283
  initial steps in synthesis of, 284
  later steps in synthesis of, 285, 286
  number of molecules formed by complete oxidation of glucose, 210
  number of molecules formed by TCA cycle, 173
  possible mechanism of formation, 155
  role in absorption of glucose, amino acids, 365
  role in starch synthesis, 214, 215
  utilization in fatty acid biosynthesis, 189
ATP-ase,
  and muscular contraction, 403–405
  and the sodium pump, 359
  in intestinal absorptive cells, 363
  in the terminal web, 363
  of erythrocytes, 359
  of myosin, 403–405
Average chain length of amylopectin, and glycogen, 78
Avogardo's number, 128
Axons, 387, 388
Azotobacter, 263

Bacteria, 6
  amino acid synthesis in, 291, 299
  cell walls, 36
  decarboxylases in, 314
  hydrolysis of cellulose by, 75
  mutant bacteria and the coding ratio, 346, 347
  structure, 28, 348, 349
  sulphonamide as bacteriostat, 295
  synthesis of aromatic rings by, 309
Bacteriophages, 341, 349, 351
Barfoed's test, 62, 73, 74

Base,
  heterocyclic, in nucleotides, 81, 82
  see also Purine, Pyrimidine
Base sequences in messenger and tranfer RNA, 335, 342–346
Bicarbonate ions,
  as buffers, 376–379
  effects of Donnan equilibrium, 361
  formation in erythrocytes, 380, 381 (Fig. 15.9)
  in plasma, erythrocytes and muscle cells, 358 (Fig. 15.8)
Bile in fat digestion, 366 (Fig. 15.4)
Bile acids,
  formation of taurocholic and glycocholic moieties of, 316, 317
  in absorption of fats, 366, 367
  structure, 15, 16
Bioenergetics, 111 et seq
Biotin,
  as cofactor in carboxylation of propionyl coenzyme A, 225, 226
  as cofactor in carboxylation of pyruvate, 223, 224
  as cofactor in fatty acid biosynthesis, 185, 188
  formula, 186
Biuret test, 46
Blood,
  carriage of amino acids by, 366
  carriage of carbon dioxide by, 377–381
  carriage of oxygen by, 372–376
  entry of glucose into, from intestine, 364
  plasma, 33, 352
  regulation of glucose level in, 232, 233
Blood pressure and capillary ultrafiltration, 371
Bohr effect, 380 (Fig. 15.7)
Botulinus toxin, 395
Branching, degree of, in glycogen and amylopectin, 78
Branching enzyme (Q-enzyme), 216, 217
"Bridge" compounds as biological intermediates, 275, 283, 286
Brush border, 363
Buffering,
  action of blood, 377–381
  amino acids and, 38, 39, 44, 45 (Fig. 2.2)
  equation for see Henderson-Hasselbalch equation
  role of carbonic anhydrase, 378
  system in cells, 376, 377
Butyric acid, 12

Cadaverine, 314
Calcium ions,
  general occurrence and as cofactor, 107,
    109, 357, 358, 360
  release by sarcoplasmic reticulum, 402
  role in muscular contraction, 402–405
    (Fig. 16.12)
Calories (cal), 113, 121, 129
Calorimeter, 118, 120
Calvin, 235
Capillaries, 371, 374, 375
Capsomeres, 350
Carbamino-haemoglobin, 379 (Fig. 15.9),
    380
Carbamyl aspartate, 287
Carbamyl phosphate,
  formation from ammonium ions, 275, 276
  in formation of citrulline, 301
  in ornithine cycle, 279–282
  in pyrimidine biosynthesis, 287, 288
Carbohydrates,
  as a source of energy, 112
  association with proteins, 35
  breakdown by glycolysis, 193 et seq
  classification, 60–62
  formation of acetyl coenzyme A from, 137,
    161
  interconversion of carbohydrates and
    amino acids, 159
  see also mono-, di- and poly-saccharides
Carbon dioxide,
  active, 186
  as cofactor in fatty acid biosynthesis, 185,
    188
  as lower 'reservoir' of an energy fall, 114–
    116
  as unit of purine structure, 283
  effect on dissociation of oxyhaemoglobin,
    379
  fixation, 241–243 (Fig. 11.4)
  formation from bicarbonate, 377
  hydration, 379
  in the ornithine cycle, 280
  passage into erythrocyte, 377–381 (Fig.
    15.9)
  production by TCA cycle, 377
  reaction with amino acids, 379
  transport by blood, 377–381
Carbonic anhydrase, 378–381
Carbon monoxide, 155
Carbon skeletons of amino acids,
  connection with nitrogen excretion, 276
  metabolism of, 291 et seq

Carbon skeletons of amino acids,
  origin and fates of, 270–272, 299
Carboxylation reactions, 186, 242
Carboxypeptidase, 109, 110
Carotene,
  conversion of β-carotene to vitamin A,
    258, 407
  in photosynthesis, 10, 237, 238
  synthesis, 255–258
Carotenoids, 10, 237, 255
Carrier,
  in sodium pump, 359 (Fig. 15.2)
  of amino acids, 366
  of glucose, 365
  meaning, 21
Carrier RNA see transfer RNA
Cartilage, protein of, 35, 46
Casein, 44
Castor oil seed, 175
Catabolism, meaning, 3
Catalyst, 87
Cattle and cellulose digestion, 75
CDP-choline, 191
Cell, 3–6 (Fig. 1.1)
Cell wall,
  cellulose in, 75, 79
  general aspects of, 4, 5, 21–23
  of bacteria, 36, 349
  of plant cells, 355
  pectin in, 79
Cellobiose, 73, 74
Cell-free systems, 345, 347
Cellulose,
  as example of β-glucoside linkage, 65, 66
  glucose in, 63
  in plant cell walls, 23, 355
  properties and structure, 75, 76 (Fig. 3.7)
  synthesis, 216
Central nervous system; complexity, 392
Chaos; relation to entropy increase, 121
Chapeville, 345
Charge,
  on nerve and muscle membranes, 383
    (Fig. 16.4)
Cheese, phosphoserine in, 44
Chemical reaction, 113–116, 120, 121
Chemical transmitter, 387, 393
  acetyl choline as, 394
  nor-adrenalin as, 394
Chemical warfare, 397
Chitin,
  formula, 76 (Fig. 3.7)
  in fungal cell walls, 23

Chitin,
  in insect cuticle, 79
Chlorella, 5, 156, 235
Chloride ions, 357, 358, 361
Chlorophyll, 5, 28, 29, 237–240
Chloroplast, 5, 28, 29, 236-238 (Fig. 11.2)
Cholesterol,
  in chylomicra, 369
  structure and functions, 15, 16, 19, 20 (Fig. 1.6)
  synthesis, 255–258
Cholic acid, 15, 16 (Fig. 1.6)
Choline,
  conversion to acetyl choline, 395 (Fig. 16.8)
  formation from acetyl choline, 397 (Fig. 16.8)
  formation from serine, 314
  formation of choline phosphate from, 191
  formation by transmethylation, 294
  structure, 13, 14 (Fig. 1.5)
Choline acetylase, 395
Choline esterase, 397
Cholinergic nerves, 394
Cholyl coenzyme A, 317
Chromosomes, 324, 328, 340
Chromatography, 34, 47, 48, 235
Chylomicron, 369 (Fig. 15.4)
Chymotrypsin, 31, 109, 110
Cilia, 21–24 (Fig. 1.9)
Cis–trans isomerism,
  configuration in peptide bond, 41
  forms of retinene, 407–409
  fumarate and maleic acid, 89
  importance of interconversion in vision, 409
Citrate,
  attachment to enzyme, 174
  in TCA cycle, 137, 162
  intramolecular rearrangement to isocitrate, 164
  method of oxidation in one turn of TCA cycle, 174
  naming, 150
  outside the mitochondrion, 184
Citrate synthase, 163
Citrulline, 279–281
Citryl coenzyme A, 164
Cleavage of C—C bonds, 166, 168, 181, 208, 250, 298
Clostridium, 263, 264
Coding ratio, 346–348
Codons, 347

Coenzymes,
  coenzyme A, 166, 181–185, 315–319
  discovery of, 106
  meaning, 92, 106, 107
  NAD, 85
  NADP, 85
Coenzyme A,
  energy of acyl derivatives of, 86
  formation of β-thioethylamine moiety, 315
  in fatty acid biosynthesis, 185
  in fatty acid oxidation spiral, 181
  in oxidative decarboxylation, 166, 167
  location in cytosol, 184
  structure, 85, 86
  synthesis of, 317–319 (Fig. 13.3)
Coenzyme F (tetrahydrofolic acid),
  formula and uses, 295 (Fig. 13.3)
  in coenzyme A formation, 318, 321
  in histidine metabolism, 312, 313
  in serine-glycine transformation, 303
  in tryptophan metabolism, 311
Coenzyme Q, 142, 151, 157
Cofactors, 7, 87
Collagen, 35, 46
Colligative properties, 355
Colloidal,
  micelles, 33
  osmotic pressure, 371
  solutions, 33, 77
Complementary bases, 331, 340–346
Conjugate proteins,
  meaning, 35
  nucleoproteins, 324 et seq
Copper ions,
  effect on pentose phosphate pathway, 253
  in cytochrome oxidase, 145
  redox potential, 152
Corpuscles see Erythrocytes
Cortisol, 15, 16, 259, 260
Coupled reactions,
  biochemical significance of, 118, 122–125
  energy changes in reversible and irreversible reactions, 134, 135
  examples of, 119, 123
  meaning, 117
  representation of, 122
  see also rerouting reactions
Covalent linkages, 123, 124
Creatine,
  conversion to creatine phosphate, 134, 316
  conversion to creatinine, 278, 316
  formation, 315, 316

Creatine,
  phosphorylation by Lohmann reaction, 134
Creatine kinase, 134, 316, 402
Creatine phosphate, 134, 315, 316, 402
Creatinine, 277, 316
Crick, F. H. C., 327, 340
Curare, 395, 396
Cuticle of insects, 79
Cyanide, 65, 104, 155, 253, 359, 365
Cyclic amino acids, 45, 309
Cystathionine, 304
Cysteine,
  as precursor of coenzyme A, 318, 321
  conversion to taurine, 317
  decarboxylation to β-thioethylamine, 315
  formation from serine and methionine, 303, 304 (Fig. 13.7)
  formula of side group, 42 (Fig. 2.4)
  residues in ferredoxin, 263
  residues in porphyrin apoenzyme, 145
  structure, 44
Cysteine desulphydratase, 305
Cystine, 43, 44 (Fig. 2.5)
Cytidine monophosphate, 80, 82 (Fig. 3.9), 288
Cytidine triphosphate, 191
Cytochrome a, 145, 151
Cytochrome b, 143, 145, 151
Cytochrome b_6, 237–240
Cytochrome c, 157
Cytochrome c_1, 144, 145
Cytochrome f, 237–240
Cytochrome p_{450}, 260, 261
Cytochrome oxidase, 104, 145
Cytosol,
  glycolytic reactions in, 196
  inorganic phosphate in, and Pasteur effect, 210
  NADH produced in, 210
  pentose phosphate pathway in, 249, 252
  some reactions occurring in, 4, 6, 8, 24, 184
Cytoplasm, 3, 22, 24, 355
Cytosine, 81, 324, 325, 329

Danielli, J. F., 17–20
Davies, R. E., and muscular contraction, 403–405
Davson, H., 17–20
Decarboxylation,
  as a step in nor-adrenalin biosynthesis, 323

Decarboxylation,
  as a step in oxidative decarboxylation, 168
  examples of, in amino acid metabolism, 314, 315
  of amino acids, 298, 314
  of pyruvate, 207–209
  role of vitamin B_6, 296
  types of, 160
Degeneracy of the genetic code, 347
Dehydrogenation, 159
Denaturation of proteins, 58, 59, 92, 93
Dendrites, 387 (Fig. 16.3)
Denitrifying bacteria, 262
Detoxication, 31, 260
Deoxyribonucleic acid see DNA
Deoxyribonucleotides, synthesis, 254, 255
2-Deoxy-D-ribose,
  formula, 66, 67
  in nucleotides, 81, 254
  in polynucleotide structure, 325
  nucleotides containing, 340
  synthesis of, 255
Dextro-rotation of glyceraldehyde, 61
Diabetes mellitus, 50, 232
Dialysis, 33 (Fig. 2.1), 87, 92, 106, 164
Differential centrifugation, 8, 33
Diffusion, 353, 361
Digestion, 15, 109, 364–366
Digitalis, 360
Diglycerides, 11, 176, 190, 367
Dihydroxyacetone (phosphate), 69, 190, 200, 201
  397, 398
Dinitrophenol, 359, 365
Dinucleotides,
  pseudo-, coenzyme A as an example, 85, 86
  structure and functions, 84–86
  see also NAD, NADP and FAD
2,3-Diphosphoglycerate, 203
1,3-Diphosphoglycerate, 202, 208
Disaccharidases, 363–365
  see also lactase, invertase, maltase
Disaccharides,
  examples of, 73–75 (Fig. 3.6)
  intestinal absorption of, 363–366
  structural relation to monosaccharides, 62
  synthesis, 216
Dihydro-orotic acid, 287
Dimethyl ethanolamine, 294
Dimethyl iso-alloxazine, 140, 146
Dinitro ortho cresol, 156

2,4-Dinitrophenol,
  as inhibitor of sodium pump, 365
  as uncoupler of oxidative phosphoryla-
    tion, 155, 156
Dithiol form of lipoic acid, 169
Dissociation,
  constant *see* pKa
  curve of oxyhaemoglobin, 373 (Fig. 15.7)
  curve of oxymyoglobin, 375
  of weak acids, 38, 39
Disulphide bridge,
  in insulin, 50
  in papain and RNA-ase, 52
  in proteins, 44, 52, 56
DNA (Deoxyribonucleic acid),
  as a primer for DNA synthesis, 341
  circular, in viruses, 351
  genetic importance of, 331
  in viruses, 349–351
  location and structure, 25, 26, 324–326,
    328
  molecular weight, 328, 330
  nucleotides in, 82, 83
  structure of double helix of, 326–331
    (Fig. 14.2)
  synthesis of, 254, 255
DNA polymerase, 340
Donnan equilibrium, 360, 361, 380
DOPA (dihydroxyphenylalanine), 323
Double helix of DNA, 326–331, 340–344
DPN, 84
Driving force of reactions, 28, 117, 120
Duodenum, 366

Effector cells, 392 (Fig. 16.5)
Electrical,
  charges across cell membrane, 358
  changes during transmission of nerve
    impulse, 398
  origin of charges, 382–386
Electrode potentials, 128, 129
Electrodes, 38, 382
Electron activator in nitrogen fixation, 266
Electron flow, analogy to waterfall *see*
    Redox reactions
Electron micrographs,
  of cells and their membranes, 4, 17, 21,
    30
  of cellulose, 79
  of intestinal absorptive cell, 363, 367
  of mitochondria, 26–28
  of muscle cells, 399
  of neurone, 387

Electron micrographs,
  of rod cells, 407
Electron transport,
  diagram of units in mitochondrial chain,
    148, 157
  in chloroplasts, 239
  in microsomes, 258–261
  in mitochondria, 136, 138
  particles, 28, 157, 349
  redox potentials of components of, 152
Electrons transferred to NAD, 84, 85
Electrophoresis, 33, 38 (Fig. 2.1)
Emulsions, 9, 10, 15, 366 (Fig. 15.4)
E.M.P. pathway, 196
Endergonic reaction,
  coupling to exergonic reactions, 118, 119,
    122–125
  in absorptive cells, 359, 363–365
  meaning, 113
End feet, 387 (Fig. 16.3), 394
Endopeptidase, 110
Endoplasmic reticulum, 4, 5, 20, 28–31
    (Fig. 1.11), 364, 367
  rough, 30, 367
  smooth, 30, 362, 364
Endothermic reactions, 117
Energy,
  associated with hydrogen and covalent
    bonds, 329
  at equilibrium, 126
  capture, in glucose catabolism, 209, 210
  changes in photosynthesis, 236–240
  changes in reversible and irreversible
    reactions, 134
  chemical and physical, 112
  conservation of, 112
  for activities in living cells, 111
  free, 119–122
  importance in active transport, 353
  in active transport, 353
  in osmoregulation and intestinal absorp-
    tion, 356, 363
  in steady state reactions, 130
  interconversion of, 112–117
  kinds of, 112, 113, 132
  potential, 112, 113–117
  squandering of, 115, 118, 124
Energy rich bonds, 131–135, 203
Enolase, 203
Enthalpy (H), 119–122
Entropy (S), 112, 115, 119–122
Environment,
  difference inside and outside cells, 352

Environment,
  speed of response to, 382
Enzyme induction (adaptive synthesis), 106, 349
Enzymes, 87 *et seq*
  as carriers in intestinal cells, 364
  as proteins, 35
  classification of, 87, 88, 90–92, Appendix 1
  commission, 91, 92, Appendix 1
  denaturation of, 58, 92, 93, 104
  enzyme-substrate interaction, 99–104 (Fig. 4.4)
  inhibition of, 104, 105
  origin of digestive, 362, 363
  serine and cysteine in active centre of, 44
  specificity of, 88–90, 410
Epimerization, 219, 245, 246
Epinephrin, 322, 323
Epithelial cells, 362
Equilibrium, 125–128, 150, 164
Equilibrium constant (K), 125–128
Equilibrium mixture, 126, 127 (Fig. 5.4)
Erythrocytes,
  ATP-ase activity, 359
  carbonic anhydrase in, 378
  chloride shift in, 380
  effect of cooling, 359
  facilitated transport of oxygen into, 353, 372
  formation of bicarbonate in, 380, 381 (Fig. 15.9)
  inorganic composition of, 358
  in tissue capillary, 381 (Fig. 15.9)
  structure, 372, 373
D-Erythrose, 66
D-Erythrose-4-P,
  as precursor of certain amino acids, 255, 292, 299, 309
  in pentose phosphate pathway, 246, 252
  in photosynthetic carbon cycle, 244–246, 248
Escherichia coli (E. coli), 5, 343, 347, 349, 351
Essential amino acids, 255, 270, 291, 299, 306
  foods rich in, 293
  table of, 272, 273
Ethanol, 207–209
Ethanolamine, 13, 14
Ethyl acetate hydrolysis, 125
Evolution, 1
Excitable cells, 382 *et seq*
Excretion,
  by tissue cells, 371

Excretion,
  of nitrogen by fish, birds and mammals, 277
Exergonic reactions,
  coupling to endergonic reactions, 118, 119, 122–125
  meaning, 113
  trapping of energy of, 113–115
Exopeptidases, 110
Exothermic reactions, 117, 118
Extracellular fluids, 352, 357, 358, 361, 372–376
  see also Plasma; Interstitial fluid
Extraction techniques for lipids, 10
Eye, photochemistry of, 406–409

Facilitated transfer, 353
  see also Transfer
FAD (H$_2$),
  as cofactor, 89, 166
  formation from TCA cycle intermediates, 137, 160
  part played in fatty acid oxidation, 179
  redox potential, 151
  structure, 146
Faraday, calculations involving the, 129
Fats,
  as efficient energy store, 175
  chemistry of, 8–13
  constant composition of, in some organisms, 177
  digestion of, 90, 366–369
  emulsification of, 366 (Fig. 15.4)
  formation of acetyl coenzyme A from, 137, 161
  percentage by weight in man, 175
  structure and functions, 9, 10
  synthesis, 190
Fatty acids,
  biosynthesis of, 185–189 (Fig. 8.5)
  chemistry, 11–13 (Table 1.3)
  cofactors required for biosynthesis of, 186–188
  efficiency of energy capture on oxidation, 182
  energy formed by oxidation of, 177
  energy involved in synthesis of, 189
  formation during digestion, 366, 367
  lengthening and shortening the chain of, 183
  location of catabolic processes, 182, 183
  oxidation spiral, 178, 180, 182 (Fig. 8.2)
  short chain, 369

Fatty acids,
    steps in catabolic spiral, 180, 181
    table of, 12
Fatty acyl-ACP complexes, 186, 188
Fatty acyl-AMP complexes,
    as primers in fatty acid oxidation, 179, 180
    comparison with amino acid activation, 337
Fatty acyl coenzyme A, 169, 179, 180, 337, 339, 367
Fatty acyl coenzyme A dehydrogenase, 180, 181
Feathers, protein in, 35
Fehling's test, 62, 73, 74
Ferredoxin,
    chemical characteristics of, 263
    in nitrogen fixation, 263, 265, 266
    in photosynthesis, 239, 240
    redox potential of, 151
Ferrocyanide, 372, (Fig. 15.6)
Ferrous ions,
    in acconitate hydratase, 164
    in haemoglobin, 373 (Fig. 15.6)
    in myoglobin, 45
    in protohaem, 373
Fibrous proteins, 35, 52
Fish,
    bony, 356
    cartilaginous, 357
    excretion of nitrogen by, 277
    gills of, 356
    osmoregulation in, 356, 357
Flaccidity of plant cells, 356
Flagellum, 21–24
Flame photometry, 384
Flavin adenine dinucleotide see FAD (H_2)
Flavin mononucleotide see FMN (H_2)
Flavoproteins, 285, 323
Fluoride,
    effect of glucose absorption, 365
    effect of glycolysis, 203
    effect on pentose phosphate pathway, 253
    effect on sodium pump, 359
Fluoroacetate, 173, 253
Fluorocitrate, 173, 365
FMN (H_2) (Flavin mononucleotide), 140, 141 (Fig. 6.4), 151
Folic acid, 294, 295
Folin and Ciocalteau test, 47 (Table 2.1)
Food as a fuel of the cell, 111
Formally endergonic reactions,
    active transport as a, 353, 363
      as coupled exergonic-endergonic changes, 118

Formally endergonic reactions,
    as re-routed reactions, 118, 122
    avoidance of, by re-routing, 336
    importance of equilibrium mixtures to, 127
    meaning, 118
Formyl kinurenine, 311
Free energy (G), 115–122
Free energy changes,
    as a function of ΔH, T and ΔS, 121
    at equilibrium, 126
    calculation of, in redox changes, 129
    definition, 115
    in coupled reactions, 118, 119, 122–125
    independence of pathway, 116
    in freely reversible reactions, 134
    in nearly irreversible reactions, 133–135
    in steady state reactions, 130
    on oxidation of glucose, 116
    overall change in coupled reactions, 123
    significance of size and sign of, 120
    standard, definition, 116
Freezing point and osmotic pressure, 355
Fresh water animals, osmoregulation, 356
D-Fructose, 62, 68, 69, 364
D-Fructose 1,6-diphosphate,
    action of aldolase on, 107, 108
    conversion to fructose-6-phosphate, 221
    formula, 71
    in glycolysis, 193, 195–199
Fructose di-phosphate phosphatase, 221–223, 252
Fructose-6-phosphate,
    formation in photosynthesis, 243
    in pentose phosphate pathway, 252
    in stage 1 of glycolysis, 197
    metabolism in green plants, 231
    reaction with transaldolase and transketolase, 244–246
Fumarate,
    formation, from aspartate, 275, 280–283, 286
    formation, from certain amino acids, 299, 309, 310
    in TCA cycle, 137, 146, 162
    naming, 147, 149
    redox potential of, 151
Fumarate hydratase, 172
Furanose form of sugars, 67–69 (Fig. 3.3)

Galactokinase, 218
D-Galactose,
    formula, 67
    intestinal absorption of, 364, 365

D-Galactose,
  metabolism of, 218–220
Galacturonic acid, 79
GDP (Guanosine di-phosphate), 170
Gene, 327
Genetic code, 346–348
Gills, 356, 372, 373, 377
Globular proteins, 56
Globulin, 33 (Fig. 2.1)
β-Globulins, 367–369
Glucagon, 232
Glucogenic amino acids, 223–229, 271, 272, 305–309
Glucogenic compounds, 220
Gluconeogenesis, 211 et seq
D-Glucose,
  α and β forms, 65 (Fig. 3.1)
  absorption of, 361, 363, 365
  conversion to galactose and lactose, 218–220
  derivatives of, 70–73
  energy yield on total oxidation of, 209
  entropy change upon total oxidation of, 121
  formation of glucose-6-phosphate from, 131
  formula, 64 (Fig. 3.1)
  intestinal absorption of, 364–365
  movement from intestine to blood, 364
  number of ATP molecules formed on total oxidation of, 131, 194
  optical rotation, 61
  passage into blood, 77, 232, 364, 371
  regulation of blood level of, 232, 233
  standard free energy change on total oxidation, 116, 131, 194 (Fig. 9.1)
  stepwise degradation of, 118, 194, 195
  structure and chemistry, 21, 61, 63–66
  synthesis see Gluconeogenesis
  uptake by liver and muscle, 365
  yield of ATP at various stages of oxidation, 194
Glucose 1,6-diphosphate, 71, 198, 203
Glucose-1-phosphate, 70, 198, 213–218, 220–223
Glucose-6-phosphatase, 232
Glucose-6-phosphate, 131, 197, 249, 250
Glucose-6-phosphate dehydrogenase, 249, 250
α-Glucosidase, 89 (Fig. 4.1)
Glucosides, 65, 66, 71, 89
Glucosyl transferase, 214–216
Glucuronic acid, 79
Glutamate semi-aldehyde, 300–302

Glutamic acid, 37, 43, 267–270, 274–276, 300, 307, 312
Glutamic acid dehydrogenase, 267–270, 276
Glutaminase, 307
Glutamine,
  formation, 274, 300, 302
  in CMP biosynthesis, 288
  in GMP biosynthesis, 286
  in uric acid formation, 282–285
  role in NAD synthesis, 322
Glutamine synthase, 274, 302
Glutarate, 149, 279
D-Glyceraldehyde, 61, 66
L-Glyceraldehyde, 61
D-Glyceraldehyde-3-phosphate,
  action of aldolase on, 107, 108
  in pentose phosphate pathway, 252
  in photosynthetic carbon cycle, 243–248
  in stage 2 of glycolysis, 200, 201
  in stage 3 of glycolysis, 201, 202, 208
Glycerides,
  absorption and digestion of, 366–369
  biosynthesis, 190
  constant proportions of, in a given sp., 177
  digestion and absorption, 366–369 (Fig. 15.4)
  hydrolysis of, 176
  mono, di and tri-, 11–13 (Fig. 1.4)
Glycerol,
  formation from di and triglycerides, 176
  structure, 11 (Fig. 1.4)
Glycine,
  as one of the phosphoglycerate family, 303
  as unit in porphyrin synthesis, 287
  as unit in purine synthesis, 282–284
  conversion to glyoxylate, 306
  formation of creatine from, 315
  formation from serine, 303
  structure, 35, 42
  use in formation of glycocholic acid, 15, 317
Glycine transamidinase, 315, 316
Glycocholic acid, 15, 16, 317, 366
Glycogen,
  as example of an α-glucoside, 66
  as precursor of fructose diphosphate, 195
  as reactant in pentose phosphate pathway, 249
  relation to blood glucose level, 233 (Fig. 10.2)
  structure, 78
  synthesis see Glycogenesis (chapter 10)
Glycogenesis, 211–233

Glycolysis,
   and the fate of pyruvate, 193 *et seq*
   as source of 3-phosphoglycerate, 303
   comparison with pentose phosphate pathway, 249, 252, 253
   four stages of, 196 (Fig. 9.2)
   importance under anaerobic conditions, 206, 207
   location in cytosol, 184
   reverse of the pathway, 221–223 (Fig. 10.8), 243, 252
   total energy produced by, 204
Glycoproteins, 35
Glycosides,
   α and β forms, 71, 73
   cellobiose as example of β-form, 73
   formation and formula, 71–73
   lactose as example of β-form, 75
   maltose as example of α-form, 71
   starch and glycogen as α-forms, 77
N-Glycosides, 82
Glyoxylate cycle, 225–229 (Fig. 10.10)
Glyoxylic acid, 225–229, 303, 306
GMP, GDP, GTP, 170, 171, 275, 284–286, 337
Goblet cells, 362 (Fig. 15.3)
Golgi bodies, 30, 31
GTP, 170, 221, 284–286, 337
Guanidine acetic acid, 316
Guanine,
   as base in polynucleotides, 324–326
   as excretory product, 277, 285
   formula and tautomerism of, 81, 82
   pairing with cytosine in DNA, 329
Guanosine diphosphate (GDP), 170, 284–286
Guanosine monophosphate (GMP), 80, 275, 284–286, 337

Haber process, 262
Haemoglobin,
   as a protein allowing facilitated transport, 353
   change of shape during oxygenation, 410
   oxygenation of, 372–376 (Fig. 15.6)
   quaternary structure of, 58
   rabbit, prepared in cell-free systems, 347
   structure, 45, 56, 372, 373
Hair, 35, 53
Harden and Young experiments in 1905, 106
Heart rate, 392
Heat, 113, 117

α-Helix,
   in globular proteins, 56
   in haemoglobin and myoglobin, 56–58
   in secondary structure of proteins, 41, 52, 53 (Fig. 2.8)
   of myosin, 53, 400 (Fig. 16.10), 404
Henderson-Hasselbalch equation, 38, 39 (Fig. 2.2), 378
Hexokinase, 197, 249
Hexose monophosphate shunt *see* Pentose phosphate pathway
Hexoses,
   definition, 61
   energy capture during oxidation of, 209, 210
   origin of carbon dioxide from, 206, 251
High energy compounds,
   active formate, 283
   acyl coenzyme A derivatives, 123, 135, 265, 337
   aminoacyl AMP, 337
   aminoacyl-transfer RNA complexes, 337
   carbamyl phosphate, 275, 287
   in sulphur bonds, 202, 205
   phosphate, 123, 131–134, 201, 203, 265
Histamine, 292, 315
L-Histidine, 42, 45, 312, 313, 372, 379
Hodgkin, A. L., 388
Homocysteine, 293, 294, 304
Homogenization, 7 (Fig. 1.2)
Homogentisic acid, 310
Homoserine, 307
Hopkins-Cole test, 47 (Table 2.1)
Hormone, local (chemical transmitters), 394, 315
Hormones *see* Adrenalin, Glucagon, Insulin, Oestrogen, Oxytocin, Testosterone, Vasopressin
Horn, 35
Huxley, A. F., 388
Huxley, H. E., 400
Hyaluronic acid, 76 (Fig. 3.7), 79
Hybrid molecules, 40, 45, 46
Hydration, 159
Hydrocarbon oil, 9
Hydrogen bonding, 41, 43, 52, 53, 327–330
Hydrogen sulphide, 104, 306
Hydrolases, 92, Appendix 1
Hydrophilic properties, 9, 15, 57, 58
Hydrophobic properties, 9, 15, 37, 57, 58, 366
L-β-Hydroxyacyl coenzyme A, 180, 181
L-β-Hydroxyacyl coenzyme A dehydrogenase, 180, 181

β-Hydroxybutyrate, 228, 229
α-Hydroxyglutaric acid, 150
β-Hydroxy-β-methyl-glutaryl coenzyme A (HMG CoA), 255, 256
L-Hydroxyproline, 46, 301, 302
Hypertonic solutions, 355, 356
Hypotonic solutions, 355, 356
Hypoxanthine, 81, 285

I-band in myofibrils, 399–401
Imidazole group, 45, 379
Imino acids, 46, 56, 301, 302, 307, 366
Imino-compounds, 282
Impulse,
    frequency of, in nerves, 392
    nature of, in nerves, 387, 391 (Fig. 16.4)
β-Indole acetic acid, 321, 322
Inhibition, 99, 104, 105, 253, 359, 365
Inhibitors,
    cyanide, 253, 365
    dinitrophenol, 365
    fluoride and fluoroacetate, 173, 203, 365
    iodoacetate, 104
    ouabain, 359
    phlorizin, 365
Inner mitochondrial membrane, 171
Inorganic ions inside and outside cells, 357, 358
Inosine monophosphate (IMP), 80, 82 (Fig. 3.9), 275, 283–286
Insecticides,
    action, by inhibiting cholinesterases, 397
    aldrin, 259, 260
    dieldrin, 259, 260
    mipafox, 397
Insulin,
    determination of structure of, 49
    α-helical configuration in, 56
    primary structure of, 50–52 (Fig. 2.7)
    regulation of blood glucose by, 232, 233
Interconversion of energy, 112, 113
Internal salt, 36
Interstitial fluid, composition, 370–372
Intestinal absorptive cells,
    absorption of sugars, amino acids, glycerides, 364–366
    origin of, 362
    structure and functions, 361–372 (Fig. 15.3)
Intestine,
    cells of, 361–372
    plasma membrane of mucosa, 20
    synthesis of vitamin A in, 407
Intracellular composition, 352, 357, 361

Invertase, 74, 75 (Fig. 3.6), 364
Iodine, reaction with starch, 77
Iodoacetate, 104
Ionic 'bridges' see Salt bridges
Ion exchange chromatography, 34, 48 (Fig. 2.1)
Ions,
    amino acids as, 37
    effect on solubility of proteins, 38
    movement into and out of cells, 357–361
Iron in porphyrins, 143
    see also Ferrous ions
Isocitrate,
    in TCA cycle, 137, 139, 162, 165
    naming, 150
    redox potential, 151
Isocitrate dehydrogenase,
    extra-mitochondrial, 184
    mitochondrial, 165
Isocitrate lyase, 225–229
Isoelectric point, 37, 38
L-Isoleucine, 43 (Fig. 2.4), 306, 307
Isomerases,
    classification of, 92, Appendix 1
    conversion of forms of retinene by, 408
    interconversion of pentose phosphates by, 246
Isomers,
    of alanine, 36
    of aldohexoses, 63
    of D-glyceraldehyde, 61
Isopentyl pyrophosphate, 255–257 (Figs. 10.9, 10.10)
Isotopes, radioactive, in study of ion movement, 358
    see also Radioactive isotopes

Joules, expressed as calories, 113

Keilin, D., 96, 99
Keratin, 35, 53
α-Keto acids,
    as precursors of amino acids, 269, 299
    formation from amino acids, 296, 299
β-Ketoacyl-acyl carrier protein complexes, 188
β-Ketoacyl-coenzyme A derivatives, 180
α-Ketobutyrate, 307
Ketogenic amino acids, 228, 229, 271, 272, 308, 309
α-Ketoglutarate,
    as precursor of glycogen, 223
    as precursor of some amino acids, 299–303 (Fig. 13.5)

α-Ketoglutarate,
   central role in nitrogen metabolism, 267–270
   family of amino acids, 300
   formation, from some amino acids, 292, 297, 299
   in TCA cycle, 137, 139, 165
   involvement in nitrogen excretion, 276
   naming, 150
   oxidatjve decarboxylation of, 165
   redox potential, 151
α-Ketoglutarate dehydrogenase, 166, 167 (Fig. 7.4)
Ketone bodies, 228, 229 (Fig. 10.11), 233
Ketoses, 61, 62, 69 (Fig. 3.4)
Kidney,
   action of vasopressin on, 50
   in osmoregulation and glucose resorption, 282, 356, 357, 365
Kinases, 197
Knoop, work on β-oxidation, 178
Krebs, H.,
   work on ornithine cycle, 279–282
   work on TCA cycle, 159 et seq
Kynurenine, 311

Lactase, 75, 363, 364
Lactate,
   buffering action of, 376, 377
   formation in exercising muscle, 195, 206–209, 211–213
   oxidation of, 88
   reconversion to glycogen, 211–213 (Fig. 10.2)
Lactate dehydrogenase, 88, 208, 213
Lactonase, 250
Lactose,
   action of lactase on, 74, 75
   formula, 74, 75 (Fig. 3.6)
   intestinal absorption of, 364
   metabolism, 218–220
   synthesis, 219 (Fig. 10.7)
Laevo amino acids, 36
Laevo-rotation of glyceraldehyde, 61
Leaf, 216, 217, 238
Lecithin,
   biosynthesis of, 191, 192 (Fig. 8.7)
   role in fat absorption, 366, 368
   structure and functions of, 9, 13, 18 (Figs. 1.5, 1.7)
Leghaemoglobin, 263
Leguminosae, root nodules, 263

L-Leucine,
   codon for, 348
   family of amino acids, 308
   formation of non-polar bridge by, 43
   intestinal absorption of, 366, 370
   side group of, 42, 43 (Fig. 2.4)
Ligases, classification, 92, Appendix 1
Light,
   action of, in eye, 406–409
   reaction in photosynthesis, 236–241 (Fig. 11.3)
   visible, 407
Lineweaver-Burk equation,
   applied to enzyme kinetics, 102–104 (Fig. 4.7)
   applied to intestinal absorption, 370
Lipase, 13, 88, 90, 176 (Fig. 8.1), 177, 366, 367 (Fig. 15.4)
Lipids, 2, 8–16, 366
Lipids, anabolism of, see Fatty acid biosynthesis
Lipids, catabolism of, 175 et seq., 182
Lipoic acid, 162, 166
Lipoprotein,
   carriage of lipids in blood as, 369
   fundamental importance of, 410
   in chylomicra, 369
   in membranes of cells, 4, 8, 15, 17–20 (Fig. 1.8), 35
   in membranes of chloroplasts, 237, 238
   in membranes of rod cells, 407 (Fig. 16.13)
Liver,
   as site of enzymes of ornithine cycle, 279
   galactose metabolism in, 218
   lactose and glycogen metabolism in, 213
   pentose phosphate pathway in, 253
   transport of amino acids to, 366
   transport of glucose to, 365
   transport of lipids to, 369
Location of cellular activities, 6 (Table 1.1)
Logarithms, conversion from base 10 to base e, 127
Lohmann reaction, 315, 316
Low-energy phosphate bonds, 131
Lowry method for protein determination, 47 (Table 2.1)
Lumi-rhodopsin, 408, 409
Lungs, 374–376 (Fig. 15.8), 377–381
Lymph, 369 (Fig. 15.3)
L-Lysine,
   absorption of, 365, 366
   an essential amino acid, 306, 307
   decarboxylation to cadaverine, 314

L-Lysine,
    formula of side chain, 42, 43, 45 (Fig. 2.4)
    in plant proteins, 273
Lysis of sea urchin eggs, 354
Lysosomes, 30, 31

Magnesium,
    as cofactor in DNA synthesis, 340
    as cofactor of enzyme reactions, 107, 197,
        203, 337
    energy change on oxidation of, 128, 129
    in cells, 357, 358, 360
    in chlorophyll, 237, 238
L-Malate,
    in T.C.A. cycle, 137, 139, 140, 162, 171
    naming of, 141, 149
    redox potential, 151, 154
Malate dehydrogenase, 139, 140, 172
Maleyl acetoacetic acid, 309, 310, 311
Malonate,
    competitive inhibitor of succinate
        dehydrogenase, 104, 173
    formula, 104, 149
Malonyl-ACP complex, 187
Malonyl coenzyme A,
    function in cell, 173
    in formation of ACP complex, 187
    role in fatty acid synthesis, 185
Maltase,
    action of, 73, 90
    in barley seedlings, 106
    in intestinal absorptive cells, 363, 364
Maltose,
    as example of a glycoside, 71, 73
    intestinal absorption of, 364
    reducing properties, 73, 74
iso-Maltose, 77
Manganous ions, 107, 165, 237
D-Mannose, 21, 67
Maximal velocity of enzyme reaction (V_{max})
        101–103 (Fig. 4.6)
Membranes, 17–20, 353
Messenger RNA,
    function, 5, 25, 26, 344
    half life, 333
    structure, 332
Metabolism, 2, 271, 359, 391
L-Methionine,
    as essential amino acid, 307
    as member of oxaloacetate family, 306
    in plant proteins, 273
    metabolism, 307
    role in transfer of methyl groups, 293, 294

L-Methionine,
    structure of side group, 42
5-Methyl cytosine, 326
Methyl malonyl coenzyme A, 225, 226
Methyl malonyl coenzyme A mutase, 225,
        226
Mevalonic acid, 256
Micelles, 33, 367, 368
Michaelis and Menton kinetics,
    applied to enzymes, 99–104 (Fig. 4.6)
    applied to transport, 369, 370
Michaelis constant (K_m), 99–104
Microorganisms,
    ability to grow on organic substrates, 212
    anaerobic respiration of, 206
    in ruminant digestive system, 75, 224, 273
    nitrogen fixation by, 262
    synthesis of amino acids by, 272
Microsomal electron transport system, 31,
        258–261
Microsomes, 8
Microvilli, 262, 263 (Fig. 15.3)
Milk,
    hormonal control of secretion, 50
    lactose in, 220
    pentose phosphate pathway in mammary
        gland, 253
    phosphoserine in, 44
Millon's test, 47
Mipafox, 398
Mitochondria,
    electron transport in, 136–158
    in amoeba, 356
    in intestinal absorptive cells, 364
    in muscle cells, 398
    oxidative decarboxylation of pyruvate in,
        184, 205
    structure, 26–28 (Fig. 1.10), 56
Mole, 37
Mole fraction, 354, 355
Molecular weight,
    of peptides and proteins, 32, 33
    of polysaccharides, 77
β-Monoglycerides,
    formation from di- and triglycerides, 176
    transacylation of, 367
Monoglyceride transacylase, 367
Mononucleotides,
    as units in DNA and RNA, 328–330
    conversion of ribo- to deoxyribonucleo-
        tides, 254, 255
    functions, 83
    nomenclature, 81 (Table 3.1)

Mononucleotides,
  steps in synthesis from PRPP, 254, 255, 338
  structure, 80 (Fig. 3.9)
  synthesis of AMP, IMP and GMP, 285, 286
Monosaccharide,
  definition, 60, 61
  derivatives of, 70–73
  examples of, (Figs. 3.2, 3.3, 3.4)
Motor end plate, 394
Mucoproteins, 35
Mucus, 23, 262
Muscle,
  contraction of, 402–405
  fibres, 398
  ionic movements in membranes of, 358, 359
  lactate formation in, 195, 206–209, 213
  myoglobin in, 375
  pentose-P pathway in, 253
  skeletal, 394
  smooth, 398
  stimulation by nerve, 394–397 (Fig. 16.7)
  striated, 398
  structure, 397–401 (Fig. 16.9)
  unstriated, 398
Mutarotation, 65
Mutation, 346–348
Myelin, 387 (Fig. 16.3)
Myofibrils, 398, 400
Myofilaments, 398 (Fig. 16.9)
Myoglobin,
  attachment of porphyrin to, 45, 46, 57
  dissociation of oxymyoglobin, 374–376 (Fig. 15.7)
  oxygenation of, 375
  structure, 56, 57 (Fig. 2.10)
Myosin,
  acting as an ATP-ase, 403, 410
  active centre of, 403
  link with actin, 401
  phosphorylation of, 403
  position in myofibril, 400 (Fig. 16.11)
  structure, 35, 400 (Fig. 16.10)

NAD (H),
  as cofactor in fatty acid oxidation, 179
  as cofactor in glycolysis, 193, 201, 202, 208
  as cofactor in TCA cycle, 137, 141, 159, 160, 165
  as cofactor of alcohol dehydrogenase, 108, 109

NAD (H),
  as cofactor of glutamate dehydrogenase, 267, 276
  energy change on oxidation of NADH, 153
  redox potential of, 151
  reduction of, 84, 85
  structure and function, 26, 84, 85
  synthesis of, 320, 321 (Fig. 13.14)
NADP (H),
  as cofactor in carotinoid and steroid synthesis, 255–258
  as cofactor in deoxyribose synthesis, 185, 254, 255
  as cofactor in fatty acid synthesis, 183, 185, 186, 188, 189
  as cofactor of glutamate dehydrogenase, 267
  extramitochondrial location of, 184
  processes requiring NADPH, 255
  reduction of, by pentose phosphate pathway, 185
  reduction of, in photosynthetic light reaction, 236, 240, 241
  relation to NAD, 85
  role in microsomal electron transport system, 258–260
  structure, 84
  synthesis, 322
NDP-glucose (nucleoside diphosphate-glucose), 214–216 (Fig. 10.3)
Nernst equation,
  for potassium diffusion potential, 383–386
  for sodium during action potential, 388
Nerve cell see Neurones
Nerve gas, 397, 398
Neurones,
  afferent and efferent, 392
  impulse carried by, 387–392
  movement of ions across membrane of, 358, 359
  structure, 386, 387 (Fig. 16.3)
Nicotinamide, 81, 139
Nicotinamide adenine dinucleotide see NAD
Nicotinamide adenine dinucleotide phosphate see NADP
Nicotinamide mononucleotide,
  formula, 80
  oxidized and reduced forms, 84, 85
Nicotinic acid, 320, 321
Nicotinic acid mononucleotide, 320, 321
Ninhydrin, 46, 48
Nitrogen bridges, 305

Nitrogen excretion,
  as urea, 279
  as uric acid, 283
  on high protein diet, 278, 279
  on low protein diet, 271, 279
  types of, in animals, 276–279 (Table 12.2)
Nitrogen fixation, 262–267 (Fig. 12.1)
Nitrogen metabolism, 262 et seq.
Nitrous acid, mutagenic effect of, 348
Non-haem iron proteins,
  in light reaction of photosynthesis, 264
  in mitochondrial electron transport sys-
    tem, 145, 147
  in nitrogen fixation, 263
Non-reducing sugar, 75, 77
Nor-adrenalin,
  as synaptic transmitter, 393
  formula, 394
  stereoisomers of, 323
  synthesis, 322, 323
Nucleic acids,
  catabolism of, 282, 285
  location and structure, 324
  see also DNA and RNA
Nucleolus, 5, 333
Nucleosides, 79–83 (Table 3.1)
Nucleotides see Mononucleotides, Dinu-
    cleotides and Polynucleotides
Nucleus, 3–6, 8, 19, 25, 26, 324, 331, 340,
    343
  lack of, 348, 372

Oestrogen, 15
Oils, 9, 10
One-carbon units, 293–295, 303, 304, 313,
    318
Opsin, 407–409
Optical rotation, 61, 63
L-Ornithine,
  absorption of, 365
  decarboxylation of, 314
  formation from glutamate, 301, 302
  in synthesis of creatine from arginine, 316
  in urea synthesis, 279–282
Ornithine cycle, 279–282 (Fig. 12.3), 301
Ornithine δ-transaminase, 301
Ornithine transcarbamylase, 280, 281
Osmoregulation in animals, 355–357
Osmosis,
  in animal cells, 353–357
  in plant cells, 355, 356
  in sea urchin eggs, 353, 354 (Fig. 15.1)
Osmotic pressure, 77, 355, 356

Ouabain, 15, 16, 359, 360
Oxaloacetate,
  as precursor of some amino acids, 299, 306
  condensation with acetyl coenzyme A, 163
  conversion to glucose-1-P, 220–223
  formation of, 291, 292
  in mitochondrion and cytosol, 184
  in TCA cycle, 137, 140, 162
  redox potential, 151
  role in nitrogen excretion, 276
  role in transamination, 269
  synthesis from acetyl coenzyme A, 225–
    229
  synthesis from amino acids, pyruvate and
    propionate, 223–225, 299
Oxidation,
  in organic compounds, 139 (Fig. 6.3)
  role of NAD in, 84, 85
β-Oxidation, 178, 180
Oxidative decarboxylation,
  cofactors involved in, 166
  definition, 160
  of α-ketoglutarate, 165
  of pyruvate, 205, 206
Oxidative phosphorylation,
  energetic aspects of, 154
  from isocitrate oxidation, 162, 165
  from α-ketoglutarate oxidation, 136, 162,
    169
  from L-malate oxidation, 171
  from pyruvate oxidation, 206
  from succinate oxidation, 162, 170
  site of, 28, 156, 157
Oxidoreductases, 92, Appendix 1
Oxygen,
  consumption by cells, 1, 5, 372
  dissociation curve for oxyhaemoglobin,
    374 (Fig. 15.7)
  lack, 155, 206–209
  reaction with haemoglobin, 373
  reaction with myoglobin, 56, 373
  transport by the blood, 372–376
Oxytocin, 49, 50 (Fig. 2.6)

Pacemaker of biochemical pathway, 105,
    130
Palmitic acid,
  energy obtained by oxidation of, 175, 181,
    182
  formula, 12
  synthesis, 185

Pancreas,
  secretion of enzymes by, 31, 109, 364, 366
  secretion of insulin by, 50
Pantothenic acid, 318
Papain, 50–52
Passive transport,
  definition, 353
  intestinal absorption of lipids as example, 367
  of oxygen into erythrocyte, 373
  of substances from cells, 371
Pasteur effect, 210
Pathway of reaction,
  energy change independent of, 116
  enzymatic significance of re-routing, 125
  re-routing of, 117, 118, 122, 123
Pauling, L., 53, 326
Pectin, 23, 76, 79
Pentose phosphate pathway,
  as supplier of amino acid precursors, 292
  comparison with other pathways of hexose catabolism, 252, 253
  reactions of, 249–252
PEP see Phosphoenol pyruvate
Pepsin, 109, 110
Peptide bond,
  enzymatic hydrolysis of, 110
  imino-acids in, 46
  reaction with cupric ions see Biuret test structure, 40, 41 (Fig. 2.3)
Peptide synthase, 337, 339, 340, 344
Peptidases see proteolytic enzymes
Peptides, 32–59
Peroxidase, 96
Perutz, M. G., 56
pH,
  definition and calculation of, 37–39
  effect of buffers on, 38, 40
  effect on enzyme action, 93, 376
  effect on ionization of amino acids, 37
  intracellular, 87, 376–377
  of water, 37
  optimal pH, 93 (Fig. 4.2)
L-Phenyl alanine,
  as precursor of DOPA, 323
  as precursor of nor-adrenalin, 323
  codon for, 348
  formula of side group, 42, 43, 45
  metabolism, 309, 310 (Fig. 13.8)
  test for, 47
Phenyl alanine hydroxylase, 309, 322
Phlorizin, 365
Phosphagen, 135

Phosphate,
  buffering by, 376, 377
  distribution in cell of, 310
  esters of carbohydrates, 62, 70, 71
  in breakdown of glycogen, 198
  in nucleotides, 81
Phosphatidic acid, 13, 14, 190 (Fig. 8.6), 192
Phosphatidyl choline see Lecithin
Phosphatidyl ethanolamine, 13, 14
Phosphatidyl serine, 13, 14
Phosphoenol pyruvate (PEP),
  as precursor of some amino acids, 299, 309
  high energy nature of, 133, 203
  in glycolysis, 203, 204
  synthesis from oxaloacetate, 220–223
Phosphoenol pyruvate carboxykinase, 221–224
Phosphofructo kinase, 197, 199, 221
Phosphogluco mutase,
  in glycolysis, 197, 198
  in pentose-P pathway, 249
  reaction of, 108, 232
6-Phosphogluconic acid, 249, 250, 255
6-Phosphogluconic acid dehydrogenase, 250
Phosphoglyceraldehyde dehydrogenase,
  in glycolysis, 202, 206
  in photosynthesis, 243
2-Phosphoglycerate, 203
3-Phosphoglycerate,
  as precursor of some amino acids, 299, 303
  conversion to fructose-6-P, 243
  conversion to phosphoenol pyruvate, 203
  formation in glycolysis, 202
  formation in photosynthesis, 241, 242
Phosphoglycerate kinase, 202
Phosphoglyceromutase, 203
Phosphohexoisomerase, 197, 199
Phosphohydroxy pyruvate, 302
Phospholipid, 9, 10, 13–15, 17, 20, 175, 369 (Fig. 1.5)
  see also lecithin
Phosphopantetheine, 35, 86, 318
Phosphopyruvate hydratase, 203, 204
Phosphoribokinase, 241, 242, 248
Phosphoribose pyrophosphate see PRPP
Phosphorylase,
  in glycolysis, 197, 198
  in pentose phosphate pathway, 249
  virtual non-reversibility of reaction, 213, 215, 232
Phosphorylation,
  oxidative, 136, 206
  photosynthetic, 173, 239, 240

Phosphorylation,
  substrate level, 138, 171, 203, 204
Phosphoserine, 13, 14, 44, 303, 304
Photoisomerization, 407
Photolysis of water, 239–241
Photosynthesis,
  ATP formed by, 119
  carbon cycle of, 241–249 (Fig. 11.7)
  energy transformation in, 3, 28, 29, 112, 116, 122
  in bacteria, 263
  light reaction of, 236–241
  overall process of, 234–236
P_i see Phosphate
Pituitary, hormones of posterior, 49
pKa (K = dissociation constant),
  and Henderson-Hasselbach equation, 38–39
  of bicarbonate (HCO_3^-/H_2CO_3), 378
  of phosphate ($HPO_4^{2-}/H_2PO_4^-$), 376
Plants,
  acetyl coenzyme A conversion to oxalo-acetate by, 212, 225, 229
  amino acids in proteins of, 273
  amino acid synthesis in, 272, 291, 299
  aromatic ring synthesis in, 309
  carbohydrate metabolism in, 229–231 (Fig. 10.12), 241–249
  cell walls of, 5, 21, 23
Plasma,
  dialysis of, 33
  electrophoresis of proteins of, 33
  inorganic ions in, 33, 358, 360
  pH of, 377
  protein content of, 33, 371
Plasmalemma,
  charge across, 358, 388
  functions, 352
  of various types of cells, 262, 387, 398, 406
  structure, 20–23
  transfer of oxygen and carbon dioxide through, 373, 378, 381
Plasmolysis, 356
Pleated sheet structure in proteins, 52, 53, 55 (Fig. 2.9)
Polar compounds, 9, 10
Polynucleotides, 324 et seq, 338
Polyoma virus, 351
Polyribosomes, 346
Polysaccharides,
  formulae and properties, 62, 75–79 (Fig. 3.7)
  synthesis, 213–218

P:O ratio, 154
Pores, in cell membranes, 20, 21
Porphobilinogen, 287, 288
Porphyrins,
  structure, 143
  synthesis, 287, 289 (Fig. 12.7), 291, 292
Potassium,
  diffusion potential, 383–386
  in intra and extracellular fluids, 357–361, 383
Potential,
  action, 387–391 (Fig. 16.4)
  of end plate, 395
  resting, 382–386
Potential energy,
  changes in physical and chemical systems, 114
  chemical and physical, 112–117, 119
  coupling of changes of, 122–125
  decrease of, in exergonic reactions, 112, 122–125
  of waterfall and electron transport compared, 138
Pressure,
  partial, of carbon dioxide (P_{CO_2}), 381
  partial, of oxygen (P_{O_2}), 374–376
Primary structure of peptides and proteins, 32, 48–52
Primers,
  DNA molecules as, 340
  glycogen and starch as, 343
Priming reactions,
  importance and meaning of, 122–125 (Fig. 5.3)
  in chemistry, 123, 124
  in fatty acid oxidation, 179, 180
  in glycolysis, 199
Proline, 46, 301, 302
  see also Imino acids
Propionate,
  conversion to oxaloacetate, 224–226 (Fig. 10.9)
  formation from amino acids, 224, 291, 292, 299, 307, 308
Propionyl coenzyme A,
  conversion to methyl malonyl coenzyme A, 186, 225, 226
  formation from isoleucine, 299, 307
Prosthetic group,
  attachment to proteins, 45
  definition, 92
  of cytochromes, 144
  of rhodopsin, 407, 410

Prosthetic group,
  of succinate dehydrogenase, 146
Proteins,
  associated with ribosomal RNA, 331
  buffering of acid by, 38, 376, 377
  chemistry of, 32–59
  denaturation of, 58, 59
  dietary requirement, 370, 371
  digestion of, 365
    see also Proteolytic enzymes
  effects of, on Donnan equilibrium, 361
  enzymes as, 87
  in cell membranes, 17
  synthesis of, 324 et seq, 343–346, 367
    (Fig. 14.7)
Proteolytic enzymes,
  action of, 109, 110 (Fig. 4.10)
  in digestion, 365
  in Golgi bodies, 31
  use in determining protein structure, 41,
    49
Protohaem,
  biosynthesis of, 289
  in cytochrome b, 143
  in haemoglobin, 56, 143, 372–376
  in myoglobin, 45, 143, 375
  structure, 144 (Fig. 6.6)
PRPP (5′-phosphoribose- 1′-pyrophosphate),
    254, 336, 338, 354
Pteridine, 295, 309
Purines,
  as source of atoms of histidine, 298
  in polynucleotides, 325
  pairing with pyrimidine bases, 327
  structure, 80–82
  synthesis, 275, 282–285 (Fig. 12.4), 291,
    292
Putrescine, 314
Pyranose forms of sugars, 63, 64, 66
Pyridine, 80, 81
Pyridoxal phosphate,
  role in decarboxylation, 296, 297
  role in transamination, 268, 296, 297
  role in various reactions, 314
  structure, 296 (Fig. 13.4)
Pyridoxamine phosphate, 268, 297
Pyrimidines,
  biosynthesis, 287
  in polynucleotides, 325
  pairing with purine bases, 327
  structure, 80–82
Pyrophosphatase,
  comparison of various functions of, 338,
    339

Pyrophosphatase,
  in CDP-choline formation, 191
  in fatty acid priming, 179
  in lecithin synthesis, 192
  in nucleotide synthesis, 336
  in polysaccharide synthesis, 215
Pyrophosphate,
  formation from ATP, 170
  formation in fatty acid priming, 179
  formation in lecithin synthesis, 191
  formation in nucleotide synthesis, 254
  formation in ornithine cycle, 280, 281
  use to draw reactions to completion, 170,
    215, 335–338 (Fig. 14.6)
    see also Pyrophosphatase
Pyrrole rings, synthesis of, 287, 288
Pyrroline carboxylic acid, 301
Pyruvate,
  anaerobic fate of, 24, 27, 208, 209 (Fig. 9.7)
  as precursor of some amino acids, 299, 308
  CO_2 formed from, 205, 206
  conversion to acetyl coenzyme A, 167,
    195, 205, 206
  conversion to lactate, 206–208
  conversion to oxaloacetate, 186, 223–225
  formation by glycolysis, 193, 195, 203, 204
  formation from amino acids, 223, 291,
    292, 308
  oxidative decarboxylation of, 167, 194
  role in nitrogen fixation, 264
Pyruvate decarboxylase, 208
Pyruvate dehydrogenase, 167 (Fig. 7.4), 184,
    205
Pyruvate kinase, 134, 138, 203, 204, 224

Quantasome, 237–238
Quarternary structure of proteins, 58
Quinolinic acid, 321

Radioactive isotopes,
  C^{14} used to study pentose phosphate
    pathway, 251
  C^{14} used to study TCA cycle, 174
  $C^{14}O_2$ used to study fatty acid synthesis,
    185
  $C^{14}O_2$ used to study photosynthesis, 235
  N^{15} used to study amino acid synthesis,
    267
  N^{15} used to study DNA, 340
  used to study movements of ions, 358
Ranvier, nodes of, 387, 391
Rate,
  of biochemical reactions, 99, 105, 130
  of passage of nerve impulse, 388, 391

Red cells *see* Erythrocytes
Redox reactions, 117, 129, 139, 151
Reducing properties of monosaccharides, 62, 63
Reductive acylation, 167
Reflex, nervous, 392
Rerouting reactions,
  selected examples, 190, 221, 224, 336
  significance of, 95, 117, 118, 125
Residues, of amino acids, 32
Respiratory quotient (R.Q.), 177, 178
Retina, 406
Retinene, 406–409
Rhodopsin, 407–410
D-Ribitol, 66, 67, 140, 146
Riboflavin phosphate *see* FMN
D-Ribose,
  formula, 66, 67
  in nucleotides, 79, 86
  in polynucleotides, 283, 284
Ribose-5-P,
  conversion in nucleotides, 254, 255
  in pentose phosphate pathway, 246, 251
  in photosynthesis, 246–248
  in purine synthesis, 283, 284
Ribosomal RNA, 25, 26, 331–333
Ribosomes,
  as site of protein synthesis, 5, 6, 29, 30, 343–345
  role in fat absorption, 367
D-Ribulose, 69
Ribulose-5-P,
  in pentose phosphate pathway, 250–252
  in photosynthesis, 241–248
RNA (Ribonucleic acid),
  location, 25, 26, 324–326
  nucleotides in, 83
  structure, 324–326
  synthesis, 341
  types of, 331–340 (Fig. 14.5)
  viruses, 349–351
RNA-ase, 31, 50, 333, 346
Rod cell, 406–409 (Fig. 16.13)
Root nodules, 263
Route of reaction, 3, 116, 117, 122
Rubber, 255–257
Ruminants, 75, 224, 273

Sakaguchi test, 47 (Table 2.1)
Salt bridge, 43, 45, 52
Sanger, F., 49, 50, 110
Sarcolemma, 398
Sarcoplasmic reticulum, 398, 402, 404
Sarin, 397, 398

Schiff base,
  in amino acid metabolism, 296, 297
  rhodopsin as a, 408, 409
Schwann cell, 387 (Fig. 16.3)
Sea urchin egg, 353, 354, 356, 357
Secondary structure of proteins, 41, 45, 46, 56
D-Sedoheptulose, 69
Sedoheptulose-7-P,
  in pentose phosphate pathway, 246, 251, 252
  in photosynthesis, 244, 246–248
Seeds,
  carbohydrate in germinating, 106, 231
  lipid, 175
Semipermeable membranes, 21, 352
Sensory cells, 393
  see also Rod cell
Sequential enzymatic reactions, 105, 130
L-Serine,
  catabolism, 305
  formula, 42, 44
  synthesis, 295, 303, 304 (Fig. 13.6)
Serine hydroxymethylase, 303, 304, 306
Side groups of amino acids, 35, 37, 41–47 (Fig. 2.4)
Silver mirror test, 62, 73
Small intestine, 361–364
Soap, 9, 13, 176
Sodium ions,
  effect on action potential, 389, 390
  in intra and extracellular fluids, 357–361, 384
  in nerve and muscle cells, 358, 390, 402
  movements of, at neuromuscular junction, 396
Sodium pump, 357–360, 383, 388, 391 (Fig. 15.2)
Soluble RNA *see* Transfer RNA
Spike *see* Action potential
Spontaneous reactions, 113, 116, 121
Squid nerve cells, 388
Standard redox potential (E_0'), 152, 153, 172
Standard states, 116, 121, 129
Starch,
  amylopectin form, 77, 78
  amylose form, 75–77
  enzymatic hydrolysis *see* Amylase
  properties, 77
  synthesis, 214–217 (Fig. 10.4)
  see also amylopectin; amylose
Starvation, 271, 273
Steady state, 129, 130
Stearic acid, 12, 177

Stereoisomers,
  of amino acids, 35, 36
  of monosaccharides, 61
Steroid glycoside, 16, 359, 360
Steroids, 10, 15, 16 (Fig. 1.6), 175, 185
Substrate, 87, 99
Substrate level phosphorylation, 138, 171, 203, 204
Succinate,
  energy change on oxidation, 151, 153
  formation from succinyl coenzyme A, 169
  in TCA cycle, 137, 146, 147, 162
  oxidation to oxaloacetate, 171
Succinate dehydrogenase,
  competitive inhibition of, 104, 173
  reaction catalyzed by, 89, 172
  site in cell of, 146, 171
Succinyl coenzyme A,
  conversion to succinate, 169
  formation from amino acids, 292, 299, 308
  formation from propionate, 225, 226
  role in TCA cycle, 162
Sucrose,
  chemistry, 74, 75
  intestinal absorption, 75, 364
  metabolism in plants, 77, 216, 231
  synthesis, 216 (Fig. 10.5)
Sugars see Carbohydrates
Sulphur group of amino acids see Cystine, Cysteine, Methionine
Sulphur test, 47
Sulphydryl groups,
  of alcohol dehydrogenase, 109
  of coenzyme A, 86
  of cysteine, 44
  of phosphoglyceraldehyde dehydrogenase, 201
  of phosphopantetheine, 187
  of reduced lipoic acid, 169
Sumner, J. B., 92
Symmetry of viruses, 350
Synapse, 393–395

Taurine, 15, 16, 292
Taurocholic acid, 15, 16, 317, 366, 368
Tautomeric forms,
  of D-glucose, 63
  of purines and pyrimidines, 81, 82
TCA cycle (Tricarboxylic acid cycle),
  energy capture by, 172, 173
  linkage with amino acid metabolism, 205, 223, 291

TCA cycle (Tricarboxylic acid cycle),
  mechanism of, 136, 137, 159 et seq
  relation to glyoxylate cycle, 225–229
Temperature,
  effect on enzyme reactions, 87, 92, 93
  effect on proteins, 58
  for preparing cell fractions, 8
  optimal, 93 (Fig. 4.2)
Templates, 338, 340, 345
Tertiary structure, 52–58, 411
Testosterone, 15, 16
Tetrahydrofolate (coenzyme F), 295
Thermodynamics,
  applied to movement of solutes, 352
  first law, 111, 112
  relation to equilibrium constant, 127
  second law, 121
Thiamine pyrophosphate see TPP
β-Thioethylamine, 318, 321
Thiokinase, 169, 170
Thiol group see Sulphydryl group
Thiolase, 180, 181
L-Threonine, 42, 44, 307
Threonine dehydratase, 307
Threshold, 392
Thymidine monophosphate (TMP), 80, 82
Thymine, 81, 324, 325, 329
Tissue fluid see Interstitial fluid
Tobacco mosaic virus, 350
Tonoplast, 4, 5, 24
TPN, 85
TPP (Thiamine pyrophosphate),
  as cofactor of transketolase, 245, 253
  in non-oxidative decarboxylation, 208, 209
  in oxidative decarboxylation, 162, 166
TPP-aldehyde complex, 167, 168, 209, 264
TPP-keto acid complex, 167, 168
Trace elements, 107, 212
Transacylase and transacylation, 187, 188
Transaldolase,
  in pentose phosphate pathway, 246, 252
  in photosynthesis, 246–248
  reactions catalyzed by, 244
Transamination,
  examples of, 91, 306, 308, 310
  role in nitrogen metabolism, 267–270, 276 (Fig. 12.2)
  role of pyridoxal phosphate in, 296, 297, 303 (Fig. 13.4)
Transcribing and translating, 331, 333, 341
Transfer RNA, 331–334, 345
Transferases, 92, Appendix 1

Transketolase,
in pentose phosphate pathway, 246, 251
in photosynthesis, 245, 246
reactions catalyzed by, 244
Transmethylation, 293, 294 (Fig. 13.2), 316, 323
Transport, active,
definition, 353
examples of, 358–360, 364, 365
Transport, facilitated,
definition, 353, 371
of disaccharides by the intestine, 364
of oxygen into erythrocyte, 372–376
Transport, passive, 353
Tricarboxylic acid cycle *see* TCA cycle
Triglycerides,
biosynthesis, 190, 191 (Fig. 8.6)
hydrolysis, 13, 90, 176
intestinal absorption, 366–369
structure, 11
Trimethylamine oxide, 277
Triose phosphate, 177, 195
Triose phosphate isomerase, 200, 201, 221
Trioses, 61
Triplet code, 347, 348
Trypsin, 31, 109, 110
L-Tryptophan,
as precursor of NAD, 320
catabolism, 311
properties, 309
side group, 42, 45
test for, 47
Tungstate, 59
Turgidity, 356
Turnover number, 104, 201
Tyrosinase, 323
L-Tyrosine,
as precursor of noradrenalin, 323
hydrogen bond formation by, 43, 45
structure, 42, 309, 310
test for, 47

Ubiquinone, 142 (Fig. 6.5), 147, 151, 256
UDP-galactose-4-epimerase, 218, 219
UDP-glucose, 215, 216, 218, 219
Ultracentrifugation, 8, 33
Ultrafiltration, 371
UMP (Uridine monophosphate), 80, 82, 288
Universality of genetic code, 347
Unsaturated fatty acids, 12, 13
Unsaturated fatty acyl-ACP hydratase, 188
Unsaturated fatty acyl coenzyme A hydratase, 180, 181

Uracil, 81, 324, 325
Urea,
as excretory product, 277
hydrolysis of, 88
synthesis in mammals and amphibia, 275, 279–282 (Fig. 12.3)
Urease, 88, 92, 279
Ureotelic animals, 278
Uric acid,
as excretory product, 277, 278
synthesis, 282–285
Uricotelic animals, 278, 282–285
Uridine monophosphate *see* UMP
Uridine triphosphate *see* UTP
Urine, 232, 356, 357
Urocanic acid, 312
Uroporphyrin III, 289
UTP (Uridine triphosphate), 215, 218, 219

Vacuole, 4, 5, 24, 30, 31, 356, 362
L-Valine,
as precursor of coenzyme A, 318, 320
non-polar properties of, 37, 42, 43
structure, 42, 308
Valonia, 357
Vasopressin, 49, 50
Villus of intestine, 362
Viruses, 347, 349–351 (Fig. 14.9)
Visual pigment *see* Rhodopsin
Vital force, 117
Vitamins,
definition, 107
requirement by microorganisms, 212
A, 257, 258, 407
B_1 *see* TPP
B_2 *see* FMN
B_6 *see* Pyridoxal-P and pyridoxamine-P
B_{12}, 225, 226, 297
C, 255
D, 15
H *see* Biotin
see also Folic acid
Volts, 113

Warburg manometer, 178
Water,
conservation during excretion, 278
movement of, into cells, 352, 353–355
Waterfall analogy to explain energy changes, 113–115, 119, 128, 138
Watson, B. E., 327, 340
Weak acids, 150, 151

X-ray crystallography, 4, 33, 52, 53, 326, 330
Xanthine, 285
Xanthoproteic test, 47
D-Xylose, 66
D-Xylulose, 69
Xylulose-5-phosphate,
  in pentose phosphate cycle, 246, 251
  in photosynthesis, 244–248

Yeast fermentation, 106, 206–209

Z-band, 400, 401
Zinc, 107, 109, 378
Zwitterion, 36
Zymogens (or proenzymes), 30, 31, 109